STATISTICS AND PROBABILITY

Proceedings of the 3rd Pannonian Symposium on Mathematical Statistics
Visegrád, Hungary, 13–18 September, 1982

Edited by

J. MOGYORÓDI
Eötvös Loránd University, Budapest

I. VINCZE
Mathematical Institute of the Hungarian Academy of Sciences, Budapest

W. WERTZ
Institute of Statistics, Technical University of Vienna

D. Reidel Publishing Company
A MEMBER OF THE KLUWER ACADEMIC PUBLISHERS GROUP
Dordrecht / Boston / Lancaster

Library of Congress Cataloging in Publication Data

CIP

Pannonian Symposium on Mathematical Statistics
(3rd: 1982 : Visegrád, Hungary)
Statistics and probability.

Includes index.
1. Mathematical statistics—Congresses. 2. Probabilities—Congresses.
I. Mogyoródi, J. II. Vincze, I. (István) III. Wertz, Wolfgang. IV. Title.
QA276.AIP36 1982 519.5 83-17615
ISBN 90-277-1675-7

Distributors for the U.S.A. and Canada
Kluwer Academic Publishers,
190 Old Derby Street, Hingham, MA 02043, U.S.A.

Distributors for Albania, Bulgaria, Chinese People's Republic,
Cuba, Czechoslovakia, German Democratic Republic, Hungary,
Korean People's Republic, Mongolia, Poland, Romania,
the U.S.S.R., Vietnam, and Yugoslavia
Kultura Hungarian Foreign Trading Company
P.O.B. 149
H-1389 Budapest 62, Hungary

Distributors for all remaining countries
Kluwer Academic Publishers Group
P.O.Box 322, 3300 AH Dordrecht, Holland

Joint edition published by D. Reidel Publishing Company
Dordrecht, Holland / Boston, U.S.A. / Lancaster, England

and

Akadémiai Kiadó, Budapest, Hungary

Printed in Hungary

CONTENTS

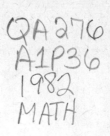

PREFACE

The Department of Probability Theory of Eötvös Loránd University, Budapest, organized the Third Pannonian Symposium on Mathematical Statistics at Visegrád, Hungary, from 13 to 18 September 1982. The preceding two Symposia took place in 1979 and 1981 at Bad Tatzmannsdorf, Austria, and were organized by the Institute of Statistics of the Technical University of Vienna. These Symposia aimed at creating an appropriate forum for statisticians living mainly in the countries which are situated on the territory of the Pannonia Provincia of the ancient Roman Empire. The collaboration of the two Universities became very fruitful and presented an excellent opportunity to exchange results and views in mathematical statistics, probability theory, and connected branches of mathematics.

The Proceedings of the preceding Symposia has already been published by D. Reidel. The present volume aims to continue this series and publish a part of contributed papers.

The participants of this Symposium came mostly from different institutes of Hungary. The 46 Hungarian research workers who participated represent the majority of the outstanding statisticians of Hungary. The 35 foreign statisticians came mainly from Austria, Czechoslovakia and Poland but there were participants from Egypt, France, the German Democratic Republic, Romania, the Soviet Union, Togo, the United States and Vietnam, too.

It was an excellent occasion to commemorate the 30th anniversary of the foundation of the Department of Probability

Theory of the Eötvös Loránd University, Budapest, and its founder, the late Professor Alfréd Rényi.

The number of presented papers was 60 and, of these, 29 were contributed by foreign statisticians.

The organizing committee is very grateful to all the Hungarian institutes who supported the organization work. Among them we have to mention the Eötvös Loránd University, Budapest, the Computing Center of the Eötvös Loránd University, the Hungarian Academy of Sciences, the Bolyai János Mathematical Society, and the Institute for Applications of Computer Technics.

Proc. of the 3rd Pannonian Symp. on Math. Stat., Visegrád, Hungary 1982
J. Mogyoródi, I. Vincze, W. Wertz, eds

LIMIT DISTRIBUTIONS FOR EHRENFEST'S URN MODEL

G. BARÓTI

Department of Probability Theory,
Eötvös Loránd University, Budapest, Hungary

The well-known Ehrenfest urn model is the following. Consider N balls, numbered 1,2,...,N, that are distributed in the urns A and B; N_0, say, in A and $N-N_0$ in B. Choose randomly an integer from 1,2,...,N. Find the ball with that number on it and transfer to the other urn. Determine the distribution of the number of balls in urn A after the nth transfer.

For fixed N, with $n\to\infty$, the limiting distribution is described in many textbooks (see for example [1]).

The aim of our paper is to determine the limit distributions if $N_0=N$ and $n,N\to\infty$ or n is fixed, with $N\to\infty$.

Let N be the initial number of balls in urn A and let $X_{n,N}$ denote the number of balls in urn A after the nth transfer. We shall prove the following theorems.

THEOREM 1. If $n,N\to\infty$, with $n^2N^{-1}\to\infty$, then the limit distribution of the standardized $X_{n,N}$ is normal.

THEOREM 2. If $n,N\to\infty$, with $n^2N^{-1}\to\lambda>0$, then the limit distribution of $(X_{n,N}-N+n)/2$ is Poisson with parameter $\lambda/2$.

THEOREM 3. If $n,N\to\infty$, with $n^2N^{-1}\to 0$, or n=const., $N\to\infty$, then

$$P(X_{n,N}=N-n)\to 1.$$

In the sense of a continuity theorem of CURTISS (see [2]) it is sufficient to show that the sequence of the moment generating functions converges to the moment generating function of the corresponding limit distribution. We shall give the moment generating functions in integral form and the asymptotic behaviour of the integrals will be obtained by the saddle point method. For the proof of Theorem 1 we shall use the following simple statement: if the variance of $X_{n,N}$ tends to infinity, var $(X_{n,N})\sim\sigma_{n,N}$ and $(E(X_{n,N})-m_{n,N})/\sigma_{n,N}=o(1)$ then $(X_{n,N}-E(X_{n,N}))/(\mathrm{var}(X_{n,N}))^{\frac{1}{2}}$ and $(X_{n,N}-m_{n,N})/\sigma_{n,N}$ have the same limiting distribution.

1. THE MOMENT GENERATING FUNCTION

If $p_k(n,N)=P(X_{n,N}=k)$, then (see for example [1])

$$p_k(n+1,N)=(1-\frac{k-1}{N})p_{k-1}(n,N)+\frac{k+1}{N}p_{k+1}(n,N). \tag{1}$$

Denote the generating function of $X_{n,N}$ by $F_{n,N}(x)$ and define the function $G_N(x,z)$ by the formula

$$G_N(x,z)=\sum_{n=0}^{\infty}\frac{(Nz)^n}{n!}F_{n,N}(x). \tag{2}$$

From (1) we obtain the following partial differential equation

$$\frac{1}{N}\cdot\frac{\partial G}{\partial z}=xG+\frac{1-x^2}{N}\cdot\frac{\partial G}{\partial x}, \tag{3}$$

where $G=G_N(x,z)$ and the initial condition is $G_N(x,0)=x^N$.

From (3) we have

$$G_N(x,z)=(\mathrm{sh}z+x\,\mathrm{ch}z)^N. \tag{4}$$

Hence by the Cauchy formula we get

$$F_{n,N}(x)=\frac{n!}{2\pi iN^n}\oint(\mathrm{sh}z+x\,\mathrm{ch}z)^N z^{-n-1}dz, \tag{5}$$

where the integration may be carried out on any circle with center 0. The moment generating function of $(X_{n,N}-m_{n,N})/\sigma_{n,N}$ is

$$M_{n,N}(t)=u^{-m_{n,N}}F_{n,N}(u), \tag{6}$$

where $u=\exp(t/\sigma_{n,N})$.

It is easy to see that the equation

$$\frac{\partial}{\partial z}\{(\mathrm{sh}z+u\,\mathrm{ch}z)^N z^{-n}\}=0 \tag{7}$$

has a unique real positive root, say z_0, which is a saddle point.

2. EXPECTATION, VARIANCE, SADDLE POINT AND ITS ASYMPTOTIC EXPANSIONS

By differentiating (4) we get

$$E(X_{n,N})=\frac{N}{2}(1+(1-\frac{2}{N})^n) \tag{8}$$

and

$$\mathrm{var}(X_{n,N})=\frac{N}{4}(1+(N-1)(1-\frac{4}{N})^n - N(1-\frac{2}{N})^{2n}) . \tag{9}$$

After some algebraic transformations, we obtain from (7),

$$z_0=\frac{\mathrm{sh}z_0+u\,\mathrm{ch}z_0}{\mathrm{ch}z_0+u\,\mathrm{sh}z_0}\cdot\alpha, \tag{10}$$

where $\alpha=nN^{-1}$.

This can be written in the following form, too

$$z_0=\alpha\left(1+\frac{se^{-2z_0}}{1+\frac{s}{2}(1-e^{-2z_0})}\right), \tag{11}$$

where $s=u-1$.

Using (8), (9) and (11) we get, by an elementary calculus, the following results

a) If $n,N\to\infty$, with $\alpha^2N\to\infty$, then

$$E(X_{n,N})=\frac{N}{2}(1+e^{-2\alpha})-\alpha e^{-2\alpha}+o(1), \tag{12}$$

$$\mathrm{var}(X_{n,N})=\frac{N}{4}(1-(1+4\alpha)e^{-4\alpha})(1+o(1)), \tag{13}$$

$$s=O(\frac{1}{\sqrt{N}}),\ \text{if } \alpha\geq\alpha_0>0 \text{ and } s=O(\frac{1}{\alpha\sqrt{N}}),\ \text{if } \alpha\to 0, \tag{14}$$

$$z_0=\alpha(1+e^{-2\alpha}s-e^{-2\alpha}(\frac{1-e^{-2\alpha}}{2}+2\alpha e^{-2\alpha})s^2)+O(\frac{1}{\alpha N\sqrt{N}}). \tag{15}$$

b) If $N\to\infty$, with $\alpha^2 N\to\lambda$ ($\lambda\in[0,+\infty)$), then

$$E(X_{n,N})=N-n+\lambda+o(1), \tag{16}$$

$$\mathrm{var}(X_{n,N})=2\lambda+o(1), \tag{17}$$

$$z_0=\alpha(\tilde{s}+1)(1-\alpha\tilde{s}(\tilde{s}+2)+O(\alpha^2)), \tag{18}$$

where $\tilde{s}=\tilde{u}-1$ and $\tilde{u}=e^t$.

3. PROOF OF THEOREM 1

In view of Curtiss' theorem it is enough to show that for any fixed real t the asymptotic relation

$$F_{n,N}(u)=u^{m_{n,N}}\cdot e^{t^2/2}(1+o(1)) \tag{19}$$

is valid, where

$$u=\exp(t/\sigma_{n,N}),\ m_{n,N}=\frac{N}{2}(1+e^{-2\alpha}),\ \sigma^2_{n,N}=\frac{N}{4}(1-(1+4\alpha)e^{-4\alpha}).$$

Substituting $z=z_0e^{iy}$ and using Stirling's formula from (5) it follows that

$$F_{n,N}(u)=e^S I_\pi(1+o(1)), \tag{20}$$

where

$$S=-n+N\log(\mathrm{sh}z_0+u\,\mathrm{ch}z_0)+n\log\alpha-n\log z_0,$$

$$I_\pi=\frac{\sqrt{n}}{\sqrt{2\pi}}\int_{-\pi}^{\pi} e^{Nf(y)-iny}dy$$

and

$$f(y)=\log(\mathrm{sh}(z_0e^{iy})+u\,\mathrm{ch}(z_0e^{iy}))-\log(\mathrm{sh}z_0+u\,\mathrm{ch}z_0).$$

Let $\varepsilon>0$ be arbitrary small, but fixed. Using the Taylor expansion and applying (10) we have

$$f(y)=\alpha iy-\frac{\alpha y^2}{2}\left(1+\frac{z_0^2-\alpha^2}{\alpha}\right)+O(\alpha y^3)\quad |y|\le\varepsilon. \tag{21}$$

To this end we note that

$$f'''(y)=i(-zh(z)-3z^2(1-h^2(z))+2z^3h(z)(1-h^2(z))),$$

where $z=z_0e^{iy}$ and

$$h(z)=\frac{chz+ushz}{shz+uchz}=1-\frac{se^{-2z}}{1+\frac{s}{2}(1+e^{-2z})}.$$

Thus

$$|1-h(z_0e^{iy})|=O(e^{-z_0}),$$

therefore

$$\max_{|y|\le\varepsilon}|f'''(y)|=O(\alpha).$$

From (15) it follows that $(z_0^2-\alpha^2)\alpha^{-1}=o(1)$ and so by (21)

$$I_\varepsilon=\frac{\sqrt{n}}{\sqrt{2\pi}}\int_{-\varepsilon}^{\varepsilon}e^{-\frac{ny^2}{2}(1+o(1)+O(y))}dy=1+o(1). \tag{22}$$

Now we consider the integral

$$I_\pi-I_\varepsilon=\frac{\sqrt{n}}{\sqrt{2\pi}}\int_{\varepsilon\le|y|\le\pi}e^{-iny}\left(\frac{sh(z_0e^{iy})+uch(z_0e^{iy})}{shz_0+uchz_0}\right)^N dy.$$

a) First suppose that $\alpha\ge\alpha_0>0$.

Introduce the following functions:

$$g(y)=|sh(z_0e^{iy})+uch(z_0e^{iy})|,\quad g_1(y)=e^{z_0\cos y}+|s|ch(z_0\cos y).$$

From the obvious inequalities

$$g(y)\le g_1(y),\quad g_1(\Pi-\varepsilon)\le g_1(\varepsilon)\le e^{z_0\cos\varepsilon}(1+|s|),\quad g(0)\ge e^{z_0}(1-|s|)$$

we have

$$\max_{\varepsilon\le|y|\le\Pi-\varepsilon}\left(\frac{g(y)}{g(0)}\right)^N\le e^{-Nz_0(1-\cos\varepsilon)}\left(\frac{1+|s|}{1-|s|}\right)^N=e^{-\beta n}, \tag{23}$$

where $\liminf\beta\ge0$.

If $\Pi-\varepsilon\le|y|\le\Pi$ then there exists a constant q $(0<q<1)$ such that

$$\frac{g_1(y)}{g(0)}\le q\,e^{-z_0(1+\cos y)}=q\,e^{-\alpha(y-\Pi)^2\beta_1} \tag{24}$$

$(\liminf\beta_1>0)$.

b) In the case $\alpha\to 0$ from (15) it follows that $z_0\to 0$ and that

$$\frac{g(y)}{g(0)}=e^{-z_0(1-\cos y)}\left|\frac{1+\frac{s}{2}(1+e^{-2z_0e^{iy}})}{1+\frac{s}{2}(1+e^{-2z_0})}\right|=e^{-z_0(1-\cos y)}(1+|s|z_0+O(z_0^2)).$$

Thus by (14), (15) we obtain

$$\max_{\varepsilon\le|y|\le\pi}\left(\frac{g(y)}{g(0)}\right)^N=O(e^{-Nz_0(1-\cos\varepsilon)+N|s|z_0})=e^{-\gamma n}, \tag{25}$$

where $\liminf\gamma>0$.

By (22)-(25) we conclude that

$$I_\pi=1+o(1). \tag{26}$$

Finally we can rewrite S in the following two forms:

$$S=N(z_0-\alpha)+(N-n)\log(1+\frac{s}{2}(1+e^{-2z_0}))+n\log(1+\frac{s}{2}(1-e^{-2z_0})), \tag{27}$$

$$S=N(z_0-\alpha)+(N-n)\log u+(N-n)\log(1-\frac{s}{2u}(1-e^{-2z_0}))+n\log(1+\frac{s}{2}(1-e^{-2z_0})). \tag{28}$$

Using the Taylor expansion from (12)-(15) it follows that

$$S=\frac{m_{n,N}t}{\sigma_{n,N}}+\frac{t^2}{2}+o(1). \tag{29}$$

(If $\alpha\ge\alpha_0>0$ then we use (27) and if $\alpha\to 0$ then we use (28).)

This completes to proof of Theorem 1.

4. PROOF OF THEOREMS 2 AND 3

The moment generating function of $(X_{n,N}-N+n)$ is

$$\tilde{M}_{n,N}(t)=\tilde{u}^{-N+n}F_{n,N}(\tilde{u}),\quad(\tilde{u}=e^t).$$

a) If $n,N\to\infty$, with $\alpha^2N\to\lambda$ ($\lambda\in[0,\infty)$) then setting $u=\tilde{u}$ in the proof of Theorem 1 we obtain from (16), (18) and (28)

$$\tilde{S}=-n+N\log(\mathrm{sh}z_0+\tilde{u}\mathrm{ch}z_0)+\tilde{u}\log\alpha-n\log z_0=(N-n)t+\frac{\lambda}{2}(e^{2t}-1)+o(1). \tag{30}$$

In this case, by Taylor expansion, using (18) we have

$$h(z_0)=\frac{\mathrm{sh}(z_0e^{iy})+\tilde{u}\mathrm{ch}(z_0e^{iy})}{\mathrm{sh}z_0+\tilde{u}\mathrm{ch}z_0}=1+\frac{z_0}{\tilde{u}}(e^{iy}-1)+O(z_0^2)=1+\alpha(e^{iy}-1)+O(\alpha^2).$$

Thus

$$|h(z_0)|^N=(1-\alpha(4\sin^2\frac{y}{2}+O(\alpha))^{\frac{N}{2}}=e^{-\delta n} \quad (\varepsilon\le|y|\le\pi),$$

where $\liminf\delta>0$.

From this it follows that

$$\tilde{I}_\pi-\tilde{I}_\varepsilon=\frac{\sqrt{n}}{\sqrt{2\pi}}\int_{\varepsilon\le|y|\le\pi} h(z_0)^N e^{-iny}dy=o(1). \tag{31}$$

Finally, the proof of (22) in this case also is valid, therefore

$$\tilde{M}_{n,N}(t)=\exp\{\frac{\lambda}{2}(e^{2t}-1)\}+o(1). \tag{32}$$

b) If n=const., $N\to\infty$, then substituting $z=z_0e^{iy}$ into (5) and using (18), we get

$$\tilde{M}_{n,N}(t)=\frac{n!}{2\pi n^n}\int_{-\pi}^{\pi}(\tilde{u}^{-1}\mathrm{sh}(z_0e^{iy})+\mathrm{ch}(z_0e^{iy}))^N e^{-iny}dy+o(1). \tag{33}$$

Since

$$\tilde{u}^{-1}\mathrm{sh}(z_0e^{iy})+\mathrm{ch}(z_0e^{iy})=1+\frac{ne^{iy}}{N}+O(\frac{1}{N^2}),$$

it follows from (33) that

$$\tilde{M}_{n,N}(t)=\frac{n!e^n}{2\pi n^n}\int_{-\pi}^{\pi}e^{n(e^{iy}-1)-niy}dy+o(1)= +o(1). \tag{34}$$

(Note that the characteristic function of the Poisson distribution with parameter n is $e^{n(e^{iy}-1)}$.)

The formulas (32) and (34) give the proof of Theorems 2,3.

Remark. Suppose that n balls are randomly placed in N urns, each urn having the same probability $\frac{1}{N}$ of assignment. Let $Y^{(r)}_{n,N}$ be the number of urns containing exactly r balls. Békéssy (see [3]) has obtained results on limiting distribution of $Y^{(r)}_{n,N}$. We have used the ideas of that paper. Denoting the number of urns containing an even number of balls by $Z_{n,N}$, it is easy to see that $X_{n,N}$ and $Z_{n,N}$ have the same distribution.

REFERENCES

1 W. FELLER: An introduction to the Probability Theory and its Applications, Vol. I. Wiley, New York 1957.

2 I.H. CURTISS: A note on the theory of moment generating functions, Ann. Math. Statist., 1942,13,3, pp.430-433.

3 A. BÉKÉSSY: On classical occupancy problems, Publ. Math. Inst. Hung. Acad. Sci., 1963,8,1-2, pp.59-71.

*Proc. of the 3rd Pannonian Symp.
on Math. Stat., Visegrád, Hungary 1982
J. Mogyoródi, I. Vincze, W. Wertz, eds*

A METHOD FOR SINGULAR VALUES DECOMPOSITION OF GENERAL REAL MATRICES

M. BOLLA, G. TUSNÁDY

Computer and Automation Institute,
Hungarian Academy of Sciences, Budapest,
Mathematical Institute,
Hungarian Academy of Sciences, Budapest, Hungary

INTRODUCTION

The singular values decomposition plays an important role in multivariate statistical analysis, mainly in linear approximation problems. The separation theorems for the singular values and the extremal value properties of them make it possible to determine the canonical correlation coefficients as the singular values of a matrix calculated from the covariance matrices of the random variables in question and to approximate normally distributed variables with special linear combinations of other variables. Some problems of this kind are discussed by Rao in [7].

Our purpose is to introduce a method for the singular values decomposition of general real matrices which is the generalization of the QR algorithm developed by Francis [3] for the spectral decomposition of square matrices. It is called QRPS algorithm and an improved version of it is discussed too. See [2].

Visually the algorithm means the subsequent orthogonalization of the orthonormal bases of the Euclidean spaces between which the linear transformation corresponding to the given matrix is mapping and leads to the eigenbasis pair of the linear transformation in question. So it can be regarded as a generalization of the power iteration. The orthogonalizations are performed by plain rotations and for bidiagonal matrices the subsequent elementary rotations are the same as those introduced by Golub and Reinsch in [4].

1. THE QRPS ALGORITHM

Let A be an m×n real matrix with singular value decomposition

$$A = V \Lambda U^T$$

where V is an m×m, U is an n×n orthogonal matrix and Λ is an m×n matrix with zeros outside of its main diagonal. The diagonal elements of Λ in non-increasing order are the singular values of A (they are unique), while the columns of V and U constitute the eigenbases of the linear transformations corresponding to A and A^T (within the so-called isotropic subspaces they are unique too).

Let us form the following series: $A_0 = A$ and if

$$A_{k-1} = Q_{k-1} R_{k-1} \quad \text{and} \quad A_{k-1}^T = P_{k-1} S_{k-1} \qquad (1)$$

are orthogonal-upper triangular decompositions, then let us take

$$A_k = R_{k-1} P_{k-1}, \text{ hence } A_k^T = S_{k-1} Q_{k-1} \ (k=1,2,\ldots). \qquad (2)$$

Let us notice that because of $A_k = Q_{k-1}^T A_{k-1} P_{k-1}$ (where Q_{k-1} and P_{k-1} are orthogonal matrices) the singular values of the subsequent elements of the series are the same.

With the notations

$$V_k = Q_0 Q_1 \ldots Q_k, \quad U_k = P_0 P_1 \ldots P_k \qquad (k=0,1,2,\ldots)$$

A_k (k=1,2,...) can be written as

$$A_k = Q_{k-1}^T A_{k-1} P_{k-1} = Q_{k-1}^T Q_{k-2}^T A_{k-2} P_{k-2} P_{k-1} =$$

$$= \ldots = V_{k-1}^T \Lambda U_{k-1} \qquad (k=0,1,2,\ldots).$$

The convergence of the series (A_k) to Λ follows from the

convergence of the series (V_k) and (U_k) to V and U, respectively.

Now it will be shown that the transformations

$$A_{2k} A_{2k}^T \rightarrow A_{2k+2} A_{2k+2}^T \qquad (k=0,1,2,\ldots)$$

and

$$A_{2k+1} A_{2k+1}^T \rightarrow A_{2k+3} A_{2k+3}^T \qquad (k=0,1,2,\ldots)$$

are QR transformations. That is, after the kth step $(k=0,1,2,\ldots)$

$$A_{2k} A_{2k}^T = Q_{2k} R_{2k} P_{2k} S_{2k} = Q_{2k} A_{2k+1} S_{2k} =$$
$$= (Q_{2k} Q_{2k+1})(R_{2k+1} S_{2k})$$

where the matrix $Q_{2k} Q_{2k+1}$ is orthogonal, while $R_{2k+1} S_{2k}$ is upper triangular. So they are the orthogonal and upper triangular factors of the orthogonal-upper triangular decomposition of $A_{2k} A_{2k}^T$. Multiplying them in reversed order:

$$R_{2k+1} S_{2k} Q_{2k} Q_{2k+1} = R_{2k+1} A_{2k+1}^T Q_{2k+1} =$$

$$= R_{2k+1} P_{2k+1} S_{2k+1} Q_{2k+1} = A_{2k+2} A_{2k+2}^T$$

which is just the matrix obtained by the QR transformation of $A_{2k} A_{2k}^T$.

Similarly, the orthogonal-upper triangular decomposition of $A_{2k+1} A_{2k+1}^T$ is

$$(Q_{2k+1} Q_{2k+2})(R_{2k+2} S_{2k+1}) \qquad (k=0,1,2,\ldots) .$$

Multiplying the matrices in brackets in reversed order $A_{2k+3} A_{2k+3}^T$ is obtained.

According to the convergence of the QR algorithm [8] applied for the matrix $A A^T$:

$$\lim_{k\to\infty} V_{2k+1} = \lim_{k\to\infty} (Q_0 Q_1)(Q_2 Q_3) \ldots (Q_{2k} Q_{2k+1}) = V.$$

Similarly, from the convergence of the QR algorithm applied for $A_1 A_1^T$:

$$\lim_{k\to\infty} (Q_1Q_2)(Q_3Q_4)\ldots(Q_{2k+1}\,Q_{2k+2}) = Q_o^T\,V.$$

As $A_1A_1^T = (Q_o^T\,V)\Lambda^2(Q_o^T\,V)^T$ is spectral decomposition, $\lim_{k\to\infty} V_{2k}=V$ follows. Hence $\lim_{k\to\infty} V_k = V$. The statement $\lim_{k\to\infty} U_k = U$ is proved in the same way, since the transformations

$$A_{2k}^T\,A_{2k} \to A_{2k+2}^T\,A_{2k+2} \qquad (k=0,1,2,\ldots)$$

and

$$A_{2k+1}^T\,A_{2k+1} \to A_{2k+3}^T\,A_{2k+3} \qquad (k=0,1,2,\ldots)$$

are also QR transformations.

As $A = V\,\Lambda\,U^T$ holds, $\lim_{k\to\infty} A_k = \Lambda$ follows from $\lim_{k\to\infty} V_k = V$ and $\lim_{k\to\infty} U_k = U$.

The procedure defined by (1), (2) is called QRPS algorithm. Let us suppose that the orthogonal-upper triangular decompositions are performed in the way, that orthogonalizing the columns of A we take the first non-zero column, and we always choose the first column which is linearly independent from the preceding ones. The remaining columns of the orthogonal factor of A constitute an orthonormal system in $\mathcal{R}(A)^{\perp}$ (the orthogonal complementary subspace of the range of the linear transformation corresponding to A in the m-dimensional Euclidean space). With these constraints the order of the limit values of the diagonal elements in A_k reflects the order of the singular values of A, that is the zero singular values are in the last positions and if the rank of A is $r<\min\{m,n\}$, then the last m-r and n-r columns of V and U can be rotated orthogonally in the subspace $\mathcal{R}(A)^{\perp}$ and $\mathcal{R}(A^T)^{\perp}$ respectively. But these elements are unique only within these isotropic subspaces.

The QRPS algorithm has a geometrical meaning too: Let us

consider the m×n matrix A as the matrix of a linear transformation mapping the n-dimensional Euclidean space into the m-dimensional one in the bases $\underline{e}_1,\ldots,\underline{e}_n \in E_n$ and $\underline{f}_1,\ldots,\underline{f}_m \in E_m$ respectively (where E_n is the n-dimensional Euclidean space). Since the orthogonal matrices P_0 and Q_0 orthogonalize the system $A\underline{e}_1,\ldots,A\underline{e}_n$ and $A^T\underline{f}_1,\ldots,A^T\underline{f}_m$ in E_m and E_n respectively, $Q_0^T A P_0$ is the matrix of the linear transformation in question in the new bases (that is the bases $\underline{e}_1,\ldots,\underline{e}_n$ and $\underline{f}_1,\ldots,\underline{f}_m$ transformed by P_0 and Q_0 respectively).

During the iteration, since $\lim_{k\to\infty} Q_0\ldots Q_k = V$ and $\lim_{k\to\infty} P_0\ldots P_k = U$ holds, the subsequent orthogonalization of the original bases in E_n and E_m leads to the eigenbases in E_n and E_m respectively, while the matrix of the transformation becomes of diagonal form. This is why the QRPS algorithm can be considered as a natural generalization of the power iteration.

2. AN IMPROVED VERSION OF THE QRPS ALGORITHM

An improved version of the QRPS algorithm is obtained if the matrix of the linear transformation corresponding to A is considered immediately in the new basis of E_m, as soon as this is constructed. In the following step the columns of the transpose of this new matrix are orthogonalized and the linear transformation in question is regarded in this new basis of E_n. That is with formulas: $A_0 = A$ and if

$$A_{k-1} = Q_{k-1} R_{k-1} \quad \text{and} \quad R_{k-1}^T = P_{k-1} S_{k-1} \tag{3}$$

are orthogonal-upper triangular decompositions, then

$$A_k = S_{k-1}^T = R_{k-1} P_{k-1} \qquad (k=1,2,\ldots) \quad . \tag{4}$$

Hence A_k (k=1,2,...) is of lower triangular form. Since

$$A_k = R_{k-1} P_{k-1} = Q_{k-1}^T A_{k-1} P_{k-1}$$

where Q_{k-1} and P_{k-1} are orthogonal matrices, the singular values of the series members are the same. Now the transformations

$$A_k A_k^T \to A_{k+1} A_{k+1}^T \qquad (k=1,2,\dots)$$

are QR transformations. That is because of

$$A_k A_k^T = Q_k R_k S_{k-1} \qquad (k=1,2,\dots)$$

the orthogonal factor of the orthogonal-upper triangular decomposition of $A_k A_k^T$ is Q_k, while the upper triangular factor is $R_k S_{k-1}$. Multiplying these two factors in reversed order:

$$R_k S_{k-1} Q_k = R_k P_k P_k^T A_k^T Q_k = A_{k+1} A_{k+1}^T \quad .$$

It is easy to see, that this version is about two times quicker than the QRPS algorithm introduced in paragraph 1, because here in every step (excluding the first one) the product of the series members with their transpose gives the QR algorithm applied for the matrix $A_1 A_1^T$. In case of the original QRPS algorithm this was true only for every second step.

Let us again introduce the following notations:

$$V_k = Q_0 Q_1 \dots Q_k, \; U_k = P_0 P_1 \dots P_k \qquad (k=1,2,\dots).$$

Since the series $(A_k A_k^T)$, $(k=1,2,\dots)$ gives the usual QR algorithm applied for the matrix $A_1 A_1^T$ and since

$$A_1 A_1^T = (Q_0^T V) \Lambda^2 (Q_0^T V)^T$$

is spectral decomposition, on the basis of the convergence of the QR algorithm:

$$\lim_{k\to\infty} A_k A_k^T = \Lambda^2 \quad \text{and} \quad \lim_{k\to\infty} Q_1 \dots Q_k = Q_0^T V \; . \qquad (5)$$

From the second relationship:

$$\lim_{k\to\infty} V_k = \lim_{k\to\infty} Q_0 Q_1 \dots Q_k = V.$$

If during the algorithm the unitary-upper triangular decompositions are performed in the way discussed in paragraph 1, then in the matrix Λ^2 only the last diagonal elements can be zeros, in case when the rank of A is $r<\min\{m,n\}$. With this restriction $\lim_{k\to\infty} A_k = \Lambda$ follows from the first part of (5). In this way the order of the limit values of the diagonal elements in A_k reflects the order of the singular values of A, as in case of the QRPS algorithm.

We have to notice that the orthogonalizations in the subsequent orthogonal-upper triangular decompositions are performed by a chain of elementary rotations in E_m and E_n, respectively. On this basis an ALGOL program for our computer CDC 3300 was developed too.

3. SOME STATISTICAL APPLICATIONS OF THE SINGULAR VALUES DECOMPOSITION

Let $\underline{\xi}$ and $\underline{\eta}$ be p- and q-dimensional random variables ($p\leq q$ can be supposed) with covariance matrices Σ_{11}, Σ_{22} and with cross-covariance matrix Σ_{12}. The singular values of the matrix

$$D = \Sigma_{11}^{-1/2}\,\Sigma_{12}\,\Sigma_{22}^{-1/2}$$

give the canonical correlation coefficients between the two variables. If the singular values decomposition of D is $V\Lambda U^T$ then $X = \Sigma_{11}^{-1/2}V$ and $Y = \Sigma_{22}^{-1/2}U$ contain the corresponding canonical vectors in their first p columns.

Let us suppose that the expectation of $\underline{\xi}$ and $\underline{\eta}$ are zero. Now the components of $\underline{\xi}$ are observable variables, while the components of $\underline{\eta}$ are either not observable or their observation is very expensive. Instead of this we have a k-dimensional channel for some fixed integer $k<p$, through which the linear combinations of the components of $\underline{\xi}$ can be transmitted. Let us denote these linear combinations by $\underline{\rho} = A^T \underline{\xi}$, where A is a $p\times k$ matrix.

Our purpose is to choose these linear combinations so that the determinant of

$$E\,(\underline{\eta} - L_{\underline{\rho}}\,\underline{\rho})\,(\underline{\eta} - L_{\underline{\rho}}\,\underline{\rho})^T \qquad (6)$$

would be minimal, where $L_{\underline{\rho}}$ is a $q\times k$ matrix giving the best linear approximation of $\underline{\eta}$ with $\underline{\rho}$.

V. J. Yohai and M. S. Garcia Ben [9] have discussed the following problem: the $p\times k$ matrix A^* containing the first k columns of X gives a particular solution of problem (6). Any other solution can be given as $A = A^*F$, where F is an arbitrary $k\times k$ orthogonal matrix. The minimum is

$$|\Sigma_{22}| \prod_{i=1}^{k} (1-\rho_i^2)$$

where ρ_i $(i=1,\dots,p)$ denote the canonical correlations.

If the covariance matrix of $\underline{\eta}$ is the identity then this statement can be strengthened:

$$\lambda_i\{E(\underline{\eta} - L_{\underline{\rho}}\,\underline{\rho})\,(\eta - L_{\underline{\rho}}\,\underline{\rho})^T\} \geq 1 - \rho_i^2 \qquad (i=1,\dots,k),$$

where the function $\lambda_i\{\cdot\}$ gives the ith eigenvalue of the matrix in the argument in descending order.

Now let $\underline{\xi}$ and $\underline{\eta}$ be both p-dimensional normally distributed random variables with covariance matrices Σ_{11} and Σ_{22}. We are looking for their joint distribution, i.e. the cross-covariance matrix Σ_{12}, so that

$$E||\underline{\xi} - \underline{\eta}||^2$$

would be minimal.

Let us take the decompositions

$$\Sigma_{11} = A\,A^T, \quad \Sigma_{22} = B\,B^T$$

where A and B are pxp matrices (since Σ_{11} and Σ_{22} are symmetrical, positive semidefinite matrices, these decompositions exist but they are not unique). Then there is a p-dimensional standard normally distributed random variable $\underline{\rho}$, so that

$$\underline{\xi} = A\underline{\rho} \quad , \quad \underline{\eta} = B\underline{\rho}$$

holds, hence $\Sigma_{12} = A\,B^T$. If

$$\Sigma_{11} = U\,\Lambda^2\,U^T \quad , \quad \Sigma_{22} = V\,S^2\,V^T$$

are spectral decompositions then A and B can be given as

$$A = U\,\Lambda\,P^T \quad , \quad B = V\,S\,Q^T \quad ,$$

where the orthogonal matrices P and Q have to be determined, so that

$$\begin{aligned} E||\underline{\xi} - \underline{\eta}||^2 &= \mathrm{tr}\{(A - B)(A - B)^T\} \\ &= \mathrm{tr}(A\,A^T) + \mathrm{tr}(B\,B^T) - 2\,\mathrm{tr}(A\,B^T) \end{aligned} \qquad (7)$$

be minimal. As $\mathrm{tr}(A\,A^T) = \mathrm{tr}\,\Sigma_{11}$ and $\mathrm{tr}(B\,B^T) = \mathrm{tr}\,\Sigma_{22}$ is given, the minimum of (7) is attained if $\mathrm{tr}(A\,B^T)$ is maximal. But $\mathrm{tr}(A\,B^T)$ is maximal if and only if Σ_{12} is symmetrical. Let us denote the singular values decomposition of $S\,V^T\,U\Lambda$ by $R_1 \Delta R^T$. Then, since

$$\mathrm{tr}(A\,B^T) = \mathrm{tr}(U\Lambda P^T\,Q\,S\,V^T) = \mathrm{tr}(S\,V^T\,U\,\Lambda\,P^T\,Q) \ ,$$

the last matrix in brackets is symmetrical - as $P^T\,Q$ is orthogonal - if and only if $Q^T\,P = R_1 R_2^T$. That is

$$\Sigma_{12} = A\,B^T = U\,\Lambda\,P^T\,Q\,S^T\,V = U\,\Lambda\,R_2 R_1^T\,S^T\,V \ ,$$

where R_1 and R_2 can be determined by the singular values decomposition of $S\,V^T\,U\Lambda$, which in turn is determined by the spectral decomposition of Σ_{11} and Σ_{22}.

REFERENCES

1 AMIR-MOÉZ, A. R. and HORN, A.: "Singular values of a matrix", Amer. Math. Monthly 65 (1958) 742-748.

2 BOLLA, M.: "The generalization of the QR algorithm for the singular values decomposition of real rectangular matrices", Alk. Mat. Lapok (1982), in Hungarian.

3 FRANCIS, J.: "The QR transformation. A unitary analogue to the LR transformation I., II.". Comput. J. 4 (1961-62) 265-271 and 332-345.

4 GOLUB, G. H. and REINSCH, G.: "Singular values decomposition and least squares solution". Numer. Math. 14 (1970), 403-420.

5 HORN, A.: "On the eigenvalues of a matrix with prescribed singular values". Proc. Amer. Math. Soc. 5 (1954) 4-7.

6 LANCZOS, C.: "Linear systems in self-adjoint form". Amer. Math. Monthly 65 (1958) 665-679.

7 RAO, C.R.: "Separation theorems for singular values of matrices and their applications in multivariate analysis". Journal of Multivariate Analysis 9 (1979) 362-377.

8 WILKINSON, J.H.: "The algebraic eigenvalue problem". Clarendon Press, Oxford (1965).

9 YOHAI V.J. and GARCIA BEN, M.S.: "Canonical variables as optimal predictors". The Annals of Stat. 8 (1980) 865-869.

Proc. of the 3rd Pannonian Symp. on Math. Stat., Visegrád, Hungary 1982
J. Mogyoródi, I. Vincze, W. Wertz, eds

ON SMALL VALUES OF THE SQUARE INTEGRAL OF A MULTIPARAMETER WIENER PROCESS

E. CSÁKI

Mathematical Institute,
Hungarian Academy of Sciences, Budapest, Hungary

1. Introduction

Let $W(t_1,\dots,t_d)$ $t_1\geq 0,\dots,t_d\geq 0$ be a d-parameter Wiener process, i.e. a continuous Gaussian process with

$$EW(t_1,\dots,t_d)=0 \tag{1.1}$$

and

$$EW(t_1,\dots,t_d)W(s_1,\dots,s_d)=\min(t_1,s_1)\dots\min(t_d,s_d). \tag{1.2}$$

A d-parameter Brownian Bridge $B(t_1,\dots,t_d)$ is defined by the following relation:

$$B(t_1,\dots,t_d)=W(t_1,\dots,t_d)-t_1\dots t_d W(1,\dots,1). \tag{1.3}$$

Put

$$I_d=\int_0^1\dots\int_0^1 W^2(t_1,\dots,t_d)dt_1,\dots,dt_d \tag{1.4}$$

and

$$I_d^o=\int_0^1\dots\int_0^1 B^2(t_1,\dots,t_d)dt_1,\dots,dt_d. \tag{1.5}$$

The characteristic functions of I_d and I_d^o have been determined by DUGUE [4], [5]:

Let

$$\psi_d(s)=E(e^{isI_d}), \tag{1.6}$$

then

$$\psi_d(s)=\prod_{n=1}^{\infty}\psi_{d-1}(s/((n-\tfrac{1}{2})^2\pi^2)) \tag{1.7}$$

with

$$\psi_1(s)=(\cos(2is)^{1/2})^{-1/2}, \tag{1.8}$$

or alternatively,

$$E(e^{-sI_d})=\prod_{j_1=1}^{\infty}\cdots\prod_{j_d=1}^{\infty}\left(1+\frac{2s}{(j_1-\frac{1}{2})^2\cdots(j_d-\frac{1}{2})^2\pi^{2d}}\right)^{-1/2}. \tag{1.9}$$

Furthermore, if

$$\psi_d^o(s)=E(e^{isI_d^o}), \tag{1.10}$$

then

$$\psi_d^o(s)=(i2^{d-1}\frac{d}{ds}C_d(s))^{-1/2}, \tag{1.11}$$

where

$$C_d(s)=\prod_{n=1}^{\infty}C_{d-1}(s/((n-\tfrac{1}{2})^2\pi^2)) \tag{1.12}$$

with

$$C_1(s)=\cos((2is)^{1/2}). \tag{1.13}$$

DURBIN [6] (for d=2), COTTERILL and CSÖRGŐ [2] (for d≥2) also give the following expression:

$$\psi_d^o(s)=\psi_d(s)\,\{i\,2^{d-1}\frac{d}{ds}\log C_d(s)\}^{-1/2}. \tag{1.14}$$

Our concern in this paper is to investigate the lower tail of the distributions of I_d and I_d^o.

From CAMERON and MARTIN [1] it follows that

$$\log P(I_1<u)\sim-\frac{1}{8u} \quad \text{as} \quad u\to 0 \tag{1.15}$$

and SYTAJA [9] showed

$$\log P(I_1^o<u)\sim-\frac{1}{8u} \quad \text{as} \quad u\to 0. \tag{1.16}$$

We generalize these results as follows.

THEOREM 1.

$$\log P(I_d<u) \sim \log P(I_d^o<u) \sim \frac{K_d^2}{u}(\log u)^{2d-2} \quad \text{as} \quad u\to 0 \tag{1.17}$$

where

$$K_d=\frac{1}{2\sqrt{2}\ \pi^{d-1}(d-1)!} \ . \tag{1.18}$$

For the maximum in the case $d=2$ RÉVÉSZ [8] proved that there are constants C_1, C_2, C_3, C_4 such that the inequality

$$C_1 \exp\{ -\frac{C_2}{u}|\log u|^{\alpha_1}\} \le$$

$$\le P(\max_{\substack{0\le t_1\le 1\\ 0\le t_2\le 1}} (W(t_1,t_2))^2<u) \le \tag{1.19}$$

$$\le C_3\{\exp -\frac{C_4}{u}|\log u|^{\alpha_2}\}$$

holds with $\alpha_1=5$, $\alpha_2=1$. It follows from Theorem 1 that we can put $\alpha_2=2$ and it is a plausible conjecture that the lower tail of the distribution of $\max|W(t_1,t_2)|^2$ has an asymptotic expression similar to (1.17).

As a corollary to Theorem 1, the following iterated logarithm law can be proved by applying the usual method.

THEOREM 2. Let

$$I_d(T)=\int_0^T\ldots\int_0^T W^2(t_1,\ldots,t_d)dt_1\ldots dt_d \ . \tag{1.20}$$

Then we have

$$\liminf_T \frac{T^{-2d}(\log\log T)I_d(T)}{(\log\log\log T)^{2d-2}}=K_d^2 \quad \text{a.s.} \tag{1.21}$$

The case $d=1$ is due to DONSKER and VARADHAN [3].

2. PROOF OF THEOREM 1.

If follows from (1.9) or from the Karhunen - Loève expansion (see DURBIN [6] for d=2) that the following representation holds:

$$I_d \overset{D}{=} \sum_{j_1=1}^{\infty} \cdots \sum_{j_d=1}^{\infty} \frac{X^2_{j_1 \ldots j_d}}{(j_1-\frac{1}{2})^2 \ldots (j_d-\frac{1}{2})^2 \pi^{2d}}, \tag{2.1}$$

where $X_{j_1 \ldots j_d}$ $(j_1=1,2,\ldots,\ \ldots,\ j_d=1,2,\ldots)$ are i.i.d. standard normal random variables and $\overset{D}{=}$ means equality in distribution. The proof of our results will be based on the following result of SYTAJA [9]:

$$P\Big(\sum_{k=1}^{\infty} \lambda_k X_k^2 < u\Big) \sim \frac{\exp\{uv-\frac{1}{2}\sum_{k=1}^{\infty} \log(1+2\lambda_k v)\}}{2\sqrt{\pi \sum_{k=1}^{\infty} \Big(\frac{\lambda_k v}{1+2\lambda_k v}\Big)^2}}, \quad (u\to 0) \tag{2.2}$$

where $X_1, X_2, \ldots$ is an i.i.d. sequence of standard normal random variables and v is the solution of the equation

$$u=\frac{1}{v}+\sum_{k=1}^{\infty} \frac{\lambda_k}{1+2\lambda_k v}. \tag{2.3}$$

Now we prove the following relations for $d\geq 2$ and $z\to\infty$

$$A(z)=\sum_{j_1=1}^{\infty} \cdots \sum_{j_d=1}^{\infty} \frac{1}{(j_1-\frac{1}{2})^2 \ldots (j_d-\frac{1}{2})^2 \pi^{2d}+z^2} = \frac{1}{2z\pi^{d-1}(d-1)!}\big((\log z)^{d-1} + O((\log z)^{d-2})\big) \tag{2.4}$$

and

$$B(z)=\sum_{j_1=1}^{\infty} \cdots \sum_{j_d=1}^{\infty} \log\Big(1+\frac{z^2}{(j_1-\frac{1}{2})^2 \ldots (j_d-\frac{1}{2})^2 \pi^{2d}}\Big) = \frac{z}{\pi^{d-1}(d-1)!}\big((\log z)^{d-1}+O((\log z)^{d-2})\big). \tag{2.5}$$

We use the following identities (see GRADSHTEYN and RYZHIK [7])

$$\sum_{k=1}^{\infty} \frac{1}{(2k-1)^2+x^2}=\frac{\pi}{4x} \text{ th } \frac{\pi x}{2} \tag{2.6}$$

and

$$\prod_{k=1}^{\infty} \left(1+\frac{x^2}{(2k-1)^2}\right)=\cosh \frac{\pi x}{2} \ . \tag{2.7}$$

Hence we get

$$A(z)=\frac{1}{2z\pi^{d-1}} \sum_{j_1=1}^{\infty} \dots \sum_{j_{d-1}=1}^{\infty} \frac{1}{(j_1-\frac{1}{2})\dots(j_{d-1}-\frac{1}{2})} \text{th}\left(\frac{z}{(j_1-\frac{1}{2})\dots(j_{d-1}-\frac{1}{2})\pi^{d-1}}\right)=$$

$$=\frac{1}{2z\,\pi^{d-1}}(\Sigma_1+\Sigma_2),$$

where the first sum runs over $(j_1,\dots,j_{d-1})$ such that $j_1\dots j_{d-1}\le z$, while Σ_2 contains the rest. By using the inequality

$$1 - e^{-x} \le \text{th } x \le 1 \tag{2.8}$$

for estimating Σ_1 and the inequality

$$0 \le \text{th } x \le x \tag{2.9}$$

for estimating Σ_2, we have

$$\Sigma_1 \le \sum_{j_1\dots j_{d-1}\le z} \dots \sum \frac{1}{(j_1-\frac{1}{2})\dots(j_{d-1}-\frac{1}{2})} = \frac{(\log z)^{d-1}}{(d-1)!} + 0((\log z)^{d-2}), \tag{2.10}$$

which can be easily seen by induction.

$$\Sigma_1 \ge \sum_{j_1\dots j_{d-1}\le z} \dots \sum \frac{1-e^{-(z/((j_1-\frac{1}{2})\dots(j_{d-1}-\frac{1}{2})\pi^{d-1}))}}{(j_1-\frac{1}{2})\dots(j_{d-1}-\frac{1}{2})} = \frac{(\log z)^{d-1}}{(d-1)!}+0((\log z)^{d-2}). \tag{2.11}$$

Furthermore it can be seen that

$$0\le\Sigma_2\le \sum_{j_1 j_2 \ldots j_{d-1}>z}\ldots\sum \frac{z}{(j_1-\frac{1}{2})^2\ldots(j_d-\frac{1}{2})^2\pi^{d-1}} = 0(1). \qquad (2.12)$$

Hence from (2.10), (2.11) and (2.12) we get (2.4).

Similarly, by using (2.7) and the inequalities

$$\frac{e^x}{2} \le \cosh x \le e^x \qquad (x>0) \qquad (2.13)$$

and

$$1 \le \cosh x \le e^{x^2} \qquad (x>0) \qquad (2.14)$$

we obtain

$$B(z)= \sum_{j_1=1}^{\infty} \ldots \sum_{j_{d-1}=1}^{\infty} \log\cosh\left(\frac{z}{(j_1-\frac{1}{2})\ldots(j_{d-1}-\frac{1}{2})\pi^{d-1}}\right)\le$$

$$\le \sum_{j_1\ldots j_{d-1}\le z} \sum \frac{z}{(j_1-\frac{1}{2})\ldots(j_{d-1}-\frac{1}{2})\pi^{d-1}} + \qquad (2.15)$$

$$+ \sum_{j_1\ldots j_{d-1}>z}\ldots\sum \frac{z}{(j_1-\frac{1}{2})^2\ldots(j_{d-1}-\frac{1}{2})^2\pi^{2d-2}} =$$

$$= z\left(\frac{(\log z)^{d-1}}{(d-1)!}+0((\log z)^{d-2})\right)$$

and

$$B(z)\ge \sum_{j_1\ldots j_{d-1}\le z}\ldots\sum \left(\frac{z}{(j_1-\frac{1}{2})\ldots(j_{d-1}-\frac{1}{2})\pi^{d-1}}-\log 2\right)=$$

$$=\frac{z}{\pi^{d-1}}\left(\frac{(\log z)^{d-1}}{(d-1)!}+0((\log z)^{d-2})\right). \qquad (2.16)$$

By putting $z=\sqrt{2v}$ in (2.4), from equation (2.3) we have

$$u=\frac{1}{v}+A(\sqrt{2v})=\frac{K_d}{\sqrt{v}}((\log \sqrt{v})^{d-1}+0((\log v)^{d-2})), \qquad (2.17)$$

i.e.

$$v=\frac{K_d^2}{u^2}((\log \sqrt{v})^{d-1}+0((\log v)^{d-2}))^2 \qquad (2.18)$$

and the exponent in (2.2) equals

$$uv-\frac{1}{2}B(\sqrt{2v})=-\frac{K_d^2}{u}((\log \sqrt{v})^{d-1}+0(\log v)^{d-2})^2 . \qquad (2.19)$$

Since from (2.17), $\log \sqrt{v} \sim -\log u$, and as easily seen, the denominator in (2.2) gives only $0(\log v)$ in the exponent, our result follows for I_d. Concerning I_d^o we note that from (1.14) we have

$$E(e^{-vI_d^o}) = E(e^{-vI_d})(2^d A \sqrt{2v}))^{-1/2} . \qquad (2.20)$$

By (2.4), as $v\to\infty$

$$\log E(e^{-vI_d^o}) = \log E(e^{-vI_d})+ 0(\log v) . \qquad (2.21)$$

Following the proof of (2.2) given in SYTAJA [9] one can see that the term $0(\log v)$ does not affect the asymptotic expression in the exponent of (2.2), hence we have the same result for I_d^o as for I_d. This completes the proof of the theorem.

REFERENCES

[1] CAMERON, R. H. and MARTIN, W. T., The Wiener measure of Hilbert neighbourhood in the space of real continuous functions, J. Math. Phys. 23 (1944), 195-209.

[2] COTTERILL, D. S. and CSÖRGŐ, M., On the limiting distribution of and critical values for the multivariate Cramér-von Mises statistic, Ann. Statistics 10 (1982), 233-244.

[3] DONSKER, M. D. and VARADHAN, S. R. S., On laws of the iterated logarithm for local times, Comm. Pure Appl. Math. 30 (1977), 707-753.

[4] DUGUE, D., Fonctions caractéristique d'integrales Browniennes, Revue Roumaine Math. Pures Appl. 12 (1967), 1207-1215.

[5] DUGUE, D., Characteristic functions of random variables connected with Brownian motion and of the von Mises multidimensional ω_n^2, Multivariate Analysis vol. 2 (P. R. Krishnaiah, ed.) 289-301. Academic Press, New York, 1969.

[6] DURBIN, J. Asymptotic distributions of some statistics based on the bivariate sample distribution function, Nonparametric Techniques in Statistical Inference (M. L. Puri, ed.) 435-451. Cambridge Univ. Press, 1970.

[7] GRADSHTEYN, I. S. and RYZHIK, I. M. Table of Integrals, Series and Products. 4th ed. Academic Press, New York, 1980.

[8] RÉVÉSZ, P., How small are the increments of a Wiener sheet? The First Pannonian Symposium on Mathematical Statistics (Bad Tatzmannsdorf, 1979), 207-219, Lecture Notes in Statist. 8, Springer, New York, 1981.

[9] SYTAJA, G. N., Certain asymptotic representations for a Gaussian measure in Hilbert space (Russian), Theory of random processes, No. 2 (Russian), 93-104, 140. Izdat. "Naukova Dumka" Kiev, 1974.

Proc. of the 3rd Pannonian Symp. on Math. Stat., Visegrád, Hungary 1982
J. Mogyoródi, I. Vincze, W. Wertz, eds

ON THE ESTIMATION OF MIXTURE PROPORTIONS*

S. CSÖRGŐ[1], H.-D. KELLER[2]

[1]Bolyai Institute, Szeged, Hungary
[2]Fachbereich IV, Angewandte Mathematik, Universität Trier, FRG

ABSTRACT
Simple estimates for the mixture proportions are proposed in a mixture of known distributions. The estimators are based on the empirical characteristic function. We derive results on consistency, rate of consistency and on asymptotic normality for these estimators.

1. INTRODUCTION.

Let $X_1,\dots,X_n$ be independent identically distributed random variables with common distribution function

$$F(x)=F(x;p_1^o,\dots,p_m^o)=\sum_{k=1}^{m+1} p_k^o F_k(x)$$

$$=F_{m+1}(x)+\sum_{k=1}^{m} p_k^o\{F_k(x)-F_{m+1}(x)\},\quad x\in\mathbb{R}, \tag{1.1}$$

where m is a fixed natural number and the component distribution functions $F_1,\dots,F_{m+1}$ are assumed to be known, and the problem is to estimate the unknown proportions $p_1^o,\dots,p_{m+1}^o$, belonging to the set

$$\mathcal{P}=\{(p_1,\dots,p_{m+1}):\ p_1,\dots,p_{m+1}\geq 0,\ \sum_{k=1}^{m+1} p_k=1\}, \tag{1.2}$$

---

*A Pannonian product: This work was done whilst the first named author was a Gastdozent of Universität Trier, Fachbereich IV, Angewandte Mathematik. He is grateful to Professor Wolfgang Sendler for the invitation and support. Both authors are grateful to Elke Henning for her prompt typing of the manuscript.

in the mixture. We assume that the model (1.1) is identifiable (TEICHER [18]), i.e., there is a unique vector $(p_1^o,\dots,p_{m+1}^o)$ to be estimated. The literature on the subject is large. Starting out, for example, from CHOI and BULGREN [1] or from the most recent significant paper by HALL [9], one can trace it back to K. PEARSON [14] who applied his method of moments for the estimation of a mixture of two normal distributions. Our estimators in Section 2 below are in certain correspondence with his moment estimators. Difficulties with maximum likelihood estimation were pointed out by many authors. Most of the work has been based either on the empirical distribution function $F_n(x)$, $x \in \mathbb{R}$, of $X_1,\dots,X_n$, and on its moments, or on density estimators. Because of space limitation we shall not comment the earlier development in more detail.

Let $C(t)$ and $C_1(t),\dots,C_{m+1}(t)$, $t \in \mathbb{R}$, be the respective characteristic functions belonging to F and $F_1,\dots,F_{m+1}$. Then the model obtained by substituting the general vector $(p_1,\dots,p_{m+1})$ for $(p_1^o,\dots,p_{m+1}^o)$ in (1.1), understanding p_{m+1} always as in (2.1) below, is equivalent to

$$C(t;p_1,\dots,p_m)=\sum_{k=1}^{m+1} p_k C_k(t)=C_{m+1}(t)+\sum_{k=1}^{m} p_k\{C_k(t)-C_{m+1}(t)\}, \quad t \in \mathbb{R}. \tag{1.3}$$

The estimators considered here are based on the empirical characteristic function

$$\hat{C}_n(t)=\int_{\mathbb{R}} e^{itx}\,dF_n(x)=\frac{1}{n}\sum_{k=1}^{n} e^{itX_k}, \quad t \in \mathbb{R}.$$

The application of transform methods for parameter estimation is not new (see, for example, PRESS [15], PAULSON, HOLCOMB and

LEITCH [13], HEATHCOTE [10], FEUERVERGER and McDUNNOUGH [6, 7]. In our present context, QUANDT and RAMSEY [16] used the empirical moment generating function for the estimation of the five parameters in a mixture of two normal distributions. Several of the discussants of the latter paper mentioned the possibility of using $\hat{C}_n$ instead. CLARKE and HEATHCOTE, among them, directed attention to the applicability of the earlier developed ([15, 13, 10]) general method of <u>integrated</u> <u>squared</u> <u>error</u> estimation. This is the line we take up in Section 3, while Section 2 will apply the general moment type method of FEUERVERGER and McDUNNOUGH [6, 7] to the mixture model. In both cases special consideration is given to the case m=1 which is perhaps the most important in practice. In this case we provide very simple explicit adaptive estimators.

2. INTEGRATED ERROR ESTIMATION

<u>2.1 General case.</u> Understanding the last proportion p_{m+1} as

$$p_{m+1}=1-\sum_{k=1}^{m} p_k, \tag{2.1}$$

consider a vector function $A(t)=(A_1(t),\ldots,A_m(t))$, where each of the components $A_k(t)$ is generally a complex-valued function of bounded variation on the whole line with $\operatorname{Re} A_k(t)=-\operatorname{Re} A_k(-t)$, $\operatorname{Im} A_k(t)=\operatorname{Im} A_k(-t)$, $t\in\mathbb{R}$, $k=1,\ldots,m$. The moment type estimators $\hat{p}_1(n),\ldots,\hat{p}_m(n)$ for $p_1^o,\ldots,p_m^o$ of FEUERVERGER and McDUNNOUGH [6, 7], termed as in the title in [4] inspired by the title of [10], solve the equation

$$\int_{-\infty}^{\infty}[\hat{C}_n(t)-C_{m+1}(t)-\sum_{k=1}^{m} p_k\{C_k(t)-C_{m+1}(t)\}]dA(t)=(0,\ldots,0)\in\mathbb{R}^m.$$

Upon introducing the quantities

$$d_{jk}=\int_{-\infty}^{\infty}[C_k(t)-C_{m+1}(t)]dA_j(t),\ j,k=1,\dots,m,$$

$$\hat{d}_j(n)=\int_{-\infty}^{\infty}[\hat{C}_n(t)-C_{m+1}]dA_j(t),\quad j=1,\dots,m,$$

we arrive at the following system of estimating equations:

$$\begin{aligned} p_1d_{11}+\dots+p_md_{1m}&=\hat{d}_1(n)\\ p_1d_{21}+\dots+p_md_{2m}&=\hat{d}_2(n)\\ \vdots \qquad\quad &\vdots\\ p_1d_{m1}+\dots+p_md_{mm}&=\hat{d}_m(n). \end{aligned} \tag{2.2}$$

Of course, this has a unique solution if and only if

$$|D|=\begin{vmatrix} d_{11}\dots d_{1m}\\ \vdots \qquad \vdots\\ d_{m1}\dots d_{mm}\end{vmatrix}\neq 0. \tag{2.3}$$

This condition excludes the case when $C_1\equiv\dots\equiv C_{m+1}$ of course, but this case has already been excluded when we assumed the identifiability of the model. Assuming (2.3) throughout in this section, let $\hat{p}_1(n),\dots,\hat{p}_m(n)$ denote the solution and let $\hat{p}_{m+1}(n)$ be defined as

$$\hat{p}_{m+1}(n)=1-\sum_{k=1}^{m}\hat{p}_k(n).$$

Letting $\hat{\Delta}_n=\max(|\hat{p}_1(n)-p_1^o|,\dots,|\hat{p}_{m+1}(n)-p_{m+1}^o|)$, for the rate of strong consistency of our estimators we have the following result.

THEOREM 2.1. $\hat{\Delta}_n=\mathcal{O}(\{(\log\log n)/n\}^{1/2})$ almost surely.

Proof. Introduce $d_j=\int_{-\infty}^{\infty}[C(t)-C_{m+1}(t)]dA_j(t)$, $j=1,\dots,m$, with $C(t)=C(t;p_1^o,\dots,p_m^o)$. Then

$$\hat{d}_j(n)-d_j=\frac{1}{n}\sum_{k=1}^{n}Y_k(j)$$

with

$$Y_k(j)=\int_{-\infty}^{\infty}[e^{itX_k}-C(t)]dA_j(t),$$

and hence (cf. Section 3 in [4] for the covariance $E\ Y_k(j)=0$, $r_{jj}(p^o)=EY_k^2(j)=\int_{-\infty}^{\infty}\int_{-\infty}^{\infty}[C(s+t)-C(s)C(t)]dA_j(t)$, $j=1,\ldots,m$, with absolutely convergent integrals. Hence the Hartmann-Winter log log law implies that $|\hat{d}_j(n)-d_j|=O(\{(\log\log n)/n\}^{1/2})$ almost surely, $j=1,\ldots,m$. The result then follows by simple manipulations on the solutions of (2.2) obtained by Cramer's rule.

Introduce now the vector notation $p=(p_1,\ldots,p_m)$ for the first m proportions, and let

$$\Delta=-D^T=-\begin{pmatrix} d_{11}\ldots d_{m1} \\ \vdots \qquad \vdots \\ d_{1m}\ldots d_{mm} \end{pmatrix}$$

be (-1) times the transpose of the matrix D in (2.3) of our system (2.2). Consider also the (row) vector functions

$$h(x;p)=r(x;p)\Delta^{-1},\ x\in\mathbb{R},$$

$$r(x;p)=\int_{-\infty}^{\infty}[e^{itx}-C(t;p)]dA(t)$$

$$=\int_{-\infty}^{\infty}[e^{itx}-C_{m+1}(t)-\sum_{k=1}^{m}p_k\{C_k(t)-C_{m+1}(t)\}]dA(t).$$

Then $N(p)=Eh(X;p)h(X;p)^T=\Delta^{-1}\Sigma_2(p)\Delta^{-1}$, where X denotes a random variable with characteristic function $C(t;p)=C(t;p_1,\ldots,p_m)$, and where

$$\Sigma_2(p)=Er(X;p)r(X;p)^T=\begin{pmatrix} r_{11}(p)\ldots r_{1m}(p) \\ \vdots \qquad \vdots \\ r_{m1}(p)\ldots r_{mm}(p) \end{pmatrix}$$

with

$$r_{jk}(p)=\int_{-\infty}^{\infty}\int_{-\infty}^{\infty}[C(s+t;p)-C(s;p)C(t;p)]dA_j(s)dA_j(t),\ j,k=1,\ldots,m.$$

Now, as a special case of Theorem 2 in [4], the following is true.

THEOREM 2.2. The distribution of the vector $n^{1/2}(\hat{p}_n - p_0) = (n^{1/2}(\hat{p}_1(n)-p_1^o),\dots,n^{1/2}(\hat{p}_m(n)-p_m^o))$ converges to the m-dimensional normal distribution with zero expectation vector and covariance matrix $N(p_0)$.

The choice of the weight functions $A(t)=(A_1(t),\dots,A_m(t))$ is crucial from the point of view of efficiency, and the qualitative information FEUERVERGER and McDUNNOUGH [6, 7] give about its efficient choice cannot really be used in practice. In most of the practical situations one would use a step function A, and the question will then be how to choose the points $t_1,\dots,t_r$ where $A(t)$ jumps, and the corresponding weights. We shall consider a version of this problem. Although it would be possible to do so in the above generality, it will be more illustrative to treat this problem for the case of a two-component mixture only.

2.2 The case m=1. In this case the above estimator $\hat{p}_n=\hat{p}_1(n)$ for $p_0=p_1^o$ in the model

$$C(t;p)=pC_1(t)+(1-p)C_2(t)=C_2(t)+p\{C_1(t)-C_2(t)\},\ t\in \mathbb{R}, \tag{2.4}$$

is

$$\hat{p}_n = \int_{-\infty}^{\infty}[\hat{C}_n(t)-C_2(t)]dA(t) / \int_{-\infty}^{\infty}[C_1(t)-C_2(t)]dA(t)$$

with $A(t)=A_1(t)$.

Since $p_0=[C(t)-C_2(t)]/[C_1(t)-C_2(t)]$ if $t\in K_0$, where

$$K_0=\{t\in \mathbb{R} : C_1(t)\neq C_2(t)\} \tag{2.5}$$

and $C(t)=C(t;p_0)$, an obvious estimator for p_0 is

$$\tilde{p}_n(t_1,\dots,t_r)=\frac{1}{r}\sum_{k=1}^{r} \mathrm{Re}\{[\hat{C}_n(t_k)-C_2(t_2)]/[C_1(t_k)-C_2(t_k)]\}, \tag{2.6}$$

where $t_1,\dots,t_r\in K_0$. For simplicity here we consider only the univariate special case

$$\tilde{P}_n(t)=\mathrm{Re}\{[\hat{C}_n(t)-C_2(t)]/[C_1(t)-C_2(t)]\}.$$

Aiming at an adaptive choice of t, we consider this functional estimator as a stochastic process in t. Since $\hat{C}_n(\cdot)$ is a uniformly consistent estimator of C almost surely on any closed interval, we have

$$\tilde{\Delta}_n=\sup_{t\in K}|\tilde{p}_n(t)-p_0|\xrightarrow{a.s.} 0 \tag{2.7}$$

where K is compact subset of K_0. The process

$$Q_n(t)=n^{1/2}\{\tilde{p}_n(t)-p_0\},\quad t\in K,$$

is a random element of the Banach space $\mathcal{C}(K)$ of continuous functions on K with the sup norm. Consider the zero mean Gaussian process

$$Q(t)=\frac{1}{|Z(t)|^2}\{Z_1(t)R(t)+Z_2(t)I(t)\},\quad t\in K,$$

where $Z(t)=C_1(t)-C_2(t)$, $Z_1(t)=\mathrm{Re}Z(t)$, $Z_2(t)=\mathrm{Im}Z(t)$, and $R(t)$ and $I(t)$ are defined as

$$R(t)+iI(t)=\int_{-\infty}^{\infty}e^{itx}dB(F(x))=Y(t),$$

where $F(x)=F(x;p_0)$ and $B(\cdot)$ is a Brownian bridge process. The process $Y(t)$, being the limit process of the empirical characteristic process $Y_n(t)=n^{1/2}\{\hat{C}_n(t)-C(t)\}$, is a random element of $\mathcal{C}(K)$ if and only if (see [2, 12, 3])

$$\int_0^1\frac{\bar{\varphi}(h)}{h(\log\frac{1}{h})^{1/2}}dh<\infty, \tag{$*$}$$

where

$$\bar{\varphi}(h)=\sup\{y:0\le y\le 1,\ \lambda\{t:|t|<1/2,\ \varphi(t)<y\}<h\},$$

with λ standing for the Lebesgue measure, is the non-decreasing rearrangement of $\varphi(t)=(1-\mathrm{Re}\,C(t))^{1/2}$. Condition ($*$) is a very-very mild tail-condition on $F(x)$ (see [2]).

THEOREM 2.3. The process $Q_n(\cdot)$ converges weakly in $\mathfrak{C}(K)$ to $Q(\cdot)$ if and only if (*) holds. Moreover, under (*), $\tilde{\Delta}_n = O(((\log\log n)/n)^{1/2})$ almost surely.

<u>Proof.</u> Since $p_0=[C(t)-C_2(t)]/Z(t)$ is real, we have

$$Q_n(t)=n^{1/2}\mathrm{Re}\{(Z(t))^{-1}(\hat{C}_n(t)-C_2(t))-(Z(t))^{-1}(C(t)-C_2(t))\}=\mathrm{Re}\{Y_n(t)/Z(t)\}$$

converging weakly if and only if (*) holds ([12, 3]) to $\mathrm{Re}\{Y(t)/Z(t)\}$ which process is easily seen to be our process Q. If instead of the weak convergence theorem just referred to we use the corresponding log log law ([3]), we obtain the second statement. Indeed, we can get this statement in the form

$$\limsup_{n\to\infty}(n/\log\log n)^{1/2}\tilde{\Delta}_n=L$$

almost surely, with a completely specified constant $L>0$ depending on the underlying distribution.

The covariance function of the limit process Q is easily computed from the cross-covariance matrix of R and I, given e.g. in [3]. Writing $C(t)=U(t)+iV(t)$, for any $s,t\in K_0$ we have

$$\begin{aligned}\sigma(s,t)&=EQ(s)Q(t)\\&=(|Z(s)||Z(t)|)^{-2}[Z_1(s)Z_1(t)\{\tfrac{1}{2}[U(s-t)+U(s+t)]-U(s)U(t)\}\\&\quad+Z_1(s)Z_2(t)\{\tfrac{1}{2}[U(t-s)+V(s+t)]-U(s)V(t)\}\\&\quad+Z_1(t)Z_2(s)\{\tfrac{1}{2}[V(s-t)+V(s+t)]-U(t)V(s)\}\\&\quad+Z_2(s)Z_2(t)\{\tfrac{1}{2}[U(s-t)-U(s+t)]-V(s)V(t)\}].\end{aligned}$$

Suppose now that the compact set K is also connected and is chosen such a way that there is a unique $t_0\in K$ which minimizes the variance function $\sigma^2(t)=\sigma(t,t)$ on K: $\sigma^2(t_0)<\sigma^2(t)$, $t\in K$, $t\neq t_0$. Let $\sigma_n^2(t)$ be the estimator of the function $\sigma^2(t)$ obtained upon replacing U and V by $\hat{U}_n$ and $\hat{V}_n$ in it, respectively, where

$\hat{C}_n(t)=\hat{U}_n(t)+i\hat{V}_n(t)$. Define the random variable t_n as the smallest value where $\sigma_n^2(t)$ is minimized on K:

$$t_n=\inf\{t \in K: \sigma_n^2(t)=\inf_{s\in K} \sigma_n^2(s)\}.$$

Clearly

$$\sup_{t\in K}|\sigma_n^2(t)-\sigma^2(t)| \xrightarrow{a.s.} 0.$$

This implies, as in Lemma 5 of [5], that $t_n \xrightarrow{a.s.} t_0$. Then for our adaptive estimator $\tilde{p}_n(t_n)$ we get, as in the proof of Theorem 4 in [5], the following consequence of Theorem 2.3.

COROLLARY. Under (*), $|\tilde{p}_n(t_n)-p_0|=0(((\log \log n)/n^{1/2}$ and

$$\lim_{n\to\infty} pr\{n^{1/2}(\tilde{p}_n(t_n)-p_0)\leq x\}=\Phi(x/\sigma(t_0)),\ x \in \mathbb{R}$$

where Φ is the standard normal distribution function.

All the statements easily generalize for the more general estimator $\tilde{p}_n(t_1,\ldots,t_r)$, but then of course we end up minimizing an r-variate estimated variance $\sigma_n^2(t_1,\ldots,t_r)$.

3. INTEGRATED SQUARED ERROR ESTIMATION

3.1 General case. These estimators were proposed and investigated in [15, 13, 10, 4] for the stable and general parametric families, and in a whole series of further papers by PAULSON and his co-authors. When applied to our model in (1.3), these estimators $p_n^*=(p_1^*(n),\ldots,p_m^*(n))$ minimize the integral

$$\begin{aligned} I_n(p)=I_n(p_1,\ldots,p_m) &= \int_{-\infty}^{\infty}|\hat{C}_n(t)-C(t,p)|^2 dH(t) \\ &= \int_{-\infty}^{\infty}|\hat{C}_n(t)-C_{m+1}(t)-\sum_{k=1}^{m} p_k\{C_k(t)-C_{m+1}(t)\}|^2 dH(t), \end{aligned} \tag{3.1}$$

where H is a known non-decreasing weight function with unit

total variation on the whole line. In a general parametric model it is hard to compute a minimum, and often sophisticated numerical methods are required (cf. [13]). First we show that, because of the linearity of the model (1.3), the present minimization problem can be reduced to the standard quadratic programming problem.

Introduce the mxm symmetric matrix Λ with (j,k)-th entry

$$\lambda_{jk}=\int_{-\infty}^{\infty}[U_j(t)U_k(t)+V_j(t)V_k(t)]dH(t),\quad j,k=1,\ldots,m,$$

the random (row) vector $h_n=(h_1(n),\ldots,h_m(n))$, with the k-th component

$$h_k(n)=\int_{-\infty}^{\infty}[\mathrm{Re}\{\hat{C}_n(t)-C_{m+1}(t)\}U_k(t)+\mathrm{Im}\{\hat{C}_n(t)-C_{m+1}(t)\}V_k(t)]dH(t),$$

and the random variable

$$\alpha_n=\int_{-\infty}^{\infty}|\hat{C}_n(t)-C_{m+1}(t)|^2dH(t),$$

where

$$U_k(t)=\mathrm{Re}\{C_k(t)-C_{m+1}(t)\},V_k(t)=\mathrm{Im}\{C_k(t)-C_{m+1}(t)\},\quad k=1,\ldots,m. \tag{3.2}$$

Then, taking the square, a simple computation yields

$$I_n(p)=\sum_{j=1}^{m}\sum_{k=1}^{m}p_jp_k\lambda_{jk}-2\sum_{k=1}^{m}p_kh_k(n)+\alpha_n$$

$$=\langle p\Lambda,p\rangle-2\langle p,h_n\rangle+\alpha_n,$$

where $\langle\cdot,\cdot\rangle$ is the inner product in $\mathbb{R}^m$. Since

$$\langle p\Lambda,p\rangle=\int_{-\infty}^{\infty}|\sum_{k=1}^{m}p_k\{C_k(t)-C_{m+1}(t)\}|^2dH(t)=\int_{-\infty}^{\infty}|C(t;p)-C_{m+1}(t)|^2dH(t)\geq 0,$$

Λ is positive semi-definite, and hence one can always compute a solution with the usual quadratic programming methods ([8]). But if in (2.1), $p_{m+1}<1$, that is, if our mixture consists of at least two components (and we are not estimating the constant zero vector), the last quantity is strictly positive. Assuming this, Λ is positive definite, and the solution of the equivalent problem maximize $-\langle p\Lambda,p\rangle+2\langle p,h_n\rangle-\alpha_n$ subject to $p_1,\ldots,p_o\geq 0$,

$$\sum_{k=1}^{m} p_k\leq 1 \tag{3.3}$$

is unique and can be obtained by standard methods ([8]). An explicit estimator will be given in the next subsection.

Let h_k^o be defined by replacing $\hat{C}_n(t)$ in $h_k(n)$ with its limit $C(t)$, $k=1,\ldots,m$, $h_o=(h_1^o,\ldots,h_m^o)$ and define similarly α_o from α_n. By similar work as in the proof of Theorem 2.1 one obtains $|h_k(n)-h_k^o|=O(((\log\log n)/n)^{1/2})$ a.s., $k=1,\ldots,m$. For the α_n, applying the triangular inequality for the $L_2(H)$ norm, we have

$$|\alpha_n^{1/2}-\alpha_o^{1/2}|\leq \left(\int_{-\infty}^{\infty}|\hat{C}_n(t)-C(t)|^2dH(t)\right)^{1/2}.$$

The upper bound is a.s. of the order of $((\log\log n)/n)^{1/2}$ by the log log law of KUELBS and KURTZ [11] in the $L_2(H)$ space, and hence $|\alpha_n-\alpha_o|=O(((\log\log n)/n)^{1/2})$ a.s. Following now the standard procedure in [8] that solves (3.2), we see that the differences in $\Delta_n^*=\max(|p_1^*(n)-p_1^o|,\ldots,|p_{m+1}^*(n)-p_{m+1}^o|)$ can be written as algebraic functions of $|\alpha_n-\alpha_o|$ and $|h_k(n)-h_k^o|$, $k=1,\ldots,m$, with bounded coefficients. Here, of course, we used the fact that p_o solves (3.3) when h_n and α_n are replaced by h_o and α_o. Hence the log log rate of consistency holds for the present estimators as well:

THEOREM 3.1. $\Delta_n^*=O(((\log\log n)/n)^{1/2})$ almost surely.

Writing $C(t;p)=U(t;p)+iV(t;p)$, introduce the function

$$q(x;p)=\int_{-\infty}^{\infty}[\cos(tx)-U(t;p)]U_k(t)dH(t)$$
$$+\int_{-\infty}^{\infty}[\sin(tx)-V(t;p)]V_k(t)dH(t),$$

with $U_k(t)=\partial U(t;p)/\partial p_k$ and $V_k(t)=\partial V(t;p)/\partial p_k$ as in (3.2), and let $\Sigma_1(p)$ be the mxm symmetric matrix whose (j,k)-th entry is $Eq_j(X;p)q_k(X;p)$. Finally, set $M(p)=\Lambda^{-1}\Sigma_1(p)\Lambda^{-1}$.

When formally applying the result of HEATHCOTE [10] on asymptotic normality, precisely formulated as Theorem C in [4], one should assume that neither of $p_1^o,\ldots,p_m^o$ takes the extreme values 0 and 1. But since we did not use this proviso in the proof of the consistency (Theorem 3.1 above), and since $p_n^*=(p_1^*(n),\ldots,p_m^*(n))$ is guaranteed to be in the closed unit cube of $\mathbb{R}^m$, a little care in checking the details in the proof of Theorem C in [4] shows that we need not exclude the extreme values 0 and 1.

THEOREM 3.2. The distribution of the vector $n^{1/2}(p_n^*-p_o)=$ $=(n^{1/2}(p_1^*(n)-p_1^o),\ldots,n^{1/2}(p_m^*(n)-p_m^o))$ converges to the m-dimensional normal distribution with zero expectation vector and covariance matrix $M(p_o)$.

3.2 General case. Explicit estimators. The function $I_n(p)$ of (3.1) may be regarded as a function on the whole space $\mathbb{R}^m$. It is convex over $\mathbb{R}^m$ and its gradient $\nabla I_n(p)$ exists for all $p\in\mathbb{R}^m$. A simple differentiation gives

$$\frac{\partial I_n(p)}{\partial p_k}=2\left\{\left(\sum_{j=1}^m p_j\lambda_{jk}\right)-h_k(n)\right\},\quad k=1,\ldots,m,$$

and hence the estimating gradient equations

$$\nabla I_n(p)=(0,\ldots,0)$$

are satisfied if and only if

$$\sum_{j=1}^m p_j\lambda_{jk}=h_k(n),\quad k=1,\ldots,n,$$

or, in vector notation, $p\Lambda = h_n$. Hence Theorem E in [17, p. 102] ensures that for the vector $p_n^{**} = h_n \Lambda^{-1}$ we have

$$I_n(p_n^{**}) = \min_{p \in \mathbb{R}^m} I_n(p). \qquad (3.4)$$

Obviously, this gradient procedure leads to estimators p_n^{**} whose components may be negative or greater than one. Asymptotically the two estimators p_n^* and p_n^{**} will be equivalent. Indeed, another simple computation gives $p_0 = h_0 \Lambda^{-1}$, and hence by the already developed log log laws for the components of $h_n - h_0$ we get

$$\Delta_n^{**} = O(((\log \log n)/n)^{1/2}) \text{ a.s.}$$

for $\Delta_n^{**} = \max(|p_1^{**}(n) - p_1^o|, \ldots, |p_{m+1}^{**}(n) - p_{m+1}^o|)$. Moreover, we see from the proof of HEATHCOTE's Theorem C in [4] that asymptotic normality requires only consistency and the fulfilment of (3.4) (if $\nabla I_n(p)$ exists everywhere in $\mathbb{R}^m$), and hence Theorem 3.2 is also true for p_n^{**} in place of p_n^*.

We emphasize that $Ep_n^{**} = p_0$, i.e., our vector estimator p_n^{**} is unbiased.

The choice of the weight function H is again crucial from the point of view of efficiency. In this regard some considerations are offered by HEATHCOTE [10].

3.3 The case m=1. In this case our explicit estimator is

$$p_n^{**} = \frac{h_n}{\lambda} = \frac{\int_{-\infty}^{\infty} [\mathrm{Re}\{\hat{C}_n(t) - C_2(t)\} U_1(t) + \mathrm{Im}\{\hat{C}_n(t) - C_2(t)\} V_1(t)] dH(t)}{\int_{-\infty}^{\infty} |C_1(t) - C_2(t)|^2 dH(t)}$$

with $U_1(t) = \mathrm{Re}\{C_1(t) - C_2(t)\}$, $V_1(t) = \mathrm{Im}\{C_1(t) - C_2(t)\}$. Now the difficulties that p_n^{**} might be outside of $[0,1]$ can be overcome easily by simple truncation via checking the possible cases separately. Thus we even get our unique quadratic programming estimator p_n^* as

$$p_n^* = \begin{cases} 0 & , \text{ if } h_n<0, \\ p_n^{**} & , \text{ if } 0\le h_n \le \lambda, \\ 1 & , \text{ if } \lambda<h_n. \end{cases}$$

It is interesting to point out that if we choose the weight function H in p_n^{**} to be the simplest one that jumps unity at the point t, then the resulting estimator is nothing else, as one would expect, but the functional estimator $\tilde{p}_n(t)$ of subsection 2.2.

Finally we note that a simple test statistic can be constructed from the functional statistics $\tilde{p}_n(t_1,\dots,t_r)$ in (2.6) for testing for the model (1.3) with unspecified proportions $p_1,\dots, p_{n+1}$. Details will be given elsewhere.

REFERENCES

[1] CHOI, K. and BULGREN, W.G. (1968): An estimation procedure for mixtures of distributions. J. Roy. Statist. Soc. Ser. B 30, pp. 444-460.

[2] CSÖRGŐ, S.(1981): Limit behaviour of the empirical characteristic process. Ann. Probability 9, pp. 130-144.

[3] CSÖRGŐ, S.(1981): Multivariate empirical characteristic functions. Z. Wahrscheinlichkeitstheorie verw. Gebiete 55, pp. 203-229.

[4] CSÖRGŐ, S. (1981): The empirical characteristic process when parameters are estimated. Contributions to Probability (E. Lukacs Festschrift; J. Gani and V.K. Rohatgi, Eds.), Academic Press, New York, pp. 215-230.

[5] CSÖRGŐ, S.(1983): The theory of functional least squares. J. Austral. Math. Soc. Ser. A. To appear.

[6] FEUERVERGER, A. and McDUNNOUGH, P. (1981): On the efficiency of empirical characteristic function procedures. J. Roy. Statist. Soc. Ser. B 43, pp. 20-27.

[7] FEUERVERGER, A. and McDUNNOUGH, P. (1981): On some Fourier methods for inference. J. Amer. Statist. Assoc. 76, pp. 379-387.

[8] HADLEY, G. (1964): Nonlinear and Dynamic Programming. Addison-Wesley, Reading, Massachusetts.

[9] HALL, P. (1981): On the nonparametric estimation of mixture proportions. J. Roy. Statist. Assoc. Ser. B 43, pp. 147-156.

[10] HEATHCOTE, C.R. (1977): The integrated squared error estimation of parameters. Biometrika 64, pp. 255-264.

[11] KUELBS, J. and KURTZ, T. (1974): Berry-Esséen estimates in Hilbert space and an application to the law of the iterated logarithm. Ann. Probability 2, pp. 387-407.

[12] MARCUS, M.B.(1981): Weak convergence of the empirical characteristic function. Ann. Probability 9, pp. 194-201.

[13] PAULSON, A.S., HOLCOMB, E.W. and LEITCH, R.A. (1975): The estimation of the parameters of the stable laws. Biometrika 62, pp. 163-170.

[14] PEARSON, K. (1894): Contributions to the mathematical theory of evolution. Phil. Trans. Roy. Soc. 185, pp. 71-110.

[15] PRESS, S.J. (1972): Estimation in univariate and multivariate stable distributions. J. Amer. Statist. Assoc. 67, pp. 842-846.

[16] QUANDT, R.E. and RAMSEY, J.B. (1978): Estimating mixtures of normal distributions and switching regressions (with discussion). J. Amer. Statist. Assoc. 73, pp. 730-752.

[17] ROBERTS, A.W. and VARBERG, D.E. (1973): Convex Functions. Academic Press, New York.

[18] TEICHER, H. (1963): Identifiability of finite mixtures. Ann. Math. Statist. 34, pp. 1265-1269.

Proc. of the 3rd Pannonian Symp. on Math. Stat., Visegrád, Hungary 1982
J. Mogyoródi, I. Vincze, W. Wertz, eds

ON RECORD TIMES ASSOCIATED WITH k-th EXTREMES

P. DEHEUVELS

École Pratique des Hautes Études
Université Paris VI, Paris, France

ABSTRACT

The record times of order k are defined on an i.i.d. sequence with continuous distribution as the times where the k-th maximum of order n increases. The case k=1 gives the usual record time sequence. In the general case, we obtain a strong approximation result for record times and inter-record times enabling us to obtain the almost sure upper and lower class for these sequences, and also an asymptotic normality of the logarithm of record times and inter-record times.

AMS 1970 Subject classification: Primary 60 F 05

Key words: Record values, maxima, extreme values, strong approximation, strong invariance principles, law of the iterated logarithm

1. INTRODUCTION

Let $X_1, X_2, \ldots$ be an i.i.d. sequence of random variables with common distribution $F(x)=P(X\leq x)$. For any $n\geq 1$, let

$$X_1^{(n)}<\ldots<X_n^{(n)} \tag{1}$$

be the order statistic corresponding to $X_1, \ldots X_n$, and consider, for a fixed $k\geq 1$, the k-th maximal sequence:

$$X_1^{(k)}\leq X_2^{(k+1)}\leq\ldots\leq X_{n-k+1}^{(n)}\leq\ldots\ . \tag{2}$$

The aim of this paper is to study the sequence of time indices where this sequence increases. We shall call these times record times of order k, and define them by

$$n_1^{(k)}=k,\ n_m^{(k)}=\mathrm{Min}\{m>n_{m-1}^{(k)};X_m>X_{m-k}^{(m-1)}\},\ m\geq 2. \tag{3}$$

For k=1, the sequence $\{n_m=n_m^{(k)},m\geq 1\}$ is the classical sequence of record times, corresponding to the times of increase of the maximum $Y_n=\mathrm{Max}\{X_1,\ldots,X_n\}=X_n^{(n)}$, and has been extensively studied, among others, by Chandler [1], 1952, Foster and Stuart [10], 1954, Rényi [17], 1962, Tata [25], 1969, Holmes and Strawderman [13], 1969, Strawderman and Holmes [24], 1970, Shorrock [20], [21], 1972, [22], 1973, Vervaat [26], 1973, [27], 1977, Williams [29], 1973, Resnick [18a], [18b], 1973, Galambos and Seneta [11], 1975, Siddiqui and Biondini [19], 1975, Westcott [28], 1977, Glick [12], 1978, Pfeifer [16], 1982.

If we define the inter-record times of order k by

$$\Delta_1^{(k)}=n_1^{(k)},\ \Delta_m^{(k)}=n_m^{(k)}-n_{m-1}^{(k)},m\geq 2, \tag{4}$$

the following results have been obtained for k=1:

In 1962, Rényi ([17]) proved that

$$\operatorname*{LimSup}_{m\to+\infty}\frac{\mathrm{Log}\ n_m-m}{\sqrt{2m\ \mathrm{Log}_2 m}}=1,\ \operatorname*{LimInf}_{n\to+\infty}\frac{\mathrm{Log}\ n_m-m}{\sqrt{2m\ \mathrm{Log}_2 m}}=-1\quad \text{a.s.} \tag{5}$$

In 1970, Strawderman and Holmes ([24)] proved that (if $\Delta_m=\Delta_m^{(1)}$)

$$\operatorname*{LimSup}_{m\to+\infty}\frac{\mathrm{Log}\ \Delta_m-m}{\sqrt{2m\ \mathrm{Log}_2 m}}=1,\ \operatorname*{LimInf}_{m\to+\infty}\frac{\mathrm{Log}\ \Delta_m-m}{\sqrt{2m\ \mathrm{Log}_2 m}}=-1\quad \text{a.s.} \tag{6}$$

In 1982, these results were precised up to a complete characterization in Deheuvels [7], by showing that, without loss of generality, there exists a Wiener process $\{W(t),t\geq 0\}$ such that

$$n_m=\exp(m+W(m)+0(\mathrm{Log}\ m))\quad \text{as } m\to\infty,\quad \text{a.s.}, \tag{7}$$

and likewise, that there exists a Wiener process $\{\tilde{W}(t), t\geq 0\}$, such that

$$\Delta_m = \exp(m+\tilde{W}(m)+O(\text{Log } m)) \quad \text{as } m\to\infty, \quad \text{a.s.} \tag{8}$$

As a consequence of (7) and (8), it follows that, if $\nu_m = n_m$, or if $\nu_m = \Delta_m$, then, for any $p\geq 4$,

$$P(\text{Log } \nu_m - m \geq \sqrt{2m\{\text{Log}_2 m+(3/2)\text{Log}_3 m+\text{Log}_4 m+\ldots+(1+\varepsilon)\text{Log}_p m\}} \text{ i.o.}) =$$

$$= P(\text{Log } \nu_m - m \leq -\sqrt{2m\{\text{Log}_2 m+(3/2)\text{Log}_3 m+\text{Log}_4 m+\ldots+(1+\varepsilon)\text{Log}_p m\}} \text{ i.o.}) =$$

$= 0$ or 1, according as $\varepsilon>0$, or $\varepsilon\leq 0$.

For $k>1$, there have been very few results available, up to a recent epoch. If instead of considering the record times, one considers the record processes:

$$\tau_t^{(k)} = \text{Min}\{n\geq 1; X_{n-k+1}^{(n)} \geq t\}, \tag{9}$$

the fact that this process has independent increments for $k=1$ was proved initially in Deheuvels [2a], 1973 (published January 22nd) and re-discovered later on by Dwass [8], 1973, and Shorrock [23], 1974. The general result for an arbitrary $k\geq 1$ has been proved in Deheuvels [2b], 1974.

In 1976, Dziubdziela and Kopocinski [9] obtained the limiting distribution of k-th record values

$$X_{n-k+1}^{(n)}, \quad n=n_m^{(k)}, \quad m\geq 1, \tag{10}$$

without giving results on record times.

In 1982, the following result was proved in Deheuvels [5]:

Without loss of generality, there exists a homogeneous Poisson process of intensity k on R, with times of arrivals

$$\ldots<z_{-1}<z_0<0<z_1<z_2<\ldots, \tag{11}$$

and an independent i.i.d. sequence $\{\omega_i, -\infty<i<+\infty\}$ of exponentially E(1) distributed random variables, such that

$$n_m^{(k)}=k+\sum_{i=1}^{m-1}\left(\left[\frac{\omega_i}{-\mathrm{Log}(1-e^{-z_i})}\right]+1\right),\quad m=1,2,\ldots. \tag{12}$$

We shall use here this result to generalize (7) and (8) to an arbitrary $k\geqq 1$.

2. STRONG APPROXIMATION OF RECORD TIMES OF ORDER k

We shall use the methods of Deheuvels [5] and [6], 1982 (see also [3]).

Consider first the inter-record times:

$$\Delta_m^{(k)}=\left[\frac{\omega_{m-1}}{-\mathrm{Log}(1-e^{-z_{m-1}})}\right]+1 \approx \omega_{m-1}e^{z_{m-1}} \tag{13}$$

It can be easily proved that, almost surely as $m\to\infty$,

$$\Delta_m^{(k)}=\exp(z_{m-1}+0(\mathrm{Log}\ m))=\exp\left(\frac{m}{k}+\frac{1}{k}W(m)+0(\mathrm{Log}\ m)\right) \tag{14}$$

where $\{W(t),t\geqq 0\}$ is a standard Wiener process. This follows from the classical Komlós-Major-Tusnády theorem of strong approximation applied to z_{m-1}.

Consider now

$$\theta_m^{(k)}=\sum_{-\infty}^{m-1}\omega_i e^{z_i}. \tag{15}$$

It is classical in the case k=1 that the point process $\{\mathrm{Log}\ \theta_m^{(1)}\}$ is a homogeneous Poisson process with intensity one on R. Indeed, we have there the points of jumps for an extremal process (see Deheuvels [7], 1982).

By using the same proof as in Deheuvels [7], 1982, one can see likewise that the point process $\{\mathrm{Log}\ \theta_m^{(k)}\}$ is a homogeneous Poisson process with intensity k on R. It follows that, almost surely as $m\to\infty$,

$$n_m^{(k)} \approx \theta_m^{(k)} = \exp\left(\frac{m}{k} + \frac{1}{k}\tilde{W}(m) + 0(\mathrm{Log}\ m)\right). \tag{16}$$

Finally, we obtain our main results:

Theorem 1. Without loss of generality, there exists a Wiener process $\{W(t), t \geq 0\}$ and a Wiener process $\{\tilde{W}(t), t \geq 0\}$, such that, almost surely as $m \to \infty$,

$$n_m^{(k)} = \exp\left(\frac{m}{k} + \frac{1}{k}\tilde{W}(m) + 0(\mathrm{Log}\ m)\right), \tag{17}$$

and

$$\Delta_m^{(k)} = \exp\left(\frac{m}{k} + \frac{1}{k}W(m) + 0(\mathrm{Log}\ m)\right). \tag{18}$$

Theorem 2.

For any $p \geq 4$,

$$P\left(\mathrm{Log}\ n_m^{(k)} - \frac{m}{k} \geq \frac{1}{k}\sqrt{2m\{\mathrm{Log}_2 m + (3/2)\mathrm{Log}_3 m + \mathrm{Log}_4 m + \ldots + (1+\varepsilon)\mathrm{Log}_p m\}} \text{ i.o.}\right) =$$

$$= P\left(\mathrm{Log}\ n_m^{(k)} - \frac{m}{k} \leq \frac{-1}{k}\sqrt{2m\{\mathrm{Log}_2 m + (3/2)\mathrm{Log}_3 m + \mathrm{Log}_4 m + \ldots + (1+\varepsilon)\mathrm{Log}_p m\}} \text{ i.o.}\right) =$$

$$= P\left(\mathrm{Log}\ \Delta_m^{(k)} - \frac{m}{k} \geq \frac{1}{k}\sqrt{2m\{\mathrm{Log}_2 m + (3/2)\mathrm{Log}_3 m + \mathrm{Log}_4 m + \ldots + (1+\varepsilon)\mathrm{Log}_p m\}} \text{ i.o.}\right) =$$

$$= P\left(\mathrm{Log}\ \Delta_m^{(k)} - \frac{m}{k} \leq \frac{-1}{k}\sqrt{2m\{\mathrm{Log}_2 m + (3/2)\mathrm{Log}_3 m + \mathrm{Log}_4 m + \ldots + (1+\varepsilon)\mathrm{Log}_p m} \text{ i.o.}\right) =$$

$$= 0 \text{ or } 1 \text{ according as } \varepsilon > 0 \text{ or } \varepsilon \leq 0. \tag{19}$$

Theorem 3.

The upper and lower almost sure classes of $\{\pm(\mathrm{Log}\ n_m^{(k)} - \frac{m}{k})\}$ and $\{\pm\mathrm{Log}\ \Delta_m^{(k)} - \frac{m}{k})\}$ are identical. If $H(t)$ is a positive function defined in a right neighborhood of zero, and such that $H\uparrow$ and $t^{-1/2}H\downarrow$, then, $mH(1/m)/k$ belongs to the preceding upper or lower class, according as

$$I = \int_{0+} t^{-3/2} H(t) \exp(-H^2(t)/2t)\, dt \tag{20}$$

converges or diverges.

Theorem 4.

As m increases to infinity,

$$\frac{k(\operatorname{Log} n_m^{(k)} - \frac{m}{k})}{\sqrt{m}} \underset{w}{\to} N(0,1), \text{ and } \frac{k(\operatorname{Log} \Delta_m^{(k)} - \frac{m}{k})}{\sqrt{m}} \underset{w}{\to} N(0,1). \quad (21)$$

Proof. Aside from the proof of Theorem 1, which has been obtained before, we have only to deduce Theorem 2, Theorem 3, and Theorem 4 from (17) and (18). This can be done by a word by word use of the proofs in Deheuvels (7), 1982. To obtain (21), we have to consider (17) and (19), and to evaluate the limiting distributions, which is straightforward.

The result in Theorem 4 is not new for k=1. In fact, Neuts showed in 1967 that, as $m \to \infty$,

$$\frac{\operatorname{Log} \Delta_m^{(1)} - m}{\sqrt{m}} \underset{w}{\to} N(0,1), \quad (22)$$

while similar results were obtained by Rényi in 1962 ([17]) for n_m.

3. CONCLUSION

The preceding results are easily deduced from strong approximation results for extreme values, used jointly with the Komlós-Major-Tusnády 1975-1976 ([14],[15]) approximation. A direct approach for an arbitrary k seems to be quite involved. Consequently, it gives a good example of the usefulness of strong approximation techniques in extreme value theory.

4. ACKNOWLEDGEMENT

I would like to express my thanks to the organizers of the 3rd Pannonian Symposium for inviting me in Visegrád to give a lecture on the preceding subject.

REFERENCES

1. CHANDLER, K.N. (1952) The distribution and frequency of record values, J. Royal Statist. Soc. B 14, 220-228

2a. DEHEUVELS, P. (1973) Sur la convergence de sommes de minima de variables aléatoires, C.R. Acad. Sci. Paris, 276, 309-312

2b. DEHEUVELS, P. (1974) Majoration et minoration presque sûre optimale des éléments de la statistique ordonnée d'un échantillon croissant de variables aléatoires indépendantes, Rendi Conti Academia Nazionale dei Lincei, 8, Vol. 56, f.5, 707-719

3. DEHEUVELS, P. (1981) The strong approximation of extremal processes, Z. Wahrscheinlichkeit. verw. Gebiete 58, 1-6

4. DEHEUVELS, P. (1982) A construction of extremal processes, in: Probability and Statistical Inference, W. Grossmann, G.C. Pflug and W. Wertz edit., Reidel, Dordrecht

5. DEHEUVELS, P. (1982) The strong approximation of extremal processes II, Z. Wahrscheinlichkeit. verw. Gebiete (to be published)

6. DEHEUVELS, P. (1982) Strong approximation in extreme value theory and applications, in: Limit Theorems in Probability Theory and Statistics, Colloquia Math. Soc. J. Bolyai, North-Holland, Amsterdam

7. DEHEUVELS, P. (1982) The complete characterization of the upper and lower class of the record and inter-record times of an i.i.d. sequence, Z. Wahrscheinlichkeit. verw. Gebiete (to be published)

8. DWASS, M. (1973) Extremal processes III, Discussion paper 41, Northwestern Univ.

9. DZIUBDZIELA, W., KOPOCINSKI, B. (1976) Limiting properties of the k-th record values, Applicationes Math. 15, 187-190

10. FOSTER, F.G., STUART, A. (1954) Distribution free tests in time series based on the breaking of records, J. Royal Statist. Soc. B 16, 1-13

11. GALAMBOS, J., SENETA, E. (1975) Record times, Proc. Amer. Math. Soc. 50, 383-387

12. GLICK, N. (1978) Breaking records and breaking boards, Amer. Math. Monthly 85 (1), 2-26

13. HOLMES, P.T., STRAWDERMAN, W.E. (1969) A note on the waiting times between record observations, J. Appl. Probability 11, 605-608

14. KOMLÓS, J., MAJOR, P., TUSNÁDY, G. (1975) An approximation of partial sums of independent R.V.'s and the sample D.F. I, Z. Wahrscheinlichkeit. verw. Gebiete 32, 111-131

15. KOMLÓS, J., MAJOR, P., TUSNÁDY, G. (1976) An approximation of partial sums of independent R.V.'s and the sample D.F. II, Z. Wahrscheinlichkeit. verw. Gebiete 34, 33-58

16. PFEIFER, D. (1982) Characterization of exponential distributions by independent non-stationary record increments, J. Appl. Prob. 19, 127-135

17. RÉNYI, A. (1962) Théorie des éléments saillants d'une suite d'observations, Coll. Aarhus, 104-115

18a. RESNICK, S. (1973) Limit laws for record values, Stoch. Processes Appl. 1, 67-82

18b. RESNICK, S. (1973) Record values and maxima, Ann. Probability 1, 650-662

19. SIDDIQUI, M.M., BIONDINI, R.W. (1975) The joint distribution of record values and inter-record times, Ann. Probability 3, 1012-1013

20. SHORROCK, R.W. (1972) A limit theorem for inter-record times, J. Appl. Probability 9, 219-223

21. SHORROCK, R.W. (1972) On record values and record times, J. Appl. Probability 9, 316-326

22. SHORROCK, R.W. (1973) Record values and interrecord times, J. Appl. Probability 10, 543-555

23. SHORROCK, R.W. (1974) On discrete time extremal processes, Adv. Appl. Probability 6, 580-592

24. STRAWDERMAN, W.E., Holmes, P.T. (1970) On the law of the iterated logarithm for inter-record times, J. Appl. Probability 7, 432-439

25. TATA, N.M. (1969) On outstanding values in a sequence of random variables, Z. Wahrscheinlichkeit. verw. Gebiete 12, 9-20

26. VERVAAT, W. (1973) Limit theorems for record from discrete distributions, Stoch. Processes Appl. 1, 317-334

27. VERVAAT, W. (1977) On record, maxima, and a stochastic difference equation, Report 7702, Univ. Nijmegen

28. WESTCOTT, M. (1977) The random record model, Proc. Royal Soc. London A 356, 529-547

29. WILLIAMS, D. (1973) On Rényi's record problem and Engel's series, Bull. London Math. Soc. 5, 235-237

Proc. of the 3rd Pannonian Symp. on Math. Stat., Visegrád, Hungary 1982
J. Mogyoródi, I. Vincze, W. Wertz, eds

MARCINKIEWICZ STRONG LAW OF LARGE NUMBERS FOR B-VALUED RANDOM VARIABLES WITH MULTIDIMENSIONAL INDICES

I. FAZEKAS

Kossuth Lajos University, Debrecen, Hungary

1. INTRODUCTION

Denote by N the set of positive integers. Let N^d, where $d \geq 1$ is an integer, denote the positive integer d-dimensional lattice points. If $n, k \in N^d$, then $n+k$, $n \leq k$ and $n < k$ are defined coordinatewise. If $n=(n_1,\dots,n_d)$, then $|n|$ is defined by $\prod_{i=1}^{d} n_i$ and $n \to \infty$ means that $n_i \to \infty$ for $i=1,2,\dots,d$. Let 1 denote $(1,\dots,1) \in N^d$.

Let B be a separable Banach space with norm $||.||$. Let $\{X_n, n \in N^d\}$ be a sequence of independent identically distributed (i.i.d.) B-valued random variables (r.v.'s). Let $S_n = \sum_{k \leq n} X_k$, $n \in N^d$. In this paper we prove a Marcinkiewicz strong law of large numbers, i.e., we show that if $1 \leq p < 2$, B is of Rademacher type p, $E||X_1||^p(\log^+||X_1||)^{d-1} < \infty$, $E X_1 = 0$, then $|n|^{-\frac{1}{p}} S_n \to 0$ a.s. as $n \to \infty$.

For real random variables this theorem has been proved by Smythe [9] for p=1 and by Gut [4] for $1 \leq p < 2$. This result has been obtained in [3] if p=1 and B is an arbitrary separable Banach space. If d=1 and B is a Banach space of Rademacher type n ($1 \leq p < 2$), then the Marcinkiewicz SLLN has been proved by Acosta [1]. In this paper we use the method of Acosta [1].

2. PRELIMINARIES

It is obvious that $(N^d, \leq)$ is a directed set. Let us consider the net (see e.g. Ash [2], p. 371) $\{a_n, n\in N^d\}$, where $a_n \in B$, $n\in N^d$.

Definition 2.1. We say that $a_n \to a$ strongly as $n\to\infty$ $(a\in B)$, if the following condition is satisfied:

(a) for any $\varepsilon>0$, $||a_n-a||>\varepsilon$ occurs finitely often, that is, there exists $n(\varepsilon)\in N^d$ such that $n\nleq n(\varepsilon)$ implies $||a_n-a||<\varepsilon$ (see Smythe [9]).

Remark 2.1. (1) It is easy to see that (a) is equivalent to

(a') the net $\{a_n, n\in K\}$ converges to a in the usual sense for any infinite directed subset $(K, \leq)$ of N^d.

(2) In (a), by Lemma V-1-1 of Neveu [8], it is sufficient to require the condition for infinite linearly ordered sets $(K, \leq)$.

Definition 2.2. Let Y and Y_n $(n\in N^d)$ be B-valued r.v.'s defined on the probability space $(\Omega, \mathfrak{A}, P)$. We say that $Y_n \to Y$ strongly almost surely as $n\to\infty$, if

(b) $P\{\omega: Y_n(\omega)\to Y(\omega)$ strongly as $n\to\infty\}=1$, that is, for $\varepsilon>0$

$P\{||Y_n-Y||>\varepsilon$ occurs finitely often$\}=1$.

Remark 2.2. By induction on d, it is easy to prove that (b) is equivalent to

(b') the net $\{Y_n, n\in K\}$ converges to Y a.s. for each infinite directed subset K of N^d.

Definition 2.3. Let Y and Y_n $(n\in N^d)$ be B-valued r.v.'s. We say that $Y_n \to Y$ strongly in probability as $n\to\infty$ if

(c) $\{Y_n, n\in K\}$ converges to Y in probability for each infinite directed subset K of N^d.

Remark 2.3. (1) By Lemma V-1-1 of Neveu [8], one can show that in (c) it can be supposed that K is linearly ordered.

(2) If $Y_n \to Y$ strongly a.s. as $n\to\infty$, then $Y_n \to_p Y$ strongly as $n\to\infty$.

Lemma 2.1. Let $r>0$ and let X be a B-valued r.v. Then

$$E||X||^r(\log^+||X||)^{d-1}<\infty$$

if and only if for each $\varepsilon>0$

$$\sum_{n\in N^d} P\{||X||\geq |n|^{\frac{1}{r}}\varepsilon\}<\infty.$$

PROOF. See Lemma 2.1 of Gut [4].

Lemma 2.2. Let $Y_n = X\, I\{||X||\leq |n|^{\frac{1}{r}}\}, n\in N^d$, where X is a B-valued r.v., $0<r<2$ and $I\{.\}$ denotes the indicator function of the set in braces. Suppose that $E||X||^r(\log^+||X||)^{d-1}<\infty$. Then

$$\sum_{n\in N^d} \frac{E||Y_n||^2}{|n|^{\frac{2}{r}}}<\infty$$

PROOF. See Gut [4], Lemma 2.2.

Lemma 2.3. Let $\{X_n,\ n\in N^d\}$ be independent symmetric B-valued r.v.'s. Let $\{a_n,\ n\in N^d\}$ be a set of positive numbers such that $a_k\geq a_n$ if $k\geq n$ and $a_n\to\infty$ strongly. Assume that

(i) $||X_n||\leq a_n$ a.s. $(n\in N^d)$;

(ii) $S_n/a_n \to_p 0$ strongly.

Then $E(||S_n||/a_n)\to 0$ strongly as $n\to\infty$.

PROOF. Lemma 3.1 of Acosta [1] shows that the sequence $\{E(||S_n||/a_n),\ n\in K\}$ converges to 0, where K is a linearly ordered infinite subset of N^d. Thus, by (2) in Remark 2.1, $E(||S_n||/a_n)\to 0$ strongly.

Lemma 2.4. Let $x_n\in B$ $(n\in N^d)$, and suppose that $x_n\to 0$ strongly as $n\to\infty$. Let $a_{n,m}$ $(n,m\in N^d)$ be numbers. Assume that

(i) $a_{n,m} \to 0$ strongly as $n \to \infty$ for each fixed $m \in N^d$;

(ii) $\sum_{m \in N^d} |a_{n,m}| \le M < \infty$ for all $n \in N^d$.

Then $y_n = \sum_{m \in N^d} a_{n,m} x_m \to 0$ strongly as $n \to \infty$.

PROOF. This is proved just as in the case d=1 (see e.g. Stout [10], Lemma 3.2.3 (ii)).

Notations. Let $N_o = \{0,1,2,\ldots\}$ and $N_{-1} = \{-1,0,1,2,\ldots\}$. If $n=(n_1,\ldots,n_d) \in N_{-1}^d$, then $2^n = (2^{n_1},\ldots,2^{n_d}) \in N_o^d$, where 2^{-1} is defined as 0. For the sake of brevity, let $|2^{k-1}| = |2^k - 2^{k-1}|$, $k \in N_o^d$ (which is a trivial identity if $k \in N^d$). Let

$$S_k^n = \sum_{k<\ell\le n} X_\ell ,$$

where $n \in N^d$, $k \in N_o^d$, $k<n$.

Lemma 2.5. Let $\{X_n,\ n \in N^d\}$ be independent symmetric B-valued r.v.'s, $1 \ge r > 0$. Assume that

(i) for all $\varepsilon > 0$

$$\sum_{n \in N_o^d} P\left\{ \left| \frac{||S_{2^{n-1}}^{2^n}|| - E||S_{2^{n-1}}^{2^n}||}{|2^{n-1}|^r} \right| > \varepsilon \right\} < \infty;$$

and

(ii) $E(||S_n||/|n|^r) \to 0$ strongly as $n \to \infty$.

Then $S_n/|n|^r \to 0$ strongly a.s. as $n \to \infty$.

PROOF. By the help of the Borel-Cantelli lemma it follows from condition (i) that

$$\frac{||S_{2^{n-1}}^{2^n}|| - E||S_{2^{n-1}}^{2^n}||}{|2^{n-1}|^r} \to 0 \quad \text{strongly a.s. as } n \to \infty.$$

It follows from (ii) that $E(||S_{2^{n-1}}^{2^n}||/|2^{n-1}|^r) \to 0$ strongly as $n \to \infty$. Thus

$$\frac{||S_{2^{n-1}}^{2^n}||}{|2^{n-1}|^r} \to 0 \text{ strongly a.s. as } n \to \infty. \tag{2.1}$$

It is easy to see that

$$\frac{S_{2^n}}{|2^n|^r} = \sum_{0\le k\le n} \frac{S_{2^{k-1}}^{2^k}}{|2^{k-1}|^r} \frac{|2^{k-1}|^r}{|2^n|^r} . \tag{2.2}$$

By (2.1), (2.2) and Lemma 2.4

$$\frac{S_{2^n}}{|2^n|^r} \to 0 \text{ strongly a.s. as } n\to\infty.$$

Now let $n\in N^d$ be fixed and choose k such that $2^{k-1}<n\le 2^k$. Then

$$\frac{||S_n||}{|n|^r} \le \frac{||S_{2^{k-1}}||}{|n|^r} + \frac{1}{|n|^r} \Big|\Big| \sum_{\substack{\ell\le n\\ \ell\nleq 2^{k-1}}} X_\ell \Big|\Big| \le$$

$$\le \frac{||S_{2^{k-1}}||}{|2^{k-1}|^r} + \frac{1}{|2^{k-1}|^r} T_k ,$$

where

$$T_k = \max_{2^{k-1}<n\le 2^k} \Big|\Big| \sum_{\substack{\ell\le n\\ \ell\nleq 2^{k-1}}} X \Big|\Big| .$$

We know that $||S_{2^{k-1}}||/|2^{k-1}|^r\to 0$ strongly a.s. Thus the proof will be completed if we show that $T_k/|2^{k-1}|^r\to 0$ strongly a.s. Using the Lévy inequality (see Hoffmann-Jørgensen [5])

$$\sum_{k\in N_o^d} P\left\{\frac{T_k}{|2^{k-1}|^r} >\varepsilon\right\} \le \sum_{k\in N_o^d} 2P \left\{\frac{||S_{2^k}-S_{2^{k-1}}||}{|2^{k-1}|^r} >\varepsilon\right\}$$

$$\le \sum_{k\in N_o^d} 2(2^d-1)P \left\{\frac{||S_{2^{k-1}}^{2^k}||}{|2^{k-1}|^r} > \frac{\varepsilon}{2^{d-1}}\right\}$$

for $\varepsilon>0$. By (2.1) and the converse of the Borel-Cantelli lemma the last expression is finite. Therefore the Borel-Cantelli lemma implies

$$\frac{T_k}{|2^{k-1}|^r} \to 0 \quad \text{strongly a.s. as } k\to\infty.$$

3. MARCINKIEWICZ STRONG LAWS OF LARGE NUMBERS

Theorem 3.1. Let B be a separable Banach space, $1 \leq p < 2$. Let $\{X_n, n \in N^d\}$ be i.i.d. B-valued r.v.'s. Assume that $E||X_1||^p(\log^+||X_1||)^{d-1} < \infty$. If

$$S_n/|n|^{\frac{1}{p}} \to_p 0 \text{ strongly as } n \to \infty, \tag{3.1}$$

then $S_n/|n|^{\frac{1}{p}} \to 0$ strongly a.s. as $n \to \infty$.

PROOF. Let us first assume that X_1 is symmetric. Define $Y_n = X_n I\{||X_n|| \leq |n|^{\frac{1}{p}}\}$, then Y_n is symmetric. Let $T_n = \sum_{\ell \leq n} Y_\ell$. By Lemma 2.1

$$\sum_{n \in N^d} P\{||X_n|| > |n|^{\frac{1}{p}}\} < \infty.$$

It follows from the Borel-Cantelli lemma that

$$P\{X_n \neq Y_n \text{ occurs finitely often}\} = 1.$$

Therefore

$$\frac{S_n - T_n}{|n|^{\frac{1}{p}}} \to 0 \text{ strongly a.s. as } n \to \infty. \tag{3.2}$$

Thus, it is enough to prove $T_n/|n|^{\frac{1}{p}} \to 0$ strongly a.s. It follows from (3.1) and (3.2) that $T_n/|n|^{\frac{1}{p}} \to_p 0$ strongly. By Lemma 2.3, $E(||T_n||/|n|^{\frac{1}{p}}) \to 0$ strongly. For $n \in N_0^d$ let

$$V_n = ||T_{2^{n-1}}^{2^n}|| - E||T_{2^{n-1}}^{2^n}||.$$

According to Lemma 2.5 the proof will be complete if we show that

$$\sum_{n \in N_0^d} P\left\{\frac{|V_n|}{|2^{n-1}|^{\frac{1}{p}}} > \varepsilon\right\} < \infty$$

for any $\varepsilon > 0$.

Theorem 2.1 of Acosta [1] implies

$$\sum_{n\in N_o^d} P\left\{\frac{|V_n|}{|2^{n-1}|^{\frac{1}{p}}} > \varepsilon\right\} \le \sum_{n\in N_o^d} \frac{1}{\varepsilon^2 |2^{n-1}|^{\frac{1}{p}}} E\, V_n^2$$

$$\le \sum_{n\in N_o^d} \frac{1}{\varepsilon^2 |2^{n-1}|^{\frac{2}{p}}}\, 4 \sum_{2^{n-1}<\ell\le 2^n} E||Y_\ell||^2$$

$$\le \sum_{\ell\in N^d} \frac{4\cdot 2^{\frac{2d}{p}}}{\varepsilon^2} \frac{E||Y_\ell||^2}{|\ell|^{\frac{2}{p}}}.$$

By Lemma 2.2, this expression is finite. Hence the theorem is proved if X_1 is symmetric.

If X_1 is not symmetric, then we consider a symmetrization (see Hoffmann-Jørgensen [5]) $X_n^s=(X_n-X_n')$ of X_n. Since the function $f(x)=x^p(\log^+ x)^{d-1}$ $(1\le p<2,\ d\in N,\ x\ge 0)$ is convex and nondecreasing it follows that

$E||X_1^s||^p(\log^+||X_1^s||^{d-1}<\infty$.

Therefore $S_n^s/|n|^{\frac{1}{p}}\to 0$ strongly a.s.

As in Theorem 2 of Chapter II of Kahane [7], one can prove that there exists a non-random sequence $\{s_n,\ n\in N^d\}$ in B such that

$$\frac{S_n}{|n|^{\frac{1}{p}}} - \frac{s_n}{|n|^{\frac{1}{p}}} \to 0 \text{ strongly a.s.} \tag{3.3}$$

(3.1) and (3.3) yield $s_n/|n|^{\frac{1}{p}}\to 0$ strongly as $n\to\infty$, which together with (3.3) shows that

$S_n/|n|^{\frac{1}{p}}\to 0$ strongly a.s.

<u>Theorem 3.2</u>. Let $1\le p<2$ and let B be of Rademacher type p (see Hoffmann-Jørgensen and Pisier [6]). Let $\{X_n,\ n\in N^d\}$ be i.i.d. B-valued r.v.'s. Suppose that $E||X_1||^p(\log^+||X_1||)^{d-1}<\infty$ and $E\,X_1=0$. Then

$S_n/|n|^{\frac{1}{p}}\to 0$ strongly a.s. as $n\to\infty$.

PROOF. Theorem 4.1 of Acosta [1] implies that the sequence $\{S_n/|n|^{\frac{1}{p}}, n\in K\}$ converges to 0 in probability, where K is an arbitrary infinite linearly ordered subset of N^d. By [1] in Remark 2.3,

$$S_n/|n|^{\frac{1}{p}} \to_p 0 \text{ strongly as } n\to\infty.$$

An application of Theorem 3.1 ends the proof.

4. FINAL COMMENTS

Remark 4.1. Let B be a separable Banach space, $0<p<1$. Let $\{X_n, n\in N^d\}$ be i.i.d. B-valued r.v.'s. Assume that

$$E||X_1||^p(\log^+||X_1||)^{d-1}<\infty.$$

Then $S_n/|n|^{\frac{1}{p}}\to 0$ strongly a.s. as $n\to\infty$.

PROOF. By Theorem 3.2 of Gut [4], $S_n/|n|^{\frac{1}{p}}\to 0$ a.s. Making use of this fact, one can prove by induction on d that

$$S_n/|n|^{\frac{1}{p}}\to 0 \text{ strongly a.s. as } n\to\infty.$$

Remark 4.2. Let $\{X_n, n\in N^d\}$ be i.i.d. B-valued r.v.'s $0<p<2$. If

$$S_n/|n|^{\frac{1}{p}}\to 0 \text{ strongly a.s. as } n\to\infty, \qquad (4.1)$$

then $E||X_1||^p(\log^+||X_1||)^{d-1}<\infty$.

PROOF. If (4.1) holds, then $X_n/|n|^{\frac{1}{p}}\to 0$ strongly a.s. By the converse of the Borel-Cantelli lemma, this implies

$$\sum_{n\in N^d} P\{||X_n||\geq |n|^{\frac{1}{p}}\varepsilon\}<\infty.$$

An application of Lemma 2.1 completes the proof.

REFERENCES

1. A. DE ACOSTA, Inequalities for B-valued random vectors with applications to the strong law of large numbers, Ann. Probability, Vol. 9 (1981), 157-161.

2. R. B. ASH, Real Analysis and Probability, Academic Press, New York and London (1972).

3. I. FAZEKAS, Convergence of vector-valued martingales with multidimensional indices, Publ. Math. (Debrecen) (to be published).

4. A. GUT, Marcinkiewicz laws and convergence rates in the law of large numbers for random variables with multidimensional indices, Ann. Probability, Vol. 6 (1978), 469-482.

5. J. HOFFMANN-JØRGENSEN, Sums of independent Banach space valued random variables, Studia Math. Vol. 52 (1974), 159-186.

6. J. HOFFMANN-JØRGENSEN and G. PISIER, The law of large numbers and the central limit theorem in Banach spaces, Ann. Probability, Vol. 4 (1976), 597-599.

7. J.-P. KAHANE, Some Random Series of Functions, Heath, Lexington, Mass. (1968).

8. J. NEVEU, Discrete-Parameter Martingales, North-Holland, Amsterdam (1975).

9. R. T. SMYTHE, Strong laws of large numbers for r-dimensional arrays of random variables, Ann. Probability, Vol. 1 (1973), 164-170.

10. W. F. STOUT, Almost Sure Convergence, Academic Press, New York (1974).

Proc. of the 3rd Pannonian Symp. on Math. Stat., Visegrád, Hungary 1982
J. Mogyoródi, I. Vincze, W. Wertz, eds

A CONVERGENT SELF-TUNING PREDICTOR

L. GERENCSÉR

Computer and Automation Institute,
Hungarian Academy of Sciences, Budapest, Hungary

ABSTRACT

In this paper we shall improve the self-tuning predictor method proposed by WITTENMARK [6] in the sense that the new version has guaranteed local convergence for constant parameter processes.

1. SOLUTION OF THE PREDICTION PROBLEM

In this section we shall summarize the method of solving the k-step ahead minimumvariance prediction problem for ARMA-processes (see ASTRÖM [1], WITTENMARK [6]).

Let us consider the ARMA process defined by

$$y(t)+a_1y(t-1)+\dots+a_ny(t-n)=e(t)+c_1e(t-1)+\dots+c_ne(t-n) \quad (1.1)$$

with $y(t)=e(t)=0$ for $t\geq 0$. This assumption on the initial conditions does not restrict the applicability of our arguments but it greatly simplifies the calculations. Here $e(t)$ is a discrete white noise process.

ARMA processes are also represented in the form

$$A\,y(t)=C\cdot e(t) \quad (1.2)$$

where $A=A(q^{-1})$, $C=C(q^{-1})$ are polynomials of the forward shift operator q^{-1}, thus e.g.

$$A=1+a_1q^{-1}+\dots+a_nq^{-n}. \quad (1.3)$$

Remark 1.1. The correspondence between the ring of difference operators and the ring of real polynomials is a homomorphism, i.e. if the polynomial $A(z)$ corresponds to the operator A then we have

$$(A+B)(z)=A(z)+B(z),\quad (AB)(z)=A(z)\cdot B(z) \tag{1.4}$$

where additon and multiplication on the left hand side should be interpreted in the operator sense. We can write (1.2) formally as

$$y(t+k)=(C/A)e(t+k) \tag{1.5}$$

which should be understood as an equivalence form of (1.2). To separate the signals after and before t we shall split C/A into the form:

$$C/A=F+q^{-k}G/A \tag{1.6}$$

where F is a polynomial of degree $K-1$ and G is a polynomial of degree $n-1$, say $G=g_0+g_1q^{-1}+\ldots+g_{n-1}q^{-n+1}$.

Then we can rewrite (1.5) in the form:

$$y(t+k)=Fe(t+k)+G/Ae(t). \tag{1.7}$$

To eliminate $e(t)$ in the second term let us observe that from (1.2) we have

$$(G/A)e(t)=(G/C)y(t). \tag{1.8}$$

Thus we get the following

Theorem 1.1. The least square estimation $\hat{y}(t+k|t)$ of $y(t+k)$ on the basis of observations up to time t is given by

$$\hat{y}(t+k|t)=(G/C)y(t). \tag{1.9}$$

Remark 1.2. This expression should be understood as a recursive formula for the computation of $\hat{y}(s+k|s)$.

Remark 1.3. There is an alternative form of (1.9) which is quite useful to know. Let the prediction error be denoted by $\varepsilon(t+k)$ i.e.

$$\varepsilon(t+k)=y(t+k)-\hat{y}(t+k|t). \tag{1.10}$$

Then we have from (1.9)

$$\hat{y}(t+k|k)=(G/C)(\hat{y}(t|t-k)+\varepsilon(t)), \tag{1.11}$$

i.e.

$$(C-q^{-k}G)\hat{y}(t+k/t)=G\varepsilon(t) \tag{1.12}$$

from which (taking into account (1.6)) we get

Theorem 1.2. The k step ahead minimum variance prediction $\hat{y}(t+k/t)$ satisfies the recursion

$$\hat{y}(t+k/t)=(G/AF)\varepsilon(t). \tag{1.13}$$

2. REFORMULATION OF THE PREDICTION PROBLEM

In this section we shall reformulate the prediction problem. This will enable us to define the best predictor on the basis of input-output data only, i.e. without the explicit use of the system parameters.

Let us introduce the notations.

$$H=AF=1+h_1q^{-1}+\ldots+h_\ell q^{-\ell} \quad (\ell=n+k-1) \tag{2.1}$$

$$\Theta^x=(h_1,\ldots,h_\ell,g_0,\ldots,g_{n-1})^T. \tag{2.2}$$

If Θ^x is not known we may try some Θ to generate predictions, and this way we shall get processes $\hat{y}(t,\Theta)$, $\varepsilon(t,\Theta)$. (We shall not emphasize in the notations that the prediction is k step ahead).

Let us form the vector

$$\phi(t,\Theta)=(-\hat{y}(t+k-1,\Theta),\ldots,-\hat{y}(t+1-n\Theta), \\ \varepsilon(t,\Theta),\ldots,\varepsilon(t-n,\Theta))^T. \tag{2.3}$$

Then the predictor (1.13) can be written in the form:

$$\hat{y}(t+k,\Theta)=\Theta^T\phi(t,\Theta) \tag{2.4}$$

and

$$\varepsilon(t,\Theta)=y(t)-\hat{y}(t,\Theta). \tag{2.5}$$

We have a relationship

$$y(t+k)=\Theta^T\phi(t,\Theta)+\varepsilon(t+k,\Theta) \tag{2.6}$$

with the property that for $\Theta=\Theta^x$ the residual $\varepsilon(t+k,\Theta^x)$ is independent of all previous $\phi(s,\Theta)$ -s with $s\geqq t$. Let us define

$$f(\Theta)=\lim_t E\phi(t,\Theta)(y(t+k)-\Theta^T\phi(t,\Theta)). \tag{2.7}$$

Theorem 2.1. If the polynomials $z^nA(z), z^{k-1}F(z)$ are stable i.e. all their roots lie inside the unit circle then the function f in (2.7) is well-defined and

$$f(\Theta^*)=0. \tag{2.8}$$

This reformulation of the prediction problem enables us to estimate Θ^* on the basis of observed values of $y^{(t)}$. To find an algorithm we shall apply the theory developed by LJUNG [4] in its improved version (LIPCSEY [3]).

The first results in this direction are due to WITTENMARK [6]. He proposed an algorithm which is a combination of the recursive least square (RLSQ) method applied for the model

$$y(t+k)=\Theta^T\phi(t,\Theta^*)+\varepsilon(t+k,\Theta) \tag{2.9}$$

and an adaptation of the predictor. Convergence cannot be quaranteed for this method.

3. APPLICATION OF A STOCHASTIC NEWTON METHOD

To guarantee local convergence we propose to use a Newton method to solve (2.8). Stochastic quasi-Newton methods are almost as old as stochastic approximation methods. In the context of LJUNG's scheme they were proposed in [5] and [2]. We can represent the Jacobian matrix of f in the form

$$f_{\Theta}(\Theta)=\lim_{t} E\{\phi_{\Theta}(t,\Theta)y(t+k)-\phi_{\Theta}(t,\Theta)\Theta^{T}\phi(t,\Theta)- \\ -\phi(t,\Theta)\phi^{T}(t,\Theta)-\phi(t,\Theta)\Theta^{T}\phi_{\Theta}(t,\Theta). \tag{3.1}$$

(Multiplication is to be understood in the tensor sense.) To generate $\phi_{\Theta}(t,\Theta)$ (i.e. $\hat{y}_{\Theta}(t,\Theta)$ and $\varepsilon_{\Theta}(t,\Theta)$ let us differentiate the equalities (2.4),(2.5) with respect to Θ. We get:

$$\varepsilon_{\Theta}(t,\Theta)=-\hat{y}_{\Theta}(t,\Theta) \tag{3.2}$$

$$\hat{y}_{\Theta}(t+k,\Theta)=\phi_{\Theta}(t,\Theta)\theta+\phi(t,\Theta). \tag{3.3}$$

It is very important to observe that these processes can be generated without the knowledge of A and C, it is sufficient to observe y(t). Thus we propose the following

Algorithm 3.1. At time k let $\Theta(0)$ be an initial estimate and let $R(0)=I$. At time t+k we proceed as follows:

1. Newton-like step:

$$\Theta(t)=\Theta(t-1)-(1/t)R^{-1}(t)\phi(t)\varepsilon(t+k). \tag{3.4}$$

2. Updating the Jacobian:

$$R(t)=R(t-1)+(1/t)(\phi_{\Theta}(t)y(t+k)-\phi_{\Theta}(t)\Theta^{T}(t)\phi(t)- \\ -\phi(t)\phi^{T}(t)-\phi(t)\Theta^{T}(t)\phi_{\Theta}(t)+\delta I-R(t-1)),\quad \delta>0. \tag{3.5}$$

Here $\phi(t)$, $\phi_{\Theta}(t)$ are on-line estimates of $\phi(t,\Theta(t))$, $\phi_{\Theta}(t,\Theta(t))$ and

$$\phi(t)=(-\hat{y}(t+k-1),\dots,-\hat{y}(t+1-n),\ \varepsilon(t),\dots,\varepsilon(t-n+1))^{T} \tag{3.6}$$

$$\varepsilon(t+k)=y(t+k)-\hat{y}(t+k). \tag{3.7}$$

To generate $\phi_{\Theta}(t)$ we use the formulae

$$\varepsilon_{o}(t)=-\hat{y}_{\Theta}(t) \tag{3.8}$$

$$\hat{y}_{\Theta}(t+k)=\phi_{\Theta}(t)\Theta(t)+\phi(t). \tag{3.9}$$

Theorem 3.1. Let us suppose that the polynomials $z^n A(z^{-1})$, $z^{k-1}F(z^{-1})$ are stable. Then the sequence $\Theta(t), R(t)$ generated by the algorithm will converge to Θ^x, R^x $(R^x=f_\Theta(\Theta^x)+\delta I)$ with positive probability if only δ is sufficiently small.

Proof: The proof is based on a modified version of LJUNG's theorem (LJUNG[4], LIPCSEY [3]). We can associate a deterministic continuous time differential equation with Algorithm 3.1 which in some sense approximates the discrete process $\Theta(t)$, $R(t)$. This differential equation has to be computed as follows: fix some Θ, R and generate the processes $\hat{y}(t,\Theta)$, $\varepsilon(t,\Theta)$. Then compute the asymptotic value of the correction term in (3.4):

$$-\lim_t E\, R^{-1}\phi(t,\Theta)\varepsilon(t+k,\Theta)=-R^{-1}f(\Theta). \tag{3.10}$$

For (3.5) we get (comparing it with (3.1)) the asymptotic value

$$f_\Theta(\Theta)+\delta I-R. \tag{3.11}$$

Hence the associated differential equation will have the form:

$$\dot{\Theta}(t)=-R^{-1}(t)f(\Theta(t) \tag{3.12}$$

$$\dot{R}(t)=f_\Theta(\Theta(t))+\delta I-R(t). \tag{3.13}$$

For sufficiently small δ this differential equation will be asymptotically stable in (Θ^x, R^x), hence Theorem 3.1 is a consequence of the results of the mentioned papers (see appendix).

Remark 3.1. The above method can be extended for the case when we restrict ourselves to low order predictors. Let the vector $\phi(t)$ be redefined as

$$\phi(t)=(-y(t+k-1),\ldots,-y(t+k-p),\varepsilon(t),\ldots,\varepsilon(t-q+1))^T, \tag{3.14}$$

where p,q are some fixed integers, and let the k step ahead predictor be defined by

$$\hat{y}(t+k,\Theta)=\Theta^T\phi(t,\Theta) \tag{3.15}$$

and

$$\varepsilon(t+k,\Theta)=y(t+k)-\hat{y}(t+k,\Theta). \tag{3.16}$$

The best predictor among the predictors of order (p,q) are chosen by solving

$$\min_{\Theta} V(\Theta) \tag{3.17}$$

where

$$V(\Theta)=\lim_{t} E\varepsilon^2(t,\Theta).$$

This problem again can be solved by stochastic Newton method however second order derivatives of $\varepsilon(t,\Theta)$, $\hat{y}(t,\Theta)$ should be computed.

Remark 3.2. Algorithm 3.1 can be extended to multivariable ARMA processes without difficulty.

Remark 3.3. The proposed algorithm exhibits an important difference between identification and control. In identification problems the trial value of the unknown parameter has effect on the simulated process only, hence derivative processes can be generated. This is in contrast with control problems where the estimated value has effect on the physical plant, hence derivative processes may not be computable, unless the plant is identified.

4. APPENDIX

LJUNG's theorem in its original form is difficult to understand and apply. Therefore we shall describe a recent improvement of the theorem due to LIPCSEY [3].

Let a discrete time stochastic process $x(t)\in R^n$ be defined by a stochastic difference equation

$$x(t)=Ax(t-1)+Be(t) \qquad x(0)=0. \tag{4.1}$$

(The initial condition was chosen 0 for the sake of simplicity.)

The following conditions are assumed:

Condition 4.1. $e(t) \in R^n$ $t=1,\ldots$ is a discrete time white noise process.

Condition 4.2. The $n \times n$ matrix A is stable i.e. its spectrum is contained in the unit circle.

If these two conditions are satisfied then we say that $x(t)$ is an asymptotically elementary Gaussian process (AEG process). It has the property that its finite dimensional distributions weakly converge to a unique stationary distribution as $t \to \infty$.

Now consider the parametrized process $x(t,\Theta) \in R^n$, $\Theta \in R^p$ defined by

$$x(t,\Theta)=A(\Theta)x(t-1,\Theta)+B(\Theta)e(t). \tag{4.2}$$

It can be interpreted as a combination of a physical process plus a process simulated by a computer attached to the physical device.

Condition 4.3. $A(\Theta)$, $B(\Theta)$ are defined for $\Theta \in D \subset R^p$ where D is some open set and the functions A,B are continuously differentiable on D. Moreover $A(\Theta)$ is stable for all $\Theta \in D$.

The environment in which an algorithm has to work is described by

Condition 4.4. $x(t)$ is a state vector of the system which consists of the physical plant plus a computer attached to the plant. That is if we fix a $\Theta(t-1)$ at time $t-1$ then an $x(t)$ will be produced by the system satisfying

$$x(t)=A(\Theta(t-1))x(t-1)+B(\Theta(t-1))e(t). \tag{4.3}$$

We shall define some functional of the process which depends on Θ and the corresponding stationary distribution. Let a function $Q: R^n x D \to R^p$ be given with the following properties.

Condition 4.5. Q is continuously differentiable on D and the estimations

$$Q(x,\Theta)<c(1+|x|^k) \tag{4.4}$$

$$Q_\Theta(x,\Theta)<c(1+|x|^k) \qquad x\in R^n \tag{4.5}$$

hold with some c, k>0.

Under this condition the function

$$f(\Theta)=\lim_t E\, Q(x(t,\Theta),\Theta) \tag{4.6}$$

is well-defined. Let us assume

Condition 4.6. The nonlinear algebraic equation $f(\Theta)=0$ has a unique solution, say Θ^x in D.

To compute Θ using available information only the following algorithm is proposed

Algorithm 4.1.

$$\Theta(t)=\Theta(t-1)+(1/t)Q(x(t),\Theta(t-1)) \tag{4.7}$$

$$x(t)=A(\Theta(t-1))+B(\Theta(t-1))e(t). \tag{4.8}$$

If we do not add any measure to prevent $\Theta(t)$ from falling outside D we have the following theorem.

Theorem 4.1. Let us assume that the differential equation

$$\dot{\Theta}(s)=f(\Theta(s)) \tag{4.9}$$

is asymptotically stable in Θ^x and $f_\Theta(\Theta^x)$ is nonsingular. Then for the sequence $\Theta(t)$ defined by Algorithm 4.1 $\Theta(t)\to\Theta^x$ with positive probability.

Remark 4.1. The probability of convergence can be close to 1 if the differential equation (4.9) has a Lyapunov function in a large region.

REFERENCES

1 ASTRÖM, K.J. (1970) Introduction to Stochastic Control Theory. Academic Press, New York, London.

2 GERENCSÉR, L. (1979) Stochastic quasi-Newton methods and convergent self-tuning regulators. Computer and Automation Inst. Hung. Acad. Sci., Working Paper IV/4.

3 LIPCSEY, ZS. (1982) A generalization of Ljung's theorem. Computer and Automation Inst. Hung. Acad. Sci., Working Paper IV/20.

4 LJUNG, L. (1977) Analysis of recursive stochastic algorithms. IEEE Trans. Automatic Control, 22,4, 551-575.

5 LJUNG, L. (1981) Analysis of a general recursive prediction error identification algorithm. Automatica, 17,1, 89-99.

6 WITTENMARK, B. (1974) A self-tuning predictor. IEEE Trans. Automatic Control, 19,6, 848-851.

Proc. of the 3rd Pannonian Symp. on Math. Stat., Visegrád, Hungary 1982
J. Mogyoródi, I. Vincze, W. Wertz, eds

THE INVERSE PÓLYA DISTRIBUTION AND ITS MOMENTS DERIVED BY THE COMPOUNDING OPERATION

T. GERSTENKORN*, J. JARZĘBSKA**

*Institute of Mathematics, Łódz University,
**Higher Pedagogical School, Częstochowa, Poland

ABSTRACT

In this paper the compounding method (randomization) has been used for obtaining, in a simple way, the inverse Pólya distribution, its factorial moments, and the recurrence relations for factorial and crude moments.

The results for the inverse Pólya distribution were obtained by compounding a negative binomial distribution with a beta one. In the case of formula for moments, there were used formulae for the distribution subject to randomization, already known. Yet the simplicity of the applied method is purchased by bounding the range of the distribution parameter to positive values. The formulae presented here can find their application in practice.

INTRODUCTION

In this paper the compounding method (randomization) has been used for obtaining, in a simple way, the inverse Pólya distribution (abbr.: i.P.d.), its factorial moments, and the recurrence relations for factorial and crude moments. In [14] J.G. Skellam obtained with this method a distribution expressed by means of the binomial and the beta ones. A slight modification of the procedure ([2], VII §4, example (a); [15], II §10, example 4) leads to the Pólya distribution. The importance of the method found its reflex in handbooks (e.g. [2], II §5; [3], §5.13; [7], §7.4). In paper [1] the randomization method was used in order to obtain formulae for moments of the Poisson, Pólya and Pascal distributions. The bibliography on some compound distributions as well as the discussion of the results concerning them can be found in N.L. Johnson and S. Kotz [8], Chap. 8, pp. 183-202 and 213-215.

The results for the i.P.d. were obtained in the paper by compounding a negative binomial distribution (abbr.: n.b.d.) with a beta one (abbr.: be.d.). In the case of formulae for moments, there were used formulae for the distribution subject to randomization, already known. Yet the simplicity of the applied method is purchased by bounding the range of the distribution parameter.

1. THE COMPOUNDING METHOD

In applications we encounter again and again a random variable (abbr.: r.v.) X whose distribution depends upon some parameter λ being a value of a r.v. Λ with a given distribution. The r.v. X is then said to have a compound distribution.

We shall consider here the case when X is a discrete r.v. with distribution $P(X=x_i|\lambda)$, whereas Λ is a continuous r.v. with density $g(\lambda)$. Let (X,Λ) be a two-dimensional r.v. with distribution $P(X=x_i,\Lambda=\lambda)$. Then the compound distribution will be written down as a marginal distribution $h(x_i)$ of the r.v. X in the two-dimensional distribution of the r.v. (X,Λ)

$$h(x_i)=P(X=x_i)=\int_{-\infty}^{+\infty} g(\lambda)P(X=x_i|\lambda)d\lambda. \qquad (1.1)$$

We shall often write the above relationship in the form

$$h(x_i)=P(X=x_i|\lambda)\underset{\Lambda}{\wedge}G(\lambda) \qquad (1.2)$$

where $G(\lambda)$ is a distribution of the r.v. Λ with density $g(\lambda)$.

2. THE INVERSE PÓLYA DISTRIBUTION

The i.P.d. $P(k,p,1,a)$ can be obtained by compounding the n.b.d. $Pa(k,\lambda)$ with the beta one Be $(p/a,\ q/a)$, i.e., by assuming that the parameter λ, characterizing the probability of gaining a success in the n.b.d. (the Pascal scheme), is a value of a r.v. Λ with the be.d. in question.

In accordance with (1.2), we may write

$$P(k,p,1,a)=Pa(k,\lambda)\underset{\Lambda}{\wedge}\ Be(p/a,\ q/a) \qquad (2.1)$$

with

$$Pa(k,\lambda)=\binom{k+r-1}{r}\lambda^k(1-\lambda)^r \qquad (2.2)$$

where $0<\lambda<1$, $r=0,1,2,\ldots,k$; k is a fixed positive integer, and the density of the be. d. $Be(p/a,\ q/a)$ is given by

$$g(\lambda,p/a,q/a)= \qquad (2.3)$$

$$= \begin{cases} \frac{\Gamma(1/a)}{\Gamma(p/a)\Gamma(q/a)} \lambda^{\frac{p}{a}-1} (1-\lambda)^{\frac{q}{a}-1} & \text{for } 0<\lambda<1, \\ 0 & \text{otherwise} \end{cases}$$

where $0<p<1$, $q=1-p$, $a>0$.

In the further part of the paper we shall use the factorial polynomials of the variable x, which, with degree r and step h, are defined by

$$x^{[r,h]}=x(x-h)(x-2h)\cdot\ldots\cdot(x-(r-1)h)$$

where $r=1,2,\ldots$, and h is an arbitrary real number.

If $h=0$, then $x^{[r,0]}=x^r$; in the case when $h=1$, we write $x^{[r,1]}=x^{[r]}$. Besides, we assume that

$$x^{[0]}=1 \text{ for } x\neq 0 \text{ and } x^{[-r,h]}=x^{[r,-h]}.$$

In the proofs of the theorems given later we shall refer to the following

Lemma. If i,j are integers satisfying the conditions $p/a + j>0$, $q/a + i>0$, and p and a satisfy the assumptions of formula (2.3), then

$$\underset{\Lambda}{\lambda^j(1-\lambda)^i}\wedge Be(p/a,q/a)=p^{[j,-a]}q^{[i,-a]}/1^{[i+j,-a]}= \\ =q^{[i,-a]}(a-1)^{[i+j,-a]}/(-1)^i(a-p)^{[j,a]} \tag{2.4}$$

holds.

Proof. The first part of (2.4) is easy to obtain from (1.1) and the equality: $\Gamma(p/a+j)=\Gamma(p/a)\cdot(p/a)^{[-j]}$. The other part is obtained after applying the transformation

$$(-1)(-p/a)^{[j]}/(-1/a)^{[i+j]}=(1/a-1)^{[-j-1]}/(p/a-1)^{[-j]}.$$

Using in (2.1) the first form of (2.4), we get the i.P.d., namely,

$$P(k,p,1,a)=\binom{k+r-1}{r}p^{[k,-a]}q^{[r,-a]}/1^{[k+r,-a]} \tag{2.5}$$

where $0<p<1$, $q=1-p$, $a>0$, $r=0,1,2,\ldots,k$ (k is a fixed positive integer).

If $a=b/N$, then (2.5) will pass into

$$P(k,Np,N,b)=\binom{k+r-1}{r}(Np)^{[k,-b]}(Nq)^{[r,-b]}/N^{[k+r,-b]}, \tag{2.6}$$

$$r=0,1,2,\ldots .$$

By adopting the notation $k+r=m$, we have

$$P(k,Np,N,b)=\binom{m-1}{k-1}(Np)^{[k,-b]}(Nq)^{[m-k,-b]}/N^{[m,-b]} \tag{2.7}$$

where $m=k,k+1,\ldots$, $k=1,2,\ldots,m$.

This formula may be interpreted on the ground of the Pólya urn scheme (the urn contains N balls; in that number, for instance, M white ones, $M/N=p$) in the way that it gives the probability of carrying out m trials to obtain k white balls.

The Pólya distribution and the inverse Pólya distribution are special cases of a rather general class of discrete distributions, the so-called generalized hypergeometric distributions ([9] and [13]).

3. MOMENTS

In order to obtain moments of a compound distribution, we apply the formula

$$E(X^{\ell})=E(X_{\lambda}^{\ell})\underset{\Lambda}{\wedge}G(\lambda)$$

and the analogous formula for factorial moments: ([1],(2′),(3′), p. 206) where X is a r.v. with a distribution dependent on the parameter λ, compounded with a distribution $G(\lambda)$.

The factorial and the crude moments of the n.b.d. (2.2) will be denoted by $m_{[\ell]}(k,\lambda)$ and $m_{\ell}(k,\lambda)$, respectively, and the moments of the i.P.d. (2.5) by $\alpha_{[\ell]}(k,p)$ and $\alpha_{\ell}(k,p)$ or, shortly, by $\alpha_{[\ell]},\alpha_{\ell}$.

The ℓ-th moment of the i.P.d. (2.5) is obtained through the randomization of the parameter λ of the moment of the n.b.d. Pa(k,λ), i.e. by assuming that this parameter has a be.d. (2.3). We then have

$$\alpha_{[\ell]}(k,p)=m_{[\ell]}(k,\lambda)\underset{\Lambda}{\wedge}Be(p/a,q/a), \tag{3.1}$$

$$\alpha_{\ell}(k,p)=m_{\ell}(k,\lambda)\underset{\Lambda}{\wedge}Be(p/a,q/a). \tag{3.2}$$

Theorem 3.1. The factorial moment of order ℓ of the i.P.d.

P(k,p,1,a) of form (2.5) is expressed (if $\ell < p/a$) by

$$\alpha_{[\ell]}=(-k)^{[\ell]}q^{[\ell,-a]}/(a-p)^{[\ell,-a]}, \quad \ell=1,2,3,\dots. \tag{3.3}$$

Proof. The ℓ-th factorial moment of the n.b.d. is given by

$$m_{[\ell]}(k,\lambda)=k^{[-\ell]}[(1-\lambda)/\lambda]^{\ell}, \quad \ell=1,2,3,\dots, \tag{3.4}$$

(e.g. [10], (5.40), p. 131). We obtain (3.3) on the ground of (3.1) by making use of (2.4).

Formula (3.3) for $\alpha_{[\ell]}$ can also be obtained by direct calculation (from definition) either as a special case of formula (14) in [9] or as a special case of the so-called incomplete moments of the i.P.d. (2.5) in [6].

To illustrate the above, we give three factorial moments calculated from (3.3):

$$\alpha_{[1]}=kq/(p-a), \alpha_{[2]}=kq(k+1)(q+a)/(2a^2-3ap+p^2),$$

$$\alpha_{[3]}=kq(k^2+3k+2)(q^2+3aq+2a^2)/(p^3-6ap^2+11a^2p-6a^3).$$

From (3.3), using the relation

$$\alpha_{\ell}=\sum_{i=0}^{\ell} S_i^{\ell}\alpha_{[i]}, \quad \ell=1,2,3,\dots, \tag{3.5}$$

where S_i^{ℓ} are the Stirling numbers of the second kind ([4], (2.26), p. 29, and the references cited there; also [11],

(18.3-31)), we get the crude moment of order ℓ of the i.P.d.

4. RECURRENCE RELATIONS

If we apply operation (3.2) to a recurrence relation for crude moments of the n.b.d. (2.2), we shall obtain a recurrence relation for crude moments of the i.P.d. (2.5). The use of recurrence relations is profitable since it allows us to avoid calculations based on the Stirling numbers occurring in formula (3.6). We shall first apply the formula ([1], (67), p. 224)

$$m_{\ell+1}(k,\lambda)=k\sum_{i=0}^{\ell}\binom{\ell}{i}\frac{1-\lambda}{\lambda}\,m_i(k+1,\lambda),\quad \ell=0,1,2,\ldots,$$

which yields, by making direct use of the definition of the moment for $Pa(k+1,\lambda)$ and, next, of formula (2.4)

$$\alpha_{\ell+1}=k\sum_{i=0}^{\ell}\binom{\ell}{i}(1-\lambda)\lambda^{-1}m_i(k+1,\lambda)\underset{\Lambda}{\wedge}Be(p/a,q/a)=$$

$$=k\sum_{i=0}^{\ell}\binom{\ell}{i}\sum_{r=0}^{\infty}r^i\binom{k+r}{r}p^{[k,-a]}q^{[r+1,-a]}/1^{[k+r+1,-a]}=$$

$$=\frac{kq}{p-a}\sum_{i=0}^{\ell}\binom{\ell}{i}\sum_{r=0}^{\infty}r^i\binom{k+r}{r}(p-a)^{[k+1,-a]}(q+a)^{[r,-a]}/1^{[k+r+1,-a]}$$

that is, we get

Theorem 4.1. The crude moments of the i.P.d. $P(k,p,1,a)$ given by (2.5) are expressed in the recurrent way by

$$\alpha_{\ell+1}(k,p)= \sum_{i=0}^{\ell} \binom{\ell}{i}\alpha_i(k+1,p-a), \quad \ell=0,1,2,\ldots, \tag{4.1}$$

where $\alpha=\alpha_1=kq/(p-a)$ is the expected value of the distribution under consideration and $\alpha_i(k+1,p-a)$ is the i-th crude moment of the i.P.d. P(k+1,p-a,1,a).

For the i.P.d. we shall determine a recurrence relation for moments, in which all moments will have the same parameters (which is different than in (4.1)).

Let us take into consideration the recurrence relation for crude moments of a n.b.d. ([1], (72), p. 226; [12], p. 329, 2nd ed. p. 94; [5], p. 21)

$$m_{\ell+1}= \frac{1-\lambda}{\lambda} \sum_{i=0}^{\ell} [k\binom{\ell}{i}+\binom{\ell}{i+1}] m_{\ell-i}, \quad \ell=0,1,2,\ldots,$$

and apply operation (3.2) to it. We shall then get

$$\alpha_{\ell+1}= \sum_{i=0}^{\ell} [k\binom{\ell}{i}+\binom{\ell}{i+1}][(1-\lambda)\lambda^{-1}m_{\ell-i}(k,\lambda)\underset{\Lambda}{\wedge}Be(p/a,q/a)]. \tag{4.2}$$

Making use of the lemma, we shall calculate

$$(1-\lambda)\lambda^{-1}m_{\ell-i}(k,\lambda)\underset{\Lambda}{\wedge}Be(p/a,q/a)=$$

$$= \sum_{r=0}^{\infty} r^{\ell-i}\binom{k+r-1}{r}\lambda^{k-1}(1-\lambda)^{r+1}\underset{\Lambda}{\wedge}Be(p/a,q/a)=$$

$$= \sum_{r=0}^{\infty} r^{\ell-i}\binom{k+r-1}{r}p^{[k-1,-a]}q^{[r+1,-a]}/1^{[r+k,-a]}=$$

$$=[1/(p+(k-1)a)][q\alpha_{\ell-i}+a\alpha_{\ell-i+1}].$$

If we substitute the above expression in (4.2) and take account of the equalities

$$\sum_{i=0}^{\ell}\binom{\ell}{i}\alpha_{\ell-i+1}=\alpha_{\ell+1}+\sum_{i=0}^{\ell}\binom{\ell}{i+1}\alpha_{\ell-i},\ \sum_{i=0}^{\ell}\binom{\ell}{i+1}\alpha_{\ell-i+1}=$$

$$=\ell\alpha_{\ell+1}+\sum_{i=0}^{\ell}\binom{\ell}{i+2}\alpha_{\ell-i},$$

then we shall obtain

Theorem 4.2. The crude moments of the i.P.d. P(k,p,1,a) given by (2.5) are expressed (if k>1) in the recurrent way by

$$\alpha_{\ell+1}=[p-(\ell+1)a]^{-1}\sum_{i=0}^{\ell}[kq\binom{\ell}{i}+(ka+q)\binom{\ell}{i+1}+ +a\binom{\ell}{i+2}]\alpha_{\ell-i},\quad \ell=0,1,2,.... \tag{4.3}$$

In turn, as the basis for our further considerations we shall take a recurrence relation for crude moments of the n.b.d. (2.2)

$$m_{\ell+1}(k,\lambda)=\frac{1-\lambda}{\lambda}[km_{\ell}(k+1,\lambda)+\frac{d}{d(\frac{1-\lambda}{\lambda})}m_{\ell}(k,\lambda)],\quad \ell=0,1,2,... \tag{4.4}$$

([1], (70), p. 225), and apply operation (3.2) to it. We then have

$$\alpha_{\ell+1}(k,p)=\frac{1-\lambda}{\lambda}[km_\ell(k+1,\lambda)\underset{\Lambda}{\wedge}Be(p/a,q/a)+$$

$$+\frac{d}{d(\frac{1-\lambda}{\lambda})}m_\ell(k,\lambda)\underset{\Lambda}{\wedge}Be(p/a,q/a)]. \tag{4.5}$$

The first term in the above sum yields, after the calculations given in the proof of Theorem 4.1 have been taken into account, the expression

$$\alpha\alpha_\ell(k+1,p-a) \quad \text{where} \quad \alpha=kq/(p-a).$$

We shall now calculate the other term, taking account of (3.4), (3.5) and (2.4):

$$\frac{1-\lambda}{\lambda}\cdot\frac{d}{d(\frac{1-\lambda}{\lambda})}\sum_{i=0}^{\ell} S_i^\ell\left(\frac{1-\lambda}{\lambda}\right)^i k^{[-i]}\underset{\Lambda}{\wedge}Be(p/a,q/a)=$$

$$=\sum_{i=0}^{\ell} S_i^\ell\, k^{[-i]}i[(1-\lambda)^i\lambda^{-i}\underset{\Lambda}{\wedge}Be(p/a,q/a)=$$

$$=\sum_{i=0}^{\ell} S_i^\ell(-k)^{[i]}iq^{[i,-a]}/(a-p)^{[i,-a]}=k\Delta_k\alpha_\ell(k,p)$$

where $\Delta_k\alpha_\ell(k,p)$ denotes the first finite difference with respect to k of the moment $\alpha_\ell(k,p)$, that is,

$$\Delta_k\alpha_\ell(k,p)=\alpha_\ell(k+1,p)-\alpha_\ell(k,p). \tag{4.6}$$

So, we may write formula (4.5) in the form expressed in the following theorem:

Theorem 4.3. The crude moment of the i.P.d. $P(k,p,1,a)$ defined by (2.5) is expressed (if $\ell<p/a$) by the precedent moment as follows:

$$\alpha_{\ell+1}(k,p)=\alpha\alpha_{\ell}(k+1,p-a)+k\Delta_k\alpha_{\ell}(k,p), \quad \ell=0,1,2,..., \tag{4.7}$$

where $\alpha=kq/(p-a)$ is the expected value of the distribution considered.

Remark 1. One can obtain (4.7) by means of the compounding operation starting from a formula, other than (4.4), for moments of the n.b.d., namely,

$$m_{\ell+1}=mm_{\ell}+q\frac{dm_{\ell}}{dq}, \quad \ell=0,1,2,...,$$

where $m=kq/p$ is the expected value of the n.b.d. ([4], (2.16), p. 398). We then have

$$\alpha_{\ell+1}(k,p)=k(1-\lambda)\lambda^{-1}m_{\ell}(k,\lambda)\underset{\Lambda}{\wedge}Be(p/a,q/a)+$$

$$+(1-\lambda)\frac{d}{d(1-\lambda)}\,m_{\ell}(k,\lambda)\underset{\Lambda}{\wedge}Be(p/a,q/a).$$

For the first term, we have $kq\alpha_{\ell}(k,p-a)$ (as in the proof of Theorem 4.1). In turn, for the other term, we get on the ground of (3.4) and (2.4)

$$\sum_{i=0}^{\ell} S_i^{\ell} k^{[-i]} i(1-\lambda)^{i}\lambda^{-i} \underset{\Lambda}{\wedge} Be(p/a,q/a)+$$

$$+\sum_{i=0}^{\ell} S_i^{\ell} k^{[-i]} i(1-\lambda)^{i+1}\lambda^{-i-1} \underset{\Lambda}{\wedge} Be(p/a,q/a)=$$

$$=\sum_{i=0}^{\ell} S_i^{\ell}(-k)^{[i]} iq^{[i,-a]}/(a-p)^{[i,-a]}-\sum_{i=0}^{\ell} S_i^{\ell}(-k)^{[i]} iq^{[i+1,-a]}/(a-p)^{[i+1,-a]}=$$

$$= k\Delta_k \alpha_\ell(k,p)+\frac{kq}{p-a}\,\Delta_k\alpha_\ell(k,p-a).$$

Consequently, we have

$$\alpha_{\ell+1}(k,p)=\alpha\alpha_\ell(k,p-a)+k\Delta_k(k,p)+\alpha\Delta_k\alpha_\ell(k,p-a)$$

where $\alpha=kq/(p-a)$ is the expected value of the i.P.d. After taking account of (4.6), one obtains (4.7).

Remark 2. The proof of (4.7) can also be carried out if one disregards the compounding operation, but then makes use of (3.5). Using the well-known relationship for the Stirling numbers of the second kind

$$S_i^{r+1}=iS_i^r+S_{i-1}^r,\quad i=1,2,3,\ldots,$$

we shall obtain

$$\alpha_{\ell+1}(k,p)=\sum_{i=1}^{\ell+1}(iS_i^{\ell}+S_{i-1}^{\ell})(-k)^{[i]}q^{[i,-a]}/(a-p)^{[i,-a]}=A+B.$$

Considering the fact that $S_{\ell+1}^{\ell}=0$, we may write

$$A=k\Delta_k \alpha_\ell(k,p).$$

In turn,

$$B=\alpha \sum_{i=1}^{\ell+1} S_{i-1}^{\ell}(-k-1)^{[-i]}(q+a)^{[i-1,-a]}/(2a-p)^{[i-1,-a]}.$$

Putting i-1=n and taking into cosideration $S_i^{\ell}=0$ for $i<0$ as well as (3.5), we have

$$B=\alpha \sum_{n=0}^{\ell} S_n^{\ell}(-k-1)^{[n]}(q+a)^{[n,-a]}/(2a-p)^{[n,-a]}=$$

$$=\alpha\alpha_\ell(k+1,p-a).$$

REFERENCES

[1] DYCZKA, W.: Application of compounding distributions for the calculations of moments (in Polish), Zeszyty Nauk. Polit. Łódzkiej (Scient. Bull. Łódź Techn. Univ.), 168 (1973), Matematyka, Fasc. 3, pp. 205-230.

[2] FELLER, W.: An Introduction to Probability Theory and Its Applications, 2nd ed. vol.I, New York 1966, John Wiley and Sons.

[3] FISZ, M.: Probability Theory and Mathematical Statistics, 3rd ed., New York 1963, John Wiley and Sons.

[4] GERSTENKORN, T.: Moment recurrence relations for the inflated negative binomial, Poisson and geometric distributions, Demonstratio Mathematica, 12, 2 (1979), pp. 389-410.

[5] GERSTENKORN, T.: The recurrence relations for the moments of the discrete probability distributions, Dissert. Math. (Rozprawy Matem.), 83 (1971), pp. 1-46.

[6] GERSTENKORN, T., Jarzębska, J.: Incomplete moments of the Pólya distribution (in Polish), Matematyka Stosowana, 20 (1982).

[7] GERSTENKORN, T., ŚRÓDKA, T.: Combinatorics and Probability Theory (in Polish), 6th ed., Warsaw 1980, Polish Scientific Publishers.

[8] JOHNSON, N.L., KOTZ, S.: Distributions in Statistics-Discrete Distributions, New York 1969, John Wiley and Sons.

[9] KEMP, C.D., KEMP, A.W.: Generalized hypergeometric distributions, J. Roy. Statist. Soc. Ser. B, 18, 2 (1956), pp. 202-211.

[10] KENDALL, M.G., Stuart, A.: The Advanced Theory of Statistics, vol. 1 - Distribution Theory, London 1958, Charles Griffin.

[11] KORN, G., KORN, Th. M.: Mathematical Handbook for Scientists and Engineers, Definitions, Theorems and Formulas for Reference and Review, New York 1961, McGraw-Hill Comp.

[12] RISSER, R., TRAYNARD, C.E.: Les Principes de la Statistique Mathématique, Livre I, Séries statistiques, Paris 1933, Gauthier-Villars; 2nd ed. 1957.

[13] SARKADI, K.: Generalized hypergeometric distributions, Publ. Math. Inst. Hung. Acad. Sci., 2, 1/2 (1957), pp. 59-68.

[14] SKELLAM, J.G.: A probability distribution derived from the binomial distribution by regarding the probability of success as variable between the sets of trials, J. Roy. Statist. Soc. Ser. B, 10, 1 (1948), pp. 257-261.

[15] TAKÁCS, L.: Combinatorial Methods in the Theory of Stochastic Processes, New York 1967, John Wiley and Sons.

Proc. of the 3rd Pannonian Symp. on Math. Stat., Visegrád, Hungary 1982
J. Mogyoródi, I. Vincze, W. Wertz, eds

ON THE SUPERPONABILITY OF STRICTLY MONOTONE INCREASING CONTINUOUS PROBABILITY DISTRIBUTION FUNCTIONS

B. GYIRES

Kossuth Lajos University, Debrecen, Hungary

I. INTRODUCTION

Let R_n be the n dimensional real column vector space. If $X \in R_n$ then let the corresponding row vector be denoted by X^*. $e \in R_n$ is the vector with components one. Denote by $S_n \subset R_n$ the set of vectors with positive components, and with the sum of these components equal to one.

Let the real numbers $a<b$ be given, where $a=-\infty$, $b=\infty$ are permitted too. Denote by $E(a,b)$ the set of continuous probability distribution functions (PDFs), which are strictly monotone increasing in $[a,b]$ and have values 0 and 1 at the points a and b, respectively. The inverse of $F \in E(a,b)$ will be denoted by F^{-1}.

Definition 1.1.

We say that PDF F is superponable by the linearly independent PDFs G_j $(j=1,\dots,n)$ if the identity

$$F(x)=\alpha_1 G_1(x)+\dots+\alpha_n G_n(x), \quad x \in R_1$$

holds with a vector $\alpha=(\alpha_j) \in S_n$.

Definition 1.2.

We say that PDF F is asymptotically superponable by the sequence of linearly independent PDFs $\{G_j\}_{j=1}^{\infty}$ if F is not superponable by G_j $(j=1,\dots,n)$ for $n=1,2,\dots$, and if the identity

$$F(x)=\sum_{j=1}^{\infty} \alpha_j G_j(x), \quad x\in R_1$$

holds with

$$\alpha_j>0 \quad (j=1,2,\dots), \qquad \sum_{j=1}^{\infty} \alpha_j=1.$$

<u>Definition 1.3.</u>

Let F, G_j $(j=1,2,\dots)$ be PDFs. We say F to be superponable by the sequence

$$G(\alpha(n))=\alpha_1(n)G_1+\dots+\alpha_n(n)G_n \quad (n=1,2,\dots)$$

with $\alpha(n)=(\alpha_j(n))\in S_n$, if $G(\alpha(n))\neq F$ $(n=1,2,\dots)$, and if the sequence $\{G(\alpha(n))\}_{n=1}^{\infty}$ converges uniformly to F if $n\to\infty$.

Two questions arise related to these definitions of the superponability.

a.) Let G_j $(j=1,2,\dots)$ be linearly independent PDFs. What are the necessary and sufficient conditions for the superponability of PDF F by G_j $(j=1,\dots,n)$, and by $\{G_j\}_{j=1}^{\infty}$, respectively?

b.) Let a PDF F be given. Determine all such linear independent PDFs G_j $(j=1,\dots,n)$, and all such linear independent sequences of PDFs $\{G_j\}_{j=1}^{\infty}$ by which F is superponable, and asymptotically superponable, respectively.

The first problem was answered by the author in full generality ([2], Chapter IV, Sections 3 and 4). Related to the second problem the author gave only ad hoc solutions, and only in the special case, when the PDFs belong to the set $E(a,b)$ with given real numbers a and b.

In this paper we deal with PDFs of sets $E(a_j,b_j)$ with

given a_j and b_j ($j=1,2,\dots$). We begin with a more general definition of the superposition of given PDFs of $E(a_j,b_j)$ ($j=1,2,\dots$) by the PDFs of the same sets. Using the results of [2] we give answers to both questions a.) and b.) in full and in almost full generality, respectively.

II. PRELIMINARY

Let $F\in E(a,b)$, and let

$$F(x_{Nk})=\frac{k}{N} \qquad (k=0,1,\dots,N),$$

where

$$x_{N_0}=a<x_{N_1}<\dots<x_{NN-1}<x_{NN}=b$$

are the N-th quantiles of F. Let $G\in E(a,b)$, and let us form the expression

$$D_N(G|F)=N\sum_{k=1}^{N}[G(x_{Nk})-G(x_{Nk-1})]^2.$$

Definition 2.1.

We say

$$D(G|F)=\sup_{N\to\infty} D_N(G|F)$$

to be the discrepancy of $G\in E(a,b)$ with respect to $F\in E(a,b)$.

This discrepancy idea was investigated more generally by the author in [2] (Chapter II), and it was shown that this concept is a measure of the distance between two PDFs indeed.

It is easy to see ([2], (2.4)) that

$$D_n(G|F)\nearrow D(G|F), \quad n\to\infty. \qquad (2.1)$$

Let $H(F)\subset E(a,b)$ be the set of PDFs with finite discrepancy with respect to $F\in E(a,b)$.

The following Theorem has an importance in our investigation.

Theorem 2.1.

If $G \in H(F)$, $F \in E(a,b)$, then $G(F^{-1}(x))$ is absolute continuous in $[0,1]$, and

$$D(G|F)= \int_0^1 g^2(x,G|F)dx,$$

where

$$g(x,G|F)= \frac{d}{dx} G(F^{-1}(x)). \tag{2.2}$$

This Theorem was formulated and proved in [2] (Theorem 2.4.). Because the proof is short, we prove the Theorem here again for the sake of completeness.

The statement of the Theorem follows immediately from

Lemma 2.1.

If f is a continuous function on [0,1], and if the supremum D of the sequence

$$\left\{ r \sum_{k=1}^{r} \left[f\left(\frac{k}{r}\right) - f\left(\frac{k-1}{r}\right) \right]^2 \right\}_{r=1}^{\infty}$$

is finite, then f is absolutely continuous, and

$$\int_0^1 (f')^2 dx = D.$$

The proof of this Lemma can be obtained using the theorem of Riesz ([3]), and Lemma 1.3.1. of [1].

Let

$$F_j \in E(a_j,b_j), G_j \in H(F_j) \qquad (j=1,2),$$

and let

$$F_1(x_{Nk})=F_2(y_{Nk})= \frac{k}{N} \qquad (k=0,1,\ldots,N).$$

Thus

$$N \sum_{k=1}^{N} \left\{ \frac{1}{2}[G_1(x_{Nk})+G_2(y_{Nk})] - \frac{1}{2}[G_1(x_{Nk-1})+G_2(y_{Nk-1})] \right\}^2 =$$

$$= \frac{1}{4}D_N(G_1|F_1)+ \frac{1}{4}D_N(G_2|F_2)+ \frac{1}{2}D_N(G_1|F_1,G_2|F_2),$$

where

$$D_N(G_1|F_1,G_2|F_2)=N\sum_{k=1}^{N}[G_1(x_{Nk})-G_1(x_{Nk-1})][G_2 y_{Nk}-G_2(y_{Nk-1})].$$

Let N=n!. Then we get on the basis of (2.1) that

$$D_{n!}(G_1|F_1,G_2|F_2)\to D(G_1|F_1,G_2|F_2),\ n\to\infty$$

exists, moreover on the basis of Theorem 2.1. we have

$$D(G_1|F_1,G_2|F_2)=\frac{1}{2}\ \{\int_0^1[g(x,G_1|F_1)+g(x,G_2|F_2)]^2dx-$$

$$-\int_0^1 g^2(x,G_1|F_1)dx-\int_0^1 g^2(x,G_2|F_2)dx\}=\int_0^1 g(x,G_1|F_1)g(x,G_2|F_2)dx,$$

where

$$g(x,G_j|F_j)=\frac{d}{dx}\ G_j(F_j^{-1}(x)),x\in[0,1]\quad (j=1,2) \qquad (2.3)$$

in accordance with (2.2).

Definition 2.2.

We call

$$D(G_1|F_1,G_2|F_2)=\lim_{n\to\infty} D_{n!}(G_1|F_1,G_2|F_2)=\int_0^1 g(x,G_1|F_1)g(x,G_2|F_2)dx$$

with the notation (2.3), the common discrepancy of $G_j\in H(F_j)$ with respect to $F_j\in E(a_j,b_j)$ (j=1,2).

Using the Schwartz inequality we get the following two theorems on the basis of Theorem 2.1. and Definition 2.2.

Theorem 2.2.

Let $F\in E(a,b)$, $G\in H(F)$. Then $D(G|F)\geq 1$ with equality if and only if G=F.

Theorem 2.3.

If

$$F_j\in E(a_j,b_j),\ G_j\in H(F_j)\quad (j=1,2\quad),$$

then

$$D(G_1|F_1,G_2|F_2)\leq D^{1/2}(G_1|F_1)D^{1/2}(G_2|F_2)$$

with equality if and only if

$$G_1(F_1^{-1}(x))=G_2(F_2^{-1}(x)), \quad x\in[0,1].$$

Definition 2.3.

The matrix

$$\Gamma(G_j|F_j \ (j=1,\ldots,n))=(D(G_j|F_j,G_k|F_k))_{j,k=1}^{n}$$

is called the Gram matrix of $G_j\in H(F_j)$ $(j=1,\ldots,n)$ with respect to $F_j\in E(a_j,b_j)$ $(j=1,\ldots,n)$.

Definition 2.4.

We say that $G_j\in E(a_j,b_j)$ $(j=1,\ldots,n)$ are linearly independent (dependent) with respect to $F_j\in E(a_j,b_j)$ $(j=1,\ldots,n)$, if the PDFs

$$G_j(F_j^{-1}(x)), \ x\in[0,1] \ (j=1,\ldots,n)$$

are linearly independent (dependent) on $[0,1]$ in the usual sense.

On the basis of this Definition we get immediately the following statement.

Theorem 2.4.

PDFs $G_j\in H(F_j)$ $(j=1,\ldots,n)$ are linearly independent with respect to $F_j\in E(a_j,b_j)$ $(j=1,\ldots,n)$ if and only if $\Gamma(G_j|F_j(j=1,\ldots,n))$ is a regular matrix.

We get the following special case from here.

Corollary 2.1.

PDFs $G_j\in E(a,b)$ $(j=1,\ldots,n)$ are linearly independent if and only if $\Gamma(G_j|F(j=1,\ldots,n))$ is a regular matrix for arbitrary $F\in E(a,b)$ with $G_j\in H(F)$ $(j=1,\ldots,n)$.

III. SUPERPOSITIONS OF PDFs

1. On the basis of Definition 1.1., and of Theorem 2.2., $F\in E(a,b)$ is superponable by the linearly independent $G_j\in H(F)$

$(j=1,\dots,n)$ if and only if

$$D(\alpha_1 G_1+\dots+\alpha_n G_n|F)=\sum_{j=1}^{n}\sum_{k=1}^{n} D(G_j|F,G_k|F)\alpha_j\alpha_k =$$

$$=\int_0^1 \Big(\sum_{j=1}^{n} \alpha_j\cdot \frac{d}{dx} G_j(F^{-1}(x))\Big)^2\, dx=1$$

with a vector $\alpha=(\alpha_j)\in S_n$, i.e.

$$\sum_{j=1}^{n} \alpha_j G_j(F^{-1}(x))=x,\ x\in[0,1].$$

Thus the following definitions are extensions of definitions 1.1., 1.2., and 1.3.

Definition 3.1.

Let $G_j\in E(a_j,b_j)$ $(j=1,\dots,n)$ be linearly independent with respect to $F_j\in E(a_j,b_j)$ $(j=1,\dots,n)$.

We say F_j $(j=1,\dots,n)$ to be superponable by G_j $(j=1,\dots,n)$ if the identity

$$\sum_{j=1}^{n} \alpha_j G_j(F_j^{-1}(x))=x,\quad x\in[0,1] \tag{3.1}$$

holds with a vector $\alpha=(\alpha_j)\in S_n$.

Definition 3.2.

Let $G_j\in E(a_j,b_j)$ $(j=1,2,\dots)$ be linearly independent with respect to $F_j\in E(a_j,b_j)$ $(j=1,2,\dots)$. We say F_j $(j=1,2,\dots)$ is asymptotically superponable by G_j $(j=1,2,\dots)$, if F_j $(j=1,\dots,n)$ is not superponable by G_j $(j=1,\dots,n)$ for $n=1,2,\dots$, and if the identity (3.1)

$$\sum_{j=1}^{\infty} \alpha_j G_j(F_j^{-1}(x))=x,\quad x\in[0,1]$$

holds with $\alpha_j>0$ $(j=1,2,\dots)$, $\sum_{j=1}^{\infty} \alpha_j=1$.

Definition 3.3.

Let $G_j \in E(a_j, b_j)$ $(j=1,2,\ldots\ )$ be linearly independent with respect to $F_j \in E(a_j, b_j)$ $(j=1,2,\ldots)$. We call F_j $(j=1,2,\ldots)$ is superponable by G_j $(j=1,2,\ldots\ )$, if F_j $(j=1,\ldots,n)$ is not superponable by G_j $(j=1,\ldots,n)$, for $n=1,2,\ldots$, and if there are vectors

$$\alpha(n) = (\alpha_j(n)) \in S_n \qquad (n=1,2,\ldots\ )$$

such that

$$\lim_{n\to\infty} \sum_{j=1}^{n} \alpha_j(n) G_j(F^{-1})x)) = x \tag{3.2}$$

uniformly in x, $x \in [0,1]$.

If in definitions 3.1., 3.2., and 3.3. $a_j = a$, $b_j = b$, $F_j = F \in E(a,b)$ $(j=1,2,\ldots\ )$ we get definitions 1.1., 1.2., and 1.3. from these definitions, respectively.

We obtain easily the following Theorem which can be looked upon as providing an equivalent condition to that of Definition 3.1.

Theorem 3.1.

1. Let $G_j \in E(a_j, b_j)$ $(j=1,\ldots,n)$ be linearly independent with respect to $F_j \in E(a_j, b_j)$ $(j=1,\ldots,n)$. F_j $(j=1,\ldots,n)$ is superponable by G_j $(j=1,\ldots,n)$ if and only if the identity

$$\sum_{j=1}^{n} \alpha_j g(x, G_j | F_j) = 1, \quad x \in [0,1]$$

holds with a vector $\alpha = (\alpha_j) \in S_n$.

Similar statements hold in the case of definitions 3.2., and 3.3. under the assumption that the limit and differentiation are interchangeable in the expressions (3.1) and (3.2), respectively.

As in the Introduction, two questions arise here, too,

related to definitions 3.1., 3.2. and 3.3.

a.) Let $G_j \in E(a_j,b_j)$ $(j=1,2,\dots)$ be linearly independent with respect to $F_j \in E(a_j,b_j)$ $(j=1,2,\dots)$. What are the necessary and sufficient conditions for the superponability of F_j $(j=1,\dots,n)$ by G_j $(j=1,\dots,n)$, and for the superponability of $\{F_j\}_{j=1}^{\infty}$ by $\{G_j\}_{j=1}^{\infty}$, respectively?

b.) Let $F_j \in E(a_j,b_j)$ $(j=1,2,\dots)$ be given. Determine all such linearly independent $G_j \in E(a_j,b_j)$ $(j=1,\dots,n)$ with respect to F_j $(j=1,\dots,n)$, and all such linearly independent sequence $G_j \in E(a_j,b_j)$ $(j=1,2,\dots)$ with respect to $\{F_j\}_{j=1}^{\infty}$, by which F_j $(j=1,\dots,n)$, and $\{F_j\}_{j=1}^{\infty}$ are asymptotically superponable, respectively.

2. Let $G_j \in E(a_j,b_j)$ $(j=1,\dots,n)$ be linearly independent with respect to $F_j \in E(a_j,b_j)$ $(j=1,\dots,n)$. Let F_j $(j=1,\dots,n)$ be superponable by G_j $(j=1,\dots,n)$. Then the converse statement, i.e., that G_j $(j=1,\dots,n)$ is superponable by F_j $(j=1,\dots,n)$ holds in a weaker sense only.

Namely let

$$\Phi_j(x)=G_j(F_j^{-1}(x)),\quad \varphi_j(x)=F_j(G_j^{-1}(x)),\quad x\in[0,1].$$

It is obvious that $\Phi_j(\varphi_j(x))=x$. Let $\alpha=(\alpha_j)\in S_n$ such a vector that

$$\sum_{j=1}^{n} \alpha_j\Phi_j(x)=x,\quad x\in[0,1].$$

Using the mean-value theorem, we get

$$\sum_{j=1}^{n} \alpha_j\Phi_j(x)-x=\sum_{j=1}^{n} \alpha_j(x-\varphi_j(x))\Phi_j'(\vartheta_j(x))=0,$$

where $\vartheta_j(x)$ is a point of the open interval with end points x, $\varphi_j(x)$. From here we have

$$\sum_{j=1}^{n} \alpha_j \Phi_j'(\vartheta_j(x))\varphi_j(x)=x \sum_{j=1}^{n} \alpha_j \Phi_j'(\vartheta_j(x)).$$

Thus G_j $(j=1,\dots,n)$ is superponable by F_j $(j=1,\dots,n)$ if and only if $\Phi_j'(x)>0$ is a constant, i.e. if $\Phi_j(x)=x$, contradicting to the assumption that Φ_j $(j=1,\dots,n)$ are linearly independent.

If we say that $F_j \in E(a_j,b_j)$ $(j=1,\dots,n)$ is superponable in the weak sense by $G_j \in E(a_j,b_j)$ $(j=1,\dots,n)$ if the identity

$$\sum_{j=1}^{n} \alpha_j(x) G_j(F^{-1}(x))=x, \quad x\in[0,1]$$

holds with

$$\alpha_j(x)>0, \quad \sum_{j=1}^{n} \alpha_j(x)=1, \quad x\in[0,1]$$

then the just obtained result can be formulated as follows.

Theorem 3.2.

Let $G_j \in E(a_j,b_j)$ $(j=1,\dots,n)$ be linearly independent with respect to $F_j \in E(a_j,b_j)$ $(j=1,\dots,n)$. Let F_j $(j=1,\dots,n)$ be superponable by G_j $(j=1,\dots,n)$. Then G_j $(j=1,\dots,n)$ is superponable by F_j $(j=1,\dots,n)$ at least in the weak sense.

Definition 3.4.

We say that the sequence $G_j \in E(a_j,b_j)$ $(j=1,2,\dots)$ converges (converges uniformly) to $F_j \in E(a_j,b_j)$ $(j=1,2,\dots\ )$, if $\{G_j(F_j^{-1}(x))\}_{j=1}^{\infty}$ converges (converges uniformly) to $x\in[0,1]$.

Definition 3.5.

We say that the sequence $G_j \in H(F_j)$ $(j=1,2,\dots\ )$ converges in discrepancy to the sequence $F_j \in E(a_j,b_j)$ $(j=1,2,\dots\ )$, if $D(G_j|F_j)\to 1$, if $j\to\infty$.

Theorem 3.3.

The sequence $G_j \in H(F_j)$ $(j=1,2,\dots\ )$ converges in discrepancy to

$F_j \in E(a_j, b_j)$ $(j=1,2,\ldots\ )$ if and only if $\{G_j\}_{j=1}^{\infty}$ converges uniformly to $\{F_j\}_{j=1}^{\infty}$.

Proof.

As a consequence of Theorem 2.1. we get that $\{G_j\}_{j=1}^{\infty}$ converges in discrepancy to $\{F_j\}_{j=1}^{\infty}$ if and only if $\{G_j(F^{-1})\}_{j=1}^{\infty}$ converges in discrepancy to $x \in [0,1]$. But ([2], Theorem 4.2.) the last sequence converges in discrepancy to x if and only if the same sequence converges uniformly to $x \in [0,1]$.

Corollary 3.1.

Let

$$G_j \in H(F_j),\ F_j \in E(a_j, b_j) \quad (j=1,2,\ldots\ )$$

be given. Let $\Phi_j(x) = G_j(F_j^{-1}(x))$, $x \in [0,1]$. If $\{\Phi_j'(x)\}_{j=1}^{\infty}$ converges uniformly to 1, $x \in [0,1]$, then $\{G_j\}_{j=1}^{\infty}$ converges in discrepancy to $\{F_j\}_{j=1}^{\infty}$.

Proof.

Since $\Phi_n'(x) \to 1$ uniformly in $[0,1]$, if $n \to \infty$, we get as a consequence of a well-known result that $\{\Phi_j(x)\}_{j=1}^{\infty}$ converges uniformly to x, $x \in [0,1]$. On the basis of Theorem 3.3. the proof is complete.

Theorem 3.4.

Assume that the sequence $G_j \in E(a_j, b_j)$ $(j=1,2,\ldots)$ converges (uniformly converges) to $F_j \in E(a_j, b_j)$ $(j=1,2,\ldots\ )$. Let

$$\Phi_j(x) = G_j(F^{-1}(x)), \quad \varphi_j(x) = F_j(G_j^{-1}(x)), \quad x \in [0,1].$$

If the sequence $\{\varphi_j'(x)\}_{j=1}^{\infty}$ is bounded (uniformly bounded) then, conversely, $\{F_j\}_{j=1}^{\infty}$ converges (uniformly converges) to $\{G_j\}_{j=1}^{\infty}$.

Proof.

Under the assumption we must show that if $\{\Phi_j(x)\}_{j=1}^{\infty}$

converges (uniformly converges) to x, $x\in[0,1]$, then $\{\varphi_j(x)\}_{j=1}^{\infty}$ converges (uniformly converges) to x, $x\in[0,1]$ too. Using the identity

$$\varphi_n(\varphi_n(x))=x, \quad x\in[0,1],$$

we get on the basis of the mean-value theorem that

$$\varphi_n[x+(\Phi_n(x)-x)]=\varphi_n(x)+\varphi_n'(\vartheta_n(x))[\Phi_n(x)-x]=x, \qquad x\in[0,1], \tag{3.3}$$

where $\vartheta_n(x)$ is a point of the open interval with end points x and $\Phi_n(x)$. Since $\{\Phi_n(x)\}_{n=1}^{\infty}$ converges (uniformly converges) to x, and $\{\varphi_n'(x)\}_{n=1}^{\infty}$ is bounded (uniformly bounded), there is a function $K(x)$, $0<K(x)<\infty$ (constant $K>0$), and a large enough positive integer $N(\varepsilon,x)$ ($N(\varepsilon)$) such that the inequalities

$$\varphi_n'(x)\leq K(x) \quad (\varphi_n'(x)\leq K),$$

$$|\Phi_n(x)-x|\leq \frac{\varepsilon}{K(x)} \quad \left(|\Phi_n(x)-x|\leq \frac{\varepsilon}{K}\right)$$

hold for $n>N(\varepsilon,x)$ ($n>N(\varepsilon)$). Using these inequalities we get the statement of Theorem 3.4. on the basis of (3.3).

3. We can get now an answer to question a.) in the following manner applying Theorems 4.9. and 4.16. of the paper [2].

Denote by A( ) the adjoint matrix of a square matrix in parenthesis.

On the basis of Theorem 4.9. of [2] we get

Theorem 3.5.

Let $G_j\in H(F_j)$ $(j=1,\dots,n)$ be linearly independent with respect to $F_j\in E(a_j,b_j)$ $(j=1,\dots,n)$, and let $\Gamma=\Gamma(G_j|F_j\ (j=1,\dots,n))$. The PDFs F_j $(j=1,\dots,n)$ are superponable by G_j $(j=1,\dots,n)$ if and only if $e^*\Gamma^{-1}e=1$, and the row sums of $A(\Gamma)$ are positive. In this case

$$\frac{1}{a}\sum_{j=1}^{n} a_j G_j(F_j^{-1}(x))=x, \qquad x\in[0,1],$$

where a_j is the sum of the elements of j-th row of $A(\Gamma)$, and $a=a_1+...+a_n$.

On the basis of Theorem 3.3. and of Theorem 4.16. of [2], we get

Theorem 3.6.

Let $G_j\in H(F_j)$ $(j=1,2,...)$ be independent with respect to $F_j\in E(a_j,b_j)$ $(j=1,2,...)$. Moreover let the row sums of matrices $A(\Gamma(n))$ with

$$\Gamma(n)=\Gamma(G_j|F_j\ (j=1,...,n))\ (n=1,2,...)$$

be positive, and let $e^*\Gamma^{-1}(n)e>1$ for all positive integers n. Then $\{F_j\}_{j=1}^{\infty}$ is superponable by $\{G_j\}_{j=1}^{\infty}$ if and only if the monotone decreasing sequence $\{e^*\Gamma^{-1}(n)e\}_{n=1}^{\infty}$ has the limit 1. In this case the sequence

$$\left\{\frac{1}{a(n)}\sum_{k=1}^{n} a_k(n)G_k(F_k^{-1}(x))\right\}_{n=1}^{\infty}, \qquad x\in[0,1]$$

converges uniformly to x, where $a_k(n)$ denote the k-th row sum of the matrix $A(\Gamma(n))$ and $a_1(n)+...+a_n(n)=a(n)$.

4. The following theorems give answers to question b.).

Theorem 3.7.

The PDFs $F_j\in E(a_j,b_j)$ $(j=1,...,n)$ are superponable by $G_j\in E(a_j,b_j)$ $(j=1,...,n)$ if and only if there are such positive, integrable and linearly independent functions $p_j(x)$, $x\in[0,1]$ $(j=1,...,n)$ that the identities

$$G_j(x)=\frac{1}{\alpha_j}\int_0^{F_j(x)} f_j(x)dx\in E(a_j,b_j), \tag{3.4}$$

$x\in[a_j,b_j]$ $(j=1,...,n)$

hold, where

$$f_j(x)=\frac{p_j(x)}{p(x)}\ , \qquad p(x)=\sum_{k=1}^{n} p_j(x), \tag{3.5}$$

$$x\in[0,1] \qquad (j=1,\dots,n),$$

and $\alpha=(\alpha_j)\in S_n$ with

$$\alpha_j=\int_o^1 f_j(x)dx. \tag{3.6}$$

Proof.

We can obtain the necessity of the conditions from the Definition 3.2., and from the Theorem 2.1., respectively. But the conditions are sufficient too. On the basis of Theorem 2.1. it is enough to see that $\sum_{j=1}^{n} f_j(x)=1 \quad x\in[0,1]$, and thus

$$\frac{d}{dx} G_j(F_j^{-1}(x))=\frac{1}{\alpha_j} f_j(x) \qquad (j=1,\dots,n),$$

i.e. (3.4) holds, and $G_j\in E(a_j,b_j)$ $(j=1,\dots,n)$ are linearly independent.

On the basis of Theorem 3.1. formula (3.4) gives the general solution of the first part of problem b.).

Similar statement holds in the case of asymptotical superposition under the assumption that the limit and differentiation are interchangeable in (3.1). Generally we have the following result.

Theorem 3.8.

Let the PDFs $F_j\in E(a_j,b_j)$ $(j=1,\dots\ \ )$ be given. Suppose that the functions $p_j(x)$ $(j=1,2,\dots\ )$ are positive, integrable, and linear independent on [0,1]. Moreover let the sequence

$$\left\{\sum_{j=1}^{n} p_j(x)\right\}_{n=1}^{\infty} \tag{3.7}$$

be uniformly convergent to p(x). Then $\{F_j\}_{j=1}^{\infty}$ is asymptotically superponable by the PDFs (3.4) with density functions (3.5) for $p(x)=\sum_{j=1}^{\infty} p_j(x)$, (j=1,2,...), where the numbers in (3.6) are positive, and the sum of these is equal to 1. The convergence of the series in (3.1) is uniform.

Proof.

Under the assumptions the sequence converges uniformly to 1 in [0,1]. Thus on the basis of a well-known theorem

$$\left\{\sum_{j=1}^{n} \alpha_j G_j(F_j^{-1}(x))\right\}_{n=1}^{\infty}$$

converges uniformly to x in the same interval. This completes the proof.

On the basis of Theorem 3.8. we get the following Corollary.

Corollary 3.2.

Under the assumption and notations of the Theorem 3.8. the sequence

$$\left\{\sum_{j=1}^{n} \alpha_j(n) G_j(F_j^{-1}(x))\right\}_{n=1}^{\infty}$$

converges uniformly to x in [0,1], where the $G_j(x)$ is defined by (3.4) with density function in (3.5). Moreover $p(x)=\sum_{j=1}^{\infty} p_j(x)$, and

$$\alpha_j(n)=\frac{\alpha_j}{\alpha_1+\ldots+\alpha_n} \qquad (j=1,\ldots,n)$$

with the numbers defined in (3.6). In other words $\{F_j\}_{j=1}^{\infty}$ is superponable by $\{G_j\}_{j=1}^{\infty}$.

REFERENCES

1 CSÖRGŐ, M. and RÉVÉSZ, P. Strong approximation in probability and mathematical statistics. Akadémiai Kiadó, Budapest, 1981.

2 GYIRES, B. Contributions to the theory of linear combinations of the probability distribution functions. Studia Mathematica, in print.

3 RIESZ, F. Untersuchungen über Systeme integrierbarer Funktionen. Math. Annalen 69 (1910), 449-457.

Proc. of the 3rd Pannonian Symp. on Math. Stat., Visegrád, Hungary 1982
J. Mogyoródi, I. Vincze, W. Wertz, eds

ON NONPARAMETRIC REGRESSION WITH RANDOMLY CENSORED DATA

L. HORVÁTH

Szeged University, Hungary

1. INTRODUCTION

Let $(X,Z)=(X^1,\dots,X^k,Z^1,\dots,Z^m)$ be a random vector taking values in $R^k \times R^m$ and let $S(x|z)$, $x=(x_1,\dots,x_k)$, $z=(z_1,\dots,z_m)$ denote the probability that $X=(X^1,\dots,X^k)$ survives $x=(x_1,\dots,x_k)$ given that $Z^1=z_1,\dots,Z^m=z_m$, i.e.

$$S(x|z)=P\{X^1 \geq x_1,\dots,X^m \geq x_m | Z^1=z_1,\dots,Z^m=z_m\}.$$

The usual estimation of the conditional survival function $S(x|z)$ is based on the observations of the vector (X,Z) but in some cases we cannot observe (X,Z).

We assume that $\{X_i^1,\dots,X_i^k,Z_i^1,\dots,Z_i^m\}_{i=1}^{\infty}$ is a sequence of independent k+m dimensional random vectors with common continuous survival function $F^0(x_1,\dots,x_{k+m})=P\{X^1 \geq x_1,\dots,Z^m \geq x_{k+m}\}$. Another sequence, independent of the

$$\{X_i^1,\dots,X_i^k,Z_i^1,\dots,Z_i^m\}_{i=1}^{\infty}, \quad \{C_i^1,\dots,C_i^{k+m}\}$$

of independent k+m dimensional random vectors with common survival function $H(x_1,\dots,x_{k+m})=P\{C_1^1 \geq x_1,\dots,C_1^{k+m} \geq x_{k+m}\}$ censors on the right the preceding one, so that our observations at the n-th stage consist of

$$\{Y_i^1,\dots,Y_i^{k+m},\delta_i^1,\dots,\delta_i^{k+m}\}_{i=1}^{n},$$

where

$$Y_i^j=\min\{X_i^j,C_i^j\},$$

$$\delta_i^j=\begin{cases}1, & \text{if } Y_i^j=X_i^j\\ 0, & \text{if } Y_i^j<X_i^j \qquad i=1,\ldots,k\end{cases}$$

and

$$Y_i^{k+\ell}=\min\{Z_i^\ell,C_i^{k+\ell}\},$$

$$\delta_i^{k+\ell}=\begin{cases}1, & \text{if } Y_i^{k+\ell}=Z_i^\ell\\ 0, & \text{if } Y_i^{k+\ell}<Z_i^\ell, \quad \ell=1,2,\ldots,m.\end{cases}$$

Semiparametric regression models for censored survival data have been widely discussed in the statistical literature during the 1970s, starting with the important paper by Cox (1972). The assumption in Cox's model is that for a patient with "covariate vector" z the hazard function $\lambda(t;z)$ defined by $\lambda(t;z)=\lim_{h\to 0}\frac{1}{h}P\{X^1\le t+h\mid X_1^1>t;Z=z\}$ $t>0$ $(k=1,m\ge 1)$ has the form

$$\lambda(t;z)=\lambda_0(t)\exp(\beta^T z), \qquad (1)$$

where $\lambda_0(t)$ is an unknown non-negative function of time not depending on the covariates, and $\beta=(\beta_1,\ldots,\beta_m)$ is a vector of unknown regression coefficients.

Beran (1981) called attention to the importance of the non-parametric estimation of $S(x|z)$ without the Cox's assumption (1) and reported an estimator for $S(x|z)$. In this note we will prove the strong uniform consistency of a kernel-type estimation of $S(x|z)$ in our general censorship model without the Cox's assumption of proportional hazards.

2. THE MULTIVARIATE PRODUCT-LIMIT ESTIMATOR

Our estimation of $S(x|z)$ is based on the multivariate product-limit estimator (MPLE) of F^o. The MPLE was first considered by Campbell (1981) and Campbell and Földes (1981). We define the MPLE $\hat{F}_n^o$ of F^o by

$$\hat{F}_n^o(x_1,\ldots,x_{k+m})=\begin{cases}\prod_{\{1\le j\le k+m\}}\prod_{\left\{\substack{1\le i\le n\\ Y_i^j<x_j}\right\}}\left\{\dfrac{nF_n(x_1,\ldots,x_{j-1},Y_i^j,-\infty,\ldots,-\infty)}{nF_n(x_1,\ldots,x_{j-1},Y_i^j,-\infty,\ldots,-\infty)}\right\} & \text{if } F_n(x_1,\ldots,x_{k+m})>0\\ 0, \text{ if } F_n(x_1,\ldots,x_{k+m})=0,\end{cases}$$

where F_n is the empirical survival function of F:

$$F_n(x_1,\ldots,x_{k+m})=\frac{1}{n}\#\{1\le i\le n, Y_i^1\ge x_1,\ldots,Y_i^{k+m}\ge x_{k+m}\}.$$

Let T_G denote the support of the survival function G:

$$T_G=\{(x_1,\ldots,x_\ell): G(x_1,\ldots,x_\ell)>0\}.$$

For $T\subset R^{k+m}$ define

$$\Delta_n(T)=\sup_{x\in T}|\hat{F}_n^o(x)-F^o(x)|$$

and the maximal deviation on R^{k+m}

$$\Delta_n=\sup_{x\in R^{k+m}}|\hat{F}_n^o(x)-F^o(x)|.$$

Introduce the following sequence of sets $T_n(c)$,

$$T_n(c)=\{x: F(x)\ge c(\log\log n/(2n))^{\frac{1}{2}}\}$$

and the numerical sequence $r_n(c)$,

$$r_n(c)=\sup\{F^o(x): x\notin T_n(c)\}.$$

The following consistency properties of $\hat{F}_n^o$ were proved by Horváth (1982):

THEOREM 1. *For any* $c>1$ *we have*

$$\overline{\lim_{n\to\infty}}\, \Delta_n / r_n(c) \le f_k(c) \text{ a.s.}$$

where

$$f_k(c)=1+\frac{3k}{c-1}\left(1+\frac{3k}{c-1}\exp\left(\frac{3k}{c-1}\right)\right) .$$

(i) *If* $T_F \subset T_{F^o}$, *then*

$$\lim_{n\to\infty} \Delta_n = \sup\{F^o(x):\ x\notin T_F\}>0 \text{ a.s.}$$

(ii) *If* $T_F=T_{F^o}$, *then* MPLE *is strongly consistent on the whole* R^{k+m} *with rate* $r_n(c)\to 0$, $n\to\infty$.

(iii) *If* $d_1=\inf\{H(x):\ x\in T_{F^o}\}>0$, *then*

$$\overline{\lim_{n\to\infty}}\left(\frac{2n}{\log\log n}\right)^{\frac{1}{2}}\Delta_n \le \frac{1}{d_1}\min\{cf_k(c):c>1\} \text{ a.s.}$$

(iv) *If* T *is a set such that* $d_2=\inf\{F(x):x\in T\}>0$, *then*

$$\overline{\lim_{n\to\infty}}\left(\frac{2n}{\log\log n}\right)^{\frac{1}{2}}\Delta_n(T) \le \frac{3k}{d_2} \quad \text{a.s.}$$

The conditional survival $S(x|z)$ can be written in the form

$$S(x|z)=\frac{s(x,z)}{g(z)},\ \text{if } g(z)>0, \qquad (2)$$

where

$$s(x,z)=\frac{\partial^m F^o(x,z_1,\ldots,z_m)}{\partial z_1\ldots\partial z_m} ,$$

$$G(z)=P\{Z^1\ge z_1,\ldots,Z^m\ge z_m\},$$

$$g(z)=\frac{\partial^m G(z_1,\ldots,z_m)}{\partial z_1\ldots\partial z_m}$$

The expression (2) suggests a kernel-type estimation of $S(x|z)$ which is based on the estimations of the derivatives $s(x,z)$ and $g(z)$. If we want to use the usual kernel-type estimation of g, at first we have to estimate G. The random vectors

$\{Z_i^1,\ldots,Z_i^m\}_{i=1}^{\infty}$ are censored on the right by the random vectors $\{C_i^{k+1},\ldots,C_i^{k+m}\}_{i=1}^{\infty}$, so we can use the MPLE G_n of G based on the censored observations

$$\{Y_i^{k+1},\ldots,Y_i^{k+m},\ \delta_i^{k+1},\ldots,\delta_i^{k+m}\}_{i=1}^{n}.$$

It follows from the definitions of G and G_n, that

$$G(z)=F^{o}(-\infty,\ldots,-\infty,z_1,\ldots,z_m), \tag{3}$$

$$G_n(z)=\hat{F}_n^{o}(-\infty,\ldots,-\infty,z_1,\ldots,z_m),\ z=(z_1,\ldots,z_m). \tag{4}$$

Given (4), it is now easy to define the empirical density functions $s_n(s,z)$ and $g_n(z)$:

$$s_n(x,z)=(h(n))^{-m}\int_{R^m} K(\frac{z-u}{h(n)})d_u\hat{F}_n^{o}(x,u),$$

$$g_n(z)=(h(n))^{-m}\int_{R^m} K(\frac{z-u}{h(n)})dG_n(u),$$

where K is an arbitrary Borel-measurable function and h(n) is a sequence of real numbers going to zero. In the following section we prove that

$$S_n(x|z)=\frac{s_n(x,z)}{g_n(z)}$$

is a strongly uniformly consistent estimation of $S(x|z)$.

3. THE STRONG UNIFORM CONSISTENCY OF S_n

To formulate our result, some notations and conditions will be introduced.

Let a<b two real numbers and consider the cube $A=\{(z_1,\ldots,z_m): a\le z_1\le b,\ldots,a\le z_m\le b\}$. We suppose that the kernel $K(z)$, $z\in R^m$ is obeying the following conditions:

(i) $\int_{R^m} K(z)dz=1$, $\int_{R^m} |K(z)|dz<\infty$

(ii) $K(z)$ is vanishing outside A

(iii) $\int_A z_i K(z)dz=0$ for $i=1,2,\ldots,m$

(iv) the partial derivatives of $K(z)$, with order not exceeding m, exist on A and

$$\sum_{\ell=0}^{m-1} \sum_{\substack{j_1+\ldots j_m\geq 0 \\ j_1+\ldots j_m=\ell}} \sup_{z\in A} \left|\frac{\partial^\ell}{\partial z_1^{j_1}\ldots\partial^{j_m} z_m} K(z)\right| = M_1<\infty,$$

$$\int_A \left|\frac{\partial^m}{\partial z_1\ldots\partial z_m} K(z)\right| = M_2<\infty.$$

For a positive number ε and a set $T_2\subset R^m$ we define $T_{2,\varepsilon}$ as

$$T_{2,\varepsilon}=\{z: \inf_{\tau\in T_2} ||z-\tau||<\varepsilon\},$$

where $||z||$ is the maximum norm in R^m. The strong uniform consistency of $S_n(x|z)$ will be proved on the set $T=T_1xT_2$, $T_1\subset R^k$, $T_2\subset R^m$. We assume that s and g are obeying the following condition on T:

(v) the second order partial derivatives of $s(x,z)$ on $T_\varepsilon=T_1xT_{2,\varepsilon}$ and the second order partial derivatives of $g(z)$ on $T_{2,\varepsilon}$ exist, moreover

$$\sup_{(x,z)\in T_\varepsilon} \max\{|s(x,z)|, \left|\frac{\partial s(x,z)}{\partial z_i}\right|, \left|\frac{\partial^2 s(x,z)}{\partial z_i \partial z_j}\right|, \ i,j=1,\ldots,m\}<\infty$$

and

$$\sup_{z\in T_{2,\varepsilon}} \max\{|g(z)|, \left|\frac{\partial g(z)}{\partial z_i}\right|, \left|\frac{\partial^2 g(z)}{\partial z_i \partial z_j}\right|, \ i,j=1,\ldots,m\}<\infty$$

(vi) $\inf_{z\in T_2} |g(z)|>0.$

Let $d_n(T)$ denote the maximal deviation of $S_n(s|z)$ from $S(x|z)$

$$d_n(T)=\sup_{(x,z)\in T} |S_n(x|z)-S(x|z)|.$$

THEOREM 2. _We assume that the conditions_ (i)-(vi) _are satisfied._

(i) _If_ $T\subseteq T_F=T_{F^o}$, _then we have_

$$d_n(T)=O(\max((h(n))^2, (h(n))^{-m}(1-F^o(T_n(c))))) \text{ a.s.},$$

and especially if $h(n)=(1-F^o(T_n(c)))^{\frac{1}{m+2}}$ _then_

$$d_n(T)=O((1-F^o(T_n(c)))^{\frac{2}{m+2}} \text{ a.s.}$$

(ii) _If_ $\inf\{H(x,z):(x,z)\in T_{F^o}\}>0$, _then we have_

$$d_n(T)=O(\max((h(n))^2,(h(n))^{-m}(\log\log n/n)^{\frac{1}{2}})) \text{ a.s.}$$

and especially if $h(n)=(\log\log n/n)^{\frac{1}{2(m+2)}}$ _then_

$$d_n(T)=O((\log\log n/n)^{\frac{1}{m+2}}) \text{ a.s.}$$

(iii) _If_ $\inf\{F(x,z):(x,z)\in T_\varepsilon\}>0$ _for some_ $\varepsilon>0$, _then we have_

$$d_n(T)=O(\max((h(n))^2,(h(n))^{-m}(\log\log n/n)^{\frac{1}{2}})) \text{ a.s.},$$

and especially if $h(n)=(\log\log n/n)^{\frac{1}{2(m+2)}}$ _then_

$$d_n(T)=O((\log\log n/n)^{\frac{1}{m+2}}) \text{ a.s.}$$

Proof. It follows from the conditions (i)-(iii) and (v) that

$$\sup_{(x,z)\in T} |(h(n))^{-m}\int_{R^m} K(\frac{z-u}{h(n)})d_uF^o(x,u)-s(x,z)|\le\rho_1(h(n))^2, \tag{5}$$

$$\sup_{z\in T}|(h(n))^{-m}\int_{R^m} K(\frac{z-u}{h(n)})dG(u)-g(z)|\le\rho_2(h(n))^2, \tag{6}$$

where ρ_1 and ρ_2 are positive constants (cf. Parzen (1962), Rejtő and Révész (1973)).

Theorem 1 and (3), (4) imply that

$$\sup_{z\in R^m} |G_n(z)-G(z)| \le \sup_{(x,z)\in R^{k+m}} |\hat{F}^o_n(x,z)-F^o(x,z)| \tag{7}$$

and if $\inf\{F(x,z):(x,z)\in T_\varepsilon\}>0$ then

$$\overline{\lim_{n\to\infty}} \left(\frac{n}{\log\log n}\right)^{\frac{1}{2}} \sup_{z\in T_{2,\varepsilon}} |G_n(z)-G(z)| < \infty \text{ a.s.}$$

An elementary computation shows that

$$\left|\frac{s_n(x,z)}{g_n(z)} - \frac{s(x,z)}{g(z)}\right| \le \left|\frac{1}{g_n(z)}\right| |s_n(x,z)-s(x,z)| + \left|\frac{s(x,z)}{g_n(z)g(z)}\right| |g_n(z)-g(z)|. \tag{8}$$

It is easy to see that

$$|s_n(x,z)-s(x,z)| \le (h(n))^{-m} \left|\int_{R^m} K\left(\frac{z-u}{h(n)}\right) d_u(\hat{F}^o_n(x,u)-F^o(x,u))\right| + (h(n))^{-m} \left|\int_{R^m} K\left(\frac{z-u}{h(n)}\right) d_u F^o(x,u) - s(x,z)\right|. \tag{9}$$

Using inequality (2.6) of Csörgő (1981) we obtain that

$$\begin{aligned} &\sup_{(x,z)\in T} (h(n))^{-m} \left|\int_{R^m} K\left(\frac{z-u}{h(n)}\right) d_u(\hat{F}^o_n(x,u)-F^o(x,u))\right| \\ &= \sup_{(x,z)\in T} \left|\int_A K(u)\, d_u(\hat{F}^o_n(x,z-uh(n))-F^o(x,z-uh(n)))\right| \\ &\le \sup_{(x,z)\in T_\varepsilon} |\hat{F}^o_n(x,z)-F^o(x,z)| \{M_2+2m(b-a)^{m-1}M_1\}. \end{aligned} \tag{10}$$

We get in a quite similar way that

$$\sup_{z\in T_2} (h(n))^{-m} \left| \int_{R^m} K\left(\frac{z-u}{h(n)}\right) dG_n(u) - g(z) \right| \le \sup_{z\in T_{2,\varepsilon}} |G_n(z)-G(z)| \{M_2+2m(b-a)^{m-1}M_1\} + \rho_2 (h(n))^2. \tag{11}$$

Inequalities (5)-(11) and Theorem 1 clearly imply our theorem.

REFERENCES

BERAN, R.I. (1981) Nonparametric regression with randomly censored survival data, IMS Bulletin 10, 271 (81t-113).

CAMPBELL, G.(1981) Nonparametric bivariate estimation with randomly censored data, Biometrika 68, 417-422.

CAMPBELL, G., Földes, A.(1981) Large-sample properties of nonparametric bivariate estimators with censored data, Colloquium on Nonparametric Statistical Inference, Budapest, June 23-28, 1980.

CSÖRGŐ, S. (1981) Multivariate empirical characteristic functions, Z. Wahrscheinlichkeitstheorie verw. Gebiete 55, 203-229.

HORVÁTH, L. (1982) The rate of strong uniform consistency for the multivariate product-limit estimator, J. Multivariate Analysis 12.

PARZEN, E. (1962) On estimation of a probability density function and mode, Ann. Math. Statist. 33, 832-837.

REJTŐ, L., RÉVÉSZ, P. (1973) Density estimation and pattern classification, Problems of Control and Information Theory 2, 67-80.

Proc. of the 3rd Pannonian Symp. on Math. Stat., Visegrád, Hungary 1982
J. Mogyoródi, I. Vincze, W. Wertz, eds

ON ESTIMATION IN THE EXPONENTIAL CASE UNDER RANDOM CENSORSHIP

J. HURT

Charles University, Prague, Czechoslovakia

1. INTRODUCTION

In classical statistical problems a complete random sample of X's from a population with distribution function F is observed. In some situations, such as clinical trials or reliability testing of complex systems, the X's represent time to the occurrence of a top event (e.g. survival time after an operation or time to failure) and the experiment is often analysed before all the top events occur. In medical research some individuals die by reasons which are desirable to exclude from consideration or they are simply lost. In reliability testing some objects are removed from the experiment before they fail. In both cases the result of the experiment is an incomplete sample and in many cases the model of random censorship may be useful.

Together with the random variable X representing lifetime or time to failure let us consider another random variable T, which is the so called time censor. The results of observation of n independent identically distributed objects under random censorship are the pairs

$$(W_1, I_1), \ldots, (W_n, I_n) \tag{1}$$

where

$W_j = \min(X_j, T_j)$,

$I_j = 1$ if $W_j = X_j$ (jth observation is uncensored)

$= 0 \quad W_j = T_j$ (jth observation is censored at time T_j),

$j = 1, \ldots, n$.

Notice that the random censorship model is a particular case of the model of competing risks with two causes of death, i.e., failure and removal.

Nonparametric estimation of the distribution function from randomly censored data is based mainly on the fundamental paper of KAPLAN and MEIER in which the product limit estimator of 1 - F was constructed. We shall briefly mention this estimator in Section 2. Parametric estimates of the unknown parameters of F are usually obtained by application of the maximum likelihood method. Some asymptotic results concerning maximum likelihood estimates for the exponential case will be given in Section 3.

2. THE PRODUCT LIMIT ESTIMATOR

Suppose that the distribution function F is continuous. Let $R(x) = 1 - F(x)$ denote the reliability function, let $W_{(1)} < \ldots < W_{(n)}$ be the ordered values of $W_1, \ldots, W_n$, and let $I_{(j)}$ denote the indicator corresponding to $W_{(j)}$, $j = 1, \ldots, n$. Then the product limit estimator is

$$\text{est } R(x) = \prod_{j=1}^{k-1} \left(\frac{n-j}{n-j+1}\right)^{I_{(j)}}, \quad x \in [W_{(k-1)}, W_{(k)}), \tag{2}$$

$$= 0, \quad x > W_{(n)}.$$

Rigorous derivations of some basic properties of this estimator were given by BRESLOW and CROWLEY (1974). In this note we mention only their result about asymptotic normality. If the censoring distribution function G is continuous and $F(T_0)<1$, $G(T_0)<1$ for some T_0 then the random process

$$\{\sqrt{n}(\text{est } R(t) - R(t)),\ 0<t<T_0\}$$

converges weakly to a zero mean Gaussian process $\{Z(t),\ 0<t<T_0\}$ with covariance function

$$\operatorname{cov}(Z(s),\ Z(t))=R(s)R(t)\int_0^s [R(x)(1-G(x))]^{-2}\, dP(X<x,\ I=1), \quad s\le t. \tag{3}$$

For further results concerning the product limit estimator see recent papers of GILL, R.D., FÖLDES, A., REJTŐ, L., BURKE, M.D., CSÖRGŐ, S., HORVÁTH, L.

The product limit estimator is suitable for estimating the mean, determining the underlying distribution etc. but it fails if the behaviour of the tail is in focus.

3. MAXIMUM LIKELIHOOD ESTIMATION

Let us assume that X and T are independent random variables distributed according to absolutely continuous distribution functions $F(x;\ \underline{\Theta}_1)$ and $G(t;\ \underline{\Theta}_2)$, respectively. The vector-valued parameters $\underline{\Theta}_1$ and $\underline{\Theta}_2$ are unknown, $\underline{\Theta}_1$ being the parameter in question and $\underline{\Theta}_2$ the nuisance parameter. Let f and g be the density functions of X and T, respectively. Then the random pair $(W,I) = (\min(X,T),\ I(X<T))$ has density function

$$\begin{aligned} h(w,i;\ \underline{\Theta}_1,\underline{\Theta}_2) &= g(w;\underline{\Theta}_2)(1-F(w;\underline{\Theta}_1)),\ w>0,\ i=0,\\ &= f(w;\underline{\Theta}_1)(1-G(w;\underline{\Theta}_2)),\ w>0,\ i=1,\\ &= 0,\ w\le 0 \end{aligned} \tag{4}$$

with respect to the Lebesgue $\times$ counting product measure. Then the likelihood function of (1) is

$$L(\underline{\Theta}_1, \underline{\Theta}_2) = \prod_{j=1}^{n} h(W_j, I_j; \underline{\Theta}_1, \underline{\Theta}_2) \quad . \tag{5}$$

An alternative form of (5) is

$$L(\underline{\Theta}_1, \underline{\Theta}_2) = \prod_{j\in U} f(X_j; \underline{\Theta}_1) \prod_{j\in C} (1-F(T_j; \underline{\Theta}_1)) \prod_{j\in C} g(T_j; \underline{\Theta}_2) \prod_{j\in U} (1-G(X_j; \underline{\Theta}_2)), \tag{6}$$

where $C = \{j: I_j=0\}$ (the indices of censored observations) and $U = \{j: I_j=1\}$ (the indices of uncensored observations). If there is no functional dependence between $\underline{\Theta}_1$ and $\underline{\Theta}_2$ then the maximum likelihood estimate is obtained by maximizing

$$L(\underline{\Theta}_1) = \prod_{j\in U} f(X_j; \underline{\Theta}_1) \prod_{j\in C} (1-F(T_j; \underline{\Theta}_1)). \tag{7}$$

Thus it does not depend on the nuisance distribution G. The distribution of the estimates, however, does generally depend on G.

Regularity conditions for the asymptotic normality of the maximum likelihood estimator are not too lucid and will be omitted here. Let us only mention that in addition to usual regularity conditions on both densities f and g we have to adopt some further similar conditions on F and G.

Next we deal with a particular case of exponentially distributed X's. Suppose that $R(x) = e^{-x/\Theta}$ where Θ is an unknown parameter. The maximization of the likelihood function (7) leads to the well-known estimator

$$\hat{\Theta} = \sum_{j=1}^{n} W_j / \sum_{j=1}^{n} I_j \, , \tag{8}$$

provided $\sum_{j=1}^{n} I_j > 0$, and $\hat{\Theta}$ is undefined in the opposite case.

Theorem 1. For any censoring distribution

$$\sqrt{n}(\hat{\Theta} - \Theta) \xrightarrow{D} N(0, \frac{1}{P^2(X<T)} \operatorname{var}(W - \Theta I)) . \tag{9}$$

Proof. First note that

$$EW = \int_0^\infty P(X>w, T>w)dw = \int_0^\infty e^{-w/\Theta}(1 - G(w))dw$$

and

$$P(X<T) = 1 - P(X>T) = 1 - \int_0^\infty G(w)\Theta^{-1} e^{-w/\Theta}dw =$$

$$= \int_0^\infty (1 - G(w))\Theta^{-1} e^{-w/\Theta}dw$$

so that $\Theta = EW/P(X<T)$ for any censoring distribution G. Let $\bar{W}$ and $\bar{I}$ denote sample means of W_j's and I_j's, respectively. Then we have

$$\sqrt{n}(\hat{\Theta} - \Theta) = \sqrt{n}(W/\bar{I} - EW/EI) = \frac{\sqrt{n}}{\bar{I}EI} (\bar{W}EI - \bar{I}EW) .$$

By the central limit theorem for i.i.d. random variables with finite variance

$$\sqrt{n}(\bar{W}EI - \bar{I}EW) \xrightarrow{D} N(0, \operatorname{var}(WEI - IEW))$$

whereas

$$\frac{1}{\bar{I}EI} \xrightarrow{P} (EI)^{-2}.$$

The limiting distribution of $\sqrt{n}(\hat{\Theta} - \Theta)$ is therefore $N(0, (EI)^{-2} \operatorname{var}(W - \Theta I))$. QED

Suppose now that the censoring distribution is also exponential with parameter ω, i.e.,

$$G(t) = 1 - e^{-t/\omega}, \quad t>0$$
$$= 0 \quad \text{otherwise.}$$

In this case, W is exponentially distributed with parameter δ satisfying

$$\frac{1}{\delta} = \frac{1}{\Theta} + \frac{1}{\omega} .$$

If there is no censoring, formally let $\omega=\infty$ so that $\delta=\Theta$. Some asymptotic properties of $\hat{\Theta}$ are stated in the following theorem.

Theorem 2. We have

$$\sqrt{n}(\hat{\Theta} - \Theta) \xrightarrow{D} N(0, \Theta^3/\delta) , \tag{10}$$

$$E\,\hat{\Theta} = \Theta + \frac{\delta}{n}\left(\frac{\Theta}{\delta}\right)^2 \left(1 - \frac{\delta}{\Theta}\right) + 0(n^{-2}), \tag{11}$$

$$\text{var}\,\hat{\Theta} = \frac{1}{n}\frac{\Theta^3}{\delta} + 0(n^{-2}) . \tag{12}$$

Proof. Since X and T are independent exponentially distributed random variables, W and I are also independent. Further, $EI = \delta/\Theta$, so that

$$\text{var}(W - \Theta I) = \text{var}\,W + \Theta^2 \text{var}\,I = \delta^2 + \Theta^2 \frac{\delta}{\Theta}\left(1 - \frac{\delta}{\Theta}\right) = \Theta\delta.$$

Thus the asymptotic variance of the limit distribution in (9) is $\Theta\delta/(\delta/\Theta)^2 = \Theta^3/\delta$ which justifies (10).

Concerning the expectation we have

$$E\hat{\Theta} = E\bar{W}E(\bar{I})^{-1} = \delta E(\bar{I})^{-1}$$

because $\bar{W}$ and $\bar{I}$ are independent. Making use of Theorem 1 of HURT (1976b) we obtain by δ-method

$$E\,\bar{I}^{-1} = \frac{\Theta}{\delta} + \frac{1}{n}\left(\frac{\Theta}{\delta}\right)^3 \frac{\delta}{\Theta}\left(1 - \frac{\delta}{\Theta}\right) + 0(n^{-2})$$

since all higher moments are $0(n^{-2})$. Hence we have (11). Similarly, using Theorem 2 in HURT, loc. cit., we get the expansion (12) for the variance. Q.E.D.

Finally we apply the above results to the maximum likelihood estimator of the reliability function. For fixed real $x>0$, $R(x) = e^{-x/\Theta}$ is estimated by

$$\hat{R}(x) = e^{-x/\hat{\Theta}} \tag{13}$$

Theorem 3. We have

$$\sqrt{n}(\hat{R}(x) - R(x)) \xrightarrow{D} N\left(0, \frac{\Theta}{\delta}\left(\frac{x}{\Theta}\right)^2 R^2(x)\right) , \tag{14}$$

$$E\,\hat{R}(x) = R(x)\left(1 + \frac{1}{n}\frac{x}{\delta}\left(\frac{1}{2}\frac{x}{\Theta} - \frac{\delta}{\Theta}\right)\right) + 0(n^{-2}), \tag{15}$$

$$\text{var}\,\hat{R}(x) = \frac{1}{n}\frac{\Theta}{\delta}\left(\frac{x}{\Theta}\right)^2 R^2(x) + 0(n^{-2}) . \tag{16}$$

Proof. Put $g(t) = R(t)$. Then $g'(t) = xt^{-2}R(t)$. Using (10) and (6a.2.5) in RAO (1973) we have (14). The expansions (15) and (16) can be obtained by applying Theorem 1 of HURT (1976b) utilizing (11) and (12). QED.

Note that for $\delta=\Theta$ (uncensored observations) the formulas (14), (15), and (16) coincide with those in HURT (1976a). Also notice that up to $0(n^{-2})$ the formulas for var $\hat{R}(x)$ and the mean squared error of $\hat{R}(x)$ are the same.

We are now in the position to compare the nonparametric Kaplan-Meier estimator est R(x) with $\hat{R}(x)$. First we calculate the asymptotic variance (a.v.) of est R(x) using (3). We have

$$P(X<x,\ I = 1) = P(W<x,\ I = 1) = P(W<x)\ P(X<T) = = \frac{\delta}{\Theta}\,(1 - e^{-x/\delta}) \quad .$$

Thus

$$\text{a.v. est } R(x) = \frac{1}{n}\, R^2(x) \int_0^x (1 - F(y))^{-2} \times \tag{17}$$

$$\times (1 - G(y))^{-2}\, d\, \frac{\delta}{\Theta}\,(1 - e^{-y/\delta}) =$$

$$= \frac{1}{n}\, R^2(x) \int_0^x e^{2y/\delta}\, \frac{1}{\Theta}\, e^{-y/\delta}\, dy =$$

$$= \frac{1}{n}\, R^2(x)\, \frac{\delta}{\Theta}\,(e^{x/\delta} - 1)\ .$$

The asymptotic efficiency of the product limit estimator is then

$$\text{eff} = \frac{\text{a.v. } \hat{R}(x)}{\text{a.v. est } R(x)} = \left(\frac{x}{\delta}\right)^2 \frac{1}{e^{x/\delta} - 1} \quad . \tag{18}$$

If we put $y = x/\delta$ then

$$\text{eff} = y^2(e^y - 1)^{-1}.$$

For all $y>0$, $\text{eff}<1$, so that the product limit estimator is not asymptotically efficient. For $x/\delta\to\infty$, $\text{eff}\to 0$.

In conclusion we present some numerical illustrations. All these illustrations are based on terms standing at n^{-1} in the asymptotic expansions.

Table 1

k	1	4/5	2/3	1/2	1/3	1/5
Bias	0	Θ/4	Θ/2	Θ	2Θ	4Θ

Table 2

z \ k		1	4/5	2/3	1/2	1/3	1/5
0.5	E	-.375	-.344	-.313	-.250	-.125	.125
	V	.0920	.1150	.1380	.1839	.2759	.4598
1.0	E	-.500	-.375	-.250	.000	.500	1.500
	V	.1353	.1692	.2030	.2707	.4060	.6767
1.5	E	-.375	-.094	.187	.750	1.875	4.125
	V	.1120	.1400	.1680	.2240	.3361	.5601
2.0	E	.000	.500	1.000	2.000	4.000	8.000
	V	.0733	.0916	.1099	.1465	.2198	.3663
2.5	E	.625	1.406	2.187	3.750	6.875	13.125
	V	.0421	.0526	.0652	.0842	.1264	.2106
3.0	E	1.500	2.625	3.750	6.000	10.500	19.500
	V	.0223	.0279	.0335	.0446	.0669	.1115

Table 3

x/δ	.5	1.0	1.5	2.0	2.5	3.0
eff	.3854	.5820	.6462	.6261	.5589	.4716

Table 1 shows the bias of $\hat{\Theta}$ expressed as a function of $k = P(X<T) = \delta/\Theta$. (Note that k=1 corresponds to the case of uncensored observations.) In such a case the term at n^{-1} in (11) is $\Theta(1 - k)/k$. We can see that the greater portion of data is censored the higher is the bias.

In Table 2 we give the terms at n^{-1} in the asymptotic expansions of E $\hat{R}$ and var $\hat{R}$ as functions of $k=\delta/\Theta$ and $z = x/\Theta$. We denote the corresponding terms as E (for the expectation) and V (for the variance). It can be seen that with increasing portion of censored observations both the bias and variance increase. The bias increases rapidly with increasing values of x/Θ whereas the variance decreases.

Table 3 contains values of the asymptotic efficiency (18).

REFERENCES

[1] BRESLOW, N. and CROWLEY, J. (1974). A large sample study of the life table and product limit estimates under random censorship. Ann. Statist. 2, 437-453.

[2] HURT, J. (1976a). On estimation of reliability in the exponential case. Apl. Mat. 21, 263-272.

[3] HURT, J. (1976b). Asymptotic expansions of functions of statistics. Apl. Mat. 21, 444-456.

[4] KAPLAN, E.L. and MEIER, P. (1958). Nonparametric estimation from incomplete observations. J. Amer. Statist. Assoc. 53, 457-481.

[5] MOESCHBERGER, M.L. and DAVID, H.A. (1971). Life tests under competing causes of failure and the theory of competing risks. Biometrics 27, 909-933.

[6] RAO, C.R. (1973). Linear statistical inference and its applications. 2nd edition. Wiley, New York.

[7] BURKE, M.D., CSÖRGŐ, S. and HORVÁTH, L. (1981). Strong approximations of some biometric estimates under random censorship. Z. Wahrscheinlichkeitstheorie verw. Gebiete 56, 87-112.

Proc. of the 3rd Pannonian Symp.
on Math. Stat., Visegrád, Hungary 1982
J. Mogyoródi, I. Vincze, W. Wertz, eds

ON A CHARACTERIZATION OF THE NORMAL DISTRIBUTION

P. KOSIK, K. SARKADI

Mathematical Institute, Hungarian Academy of Sciences,
Budapest, Hungary

1. INTRODUCTION

Let $X_1,\ldots,X_n$ $(n\geq 3)$ be independent, identically distributed random variables

$$\overline{X}= X_i/n$$

$$S=\sqrt{\Sigma(X_i-\overline{X}^2)}$$

It is well known that if the distribution of X_1 is normal then the vector

$$Z=\left(\frac{X_i-\overline{X}}{S}\right)_{i=1,\ldots,n} \qquad (1)$$

is uniformly distributed on the surface of an (n-1)-dimensional sphere. Fig.1 represents the case n=3 when the sphere surface becomes a circle.

It is also known that in the case $n\geq 6$ the condition of the normality is necessary. Thus this property characterizes the normal distribution. This is the theorem of Zinger [13].

For some generalizations of Zinger's theorem, we refer to Kagan et al. [3, Theorem 13.5.9] and Bondesson [1, Theorem 9.3]. However, as far as we know, there is no extension in respect to sample size, in other words, the problem is unsolved for n=3,4,5.

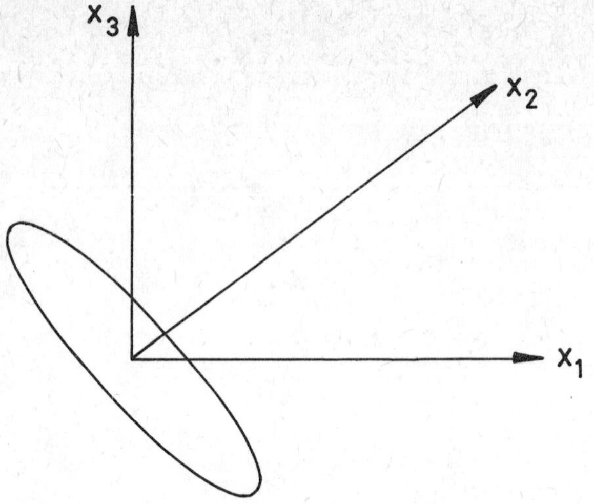

Fig. 1

It is also known that the ratio of two independent normal deviates follows the Cauchy law. This property is not characteristic for the normal distribution. First Mauldon [6] gave some counter examples; later Laha [5] gave necessary condition that the ratio of two independent, identically distributed variables follows the Cauchy law.

It is clear that if the distribution of X_i is a counter-example for the extension of Zinger's theorem for n=4 or n=5, then the distribution of X_1-X_2 is of Laha type (i.e. it has the properties described in [5, Theorem 2.1]).

Zinger's problem for $3\leq n\leq 5$ has practical aspects: it is closely connected with the characterization of the multisample tests of normality.

In the practice often occurs that we wish to assess the normality of the distribution and find that the sizes of the available samples are too low. The corresponding population parameters may differ from sample to sample. In such instances, a multisample test of normality can be used. In the heteroscedastic case (i.e. if homogeneity in variance cannot be assumed) the available such tests are [12], [11], [2, Eq. (26)], [8] and [9, Eq. (7.1)].

These tests use functions of statistics of type (1) for test statistics. If, for some initial non-normal distribution, Zinger's statistics is uniformly distributed on its image set then none of the above tests can distinguish between this distribution and the normal one at the given sample size.

Instead of trying to find an exact solution for the extension of the problem of Zinger we have sought for a practical solution. We considered the "practically important" alternatives and tested whether the empirical distribution of Z differs from the normal distribution. We have chosen the sample size 3. The reason for this choice is the following: if there is an example which disproves the extension of Zinger's theorem for n=4 or n=5 then functional relationship shows that for n=3 the case is the same.

As "practically important" alternatives those appearing in the extensive empirical Monte Carlo investigations on normality tests of Shapiro et al. [10] and Pearson et al. [7] have been accepted.

2. THE RESULTS

Let n=3. Instead of Z we considered the statistic

$$Y = \frac{X_{(2)}-X_{(1)}}{X_{(3)}-X_{(1)}}$$

where

$$X_{(1)} \leq X_{(2)} \leq X_{(3)}$$

is the rearrangement of X_1, X_2 and X_3 in the order of magnitude. Y is in one-by-one functional relationship with the set of order statistics of the components of Z; clearly, the mutual order of magnitude of these components contains no information on the distribution of Z.

Y has a truncated Cauchy distribution under the normal distribution and the theoretical probabilities can be calculated easily. The interval (0,1) has been divided into 25 equal subintervals. There are 68 non-normal distributions which appear at least in one of papers [10] and [7]. For each of these distributions, 1000 replicates of samples of size 3 have been generated and the empirical distributions were tested by χ^2-test. The results are given in Tables 1 (significant results) and 2 (non-significant results). For more accurate data of the alternatives, we refer to the sources mentioned above.

3. CONCLUSIONS AND REMARKS

1. We see that although the result of the test is significant for the majority of "practically important" alternatives a considerable number of such alternatives seems practically undistinguishable from the normal distribution via samples of size 3.

This fact suggests caution in the application of multisample tests if the sample size is 3 or near to it.

2. For sample of size 4, the empirical power of the tests of Wilk-Shapiro and Durbin is low against the uniform and double exponential alternatives [4]. This is in accordance with our present results. On the other hand, at sample of size 10 the empirical power is reasonably high against the same alternatives.

Table 1.

Significant results

Size of the test: 5%, critical value: 36.4

Code name	No. in[7]	Alternative	χ^2
BE(2,1)	38	Beta(2,1)	72.4
BI(4)	-	Binomial k=4, p=0.5	6500
BI(8)	-	" k=8 "	4249
BI(12)	-	k=12 "	3197
BI(20)	-	" k=20 "	1823
$\chi^2(1)$	55	Chi square 1 d.f.	668
$\chi^2(2)$	53	" 2 " exponential	230
$\chi^2(4)$	50	" 4 "	106
$\chi^2(10)$	46	" 10 "	51.2
$D\chi^2(-0.8)$	-	Double chi sq. β=-0.8	52.1
$D\chi^2(-0.5)$	-	" β=-0.5	39.8
$D\chi^2(1.5)$	-	" β=1.5	42.0
$D\chi^2(2)$	-	" β=2	93.8
SB(1,2)	39	Johnson SB (1,2)	36.7
LN	58	Log. normal (0,1)	279
P(1)	-	Poisson λ=1	7330
P(4)	-	" λ=4	2224
P(10)	-	" λ=10	973
T(1)	32	Student 1 d.f. (Cauchy)	129
TU(1,5)	-	Tukey (1,5)	40.9
TU(1,10)	17	" (1,10)	267
WE(0,5)	56	Weibull (0,5)	1771
SU(0,0.9)	29	Johnson SU (0, 0.9)	37.7
SU(1,1)	57	" (1,1)	129
LC(0.2,3)	41	Location contaminated normal (0.2,3)	77.5
LC(0.2,5)	42	" (0.2,5)	167
LC(0.2,7)	43	" (0.2,7)	276
LC(0.1,5)	48	" (0.1,5)	107
LC(0.1,7)	51	" (0.1,7)	157
LC(0.05,5)	52	" (0.05,5)	70.8
LC(0.05,7)	54	" (0.05,7)	66.3
SC(0.2,3)	19	Scale contaminated normal (0.2,3)	38.0
SC(0.2,5)	22	" (0.2,5)	52.5
SC(0.2,7)	23	" (0.2,7)	79.2
SC(0.1,7)	26	" (0.1,7)	47.2

Table 2.

Non-significant results

Code name	No. in[7]	Alternative	χ^2
BE(1,1)	3	Beta(1,1) (Uniform)	36.2
BE(1.1, 1.1)	-	" (1.1, 1.1)	32.5
BE(1.3, 1.3)	-	" (1.3, 1.3)	10.8
BE(1.5, 1.5)	-	" (1.5, 1.5)	31.8
BE(2,2)	7	" (2,2)	32.2
BE(3,2)	37	" (3,2)	13.5
Dχ^2(1)	18	Double exponential (Laplace)	26.2
SB(0, 0.7071)	4	Johnson SB (0, 0.7071)	24.7
SB(0.533, 0.5)	36	" (0.533, 0.5)	28.0
SB(0,2)	-	" (0,2)	23.4
SB(1,1)	40	" (1,1)	30.2
L	15	Logistic(0,2)	32.1
NCχ^2	-	Non-central χ^2 (1,16)	18.0
T(2)	31	Student 2 d.f.	31.6
T(4)	30	" 4 "	14.5
T(10)	14	" 10 "	20.5
WE(2)	44	Weibull(2)	36.3
TU(1, 0.1)	-	Tukey (1, 0.1)	22.4
TU(1, 0.2)	-	" (1, 0.2)	29.6
TU(1, 0.7)	5	" (1, 0.7)	33.4
TU(1, 1.5)	2	" (1, 1.5)	31.3
TU(1, 3)	6	" (1, 3)	19.6
SU(0,3)	13	Johnson SU (0,3)	29.0
SU(0,2)	16	" (0,2)	19.9
SU(0,1)	28	" (0,1)	30.4
SU(-1,2)	49	" (-1,2)	24.5
LC(0.1,3)	45	Location contaminated normal (0.1,3)	31.6
LC(0.05,3)	47	" (0.05,3)	25.7
SC(0.1,3)	21	Scale contaminated normal (0.1,3)	13.2
SC(0.1,5)	24	" (0.1,5)	29.8
SC(0.05,3)	20	" (0.05,3)	31.5
SC(0.05,5)	25	" (0.05,5)	34.5
SC(0.05,7)	27	" (0.05,7)	33.8

It would be suitable to extend the experiments for an intermediate sample size.

3. If the assumption of homoscedasticity is applicable then it is better to apply the method suggested in [4, Eq. (2.1)] if the sample size is low. In this case, Cramér's theorem assures the characterization. For the empirical power of this test, we refer to [4].

4. For the majority of the alternatives listed in Table 2, the experiments of Shapiro et al. [10] yield small or moderate empirical power for the sample size 10.

5. Although the result for the uniform distribution is not significant, easy calculation shows that the exact distribution of Z under this alternative differs from its null distribution.

6. The distribution of nonsignificant results (a third quarter of them is above the median of the χ^2-distribution) suggests that the number of significant results would certainly increase by increasing the number of replications in the experiments.

7. For some selected alternatives the histograms of the empirical distributions of Y are represented in the Appendix. In case of symmetrical initial distributions the histograms have been symmetrized. The last figure represents the density function under the null hypothesis.

APPENDIX

Histograms of the empirical distributions of Y for selected alternatives.

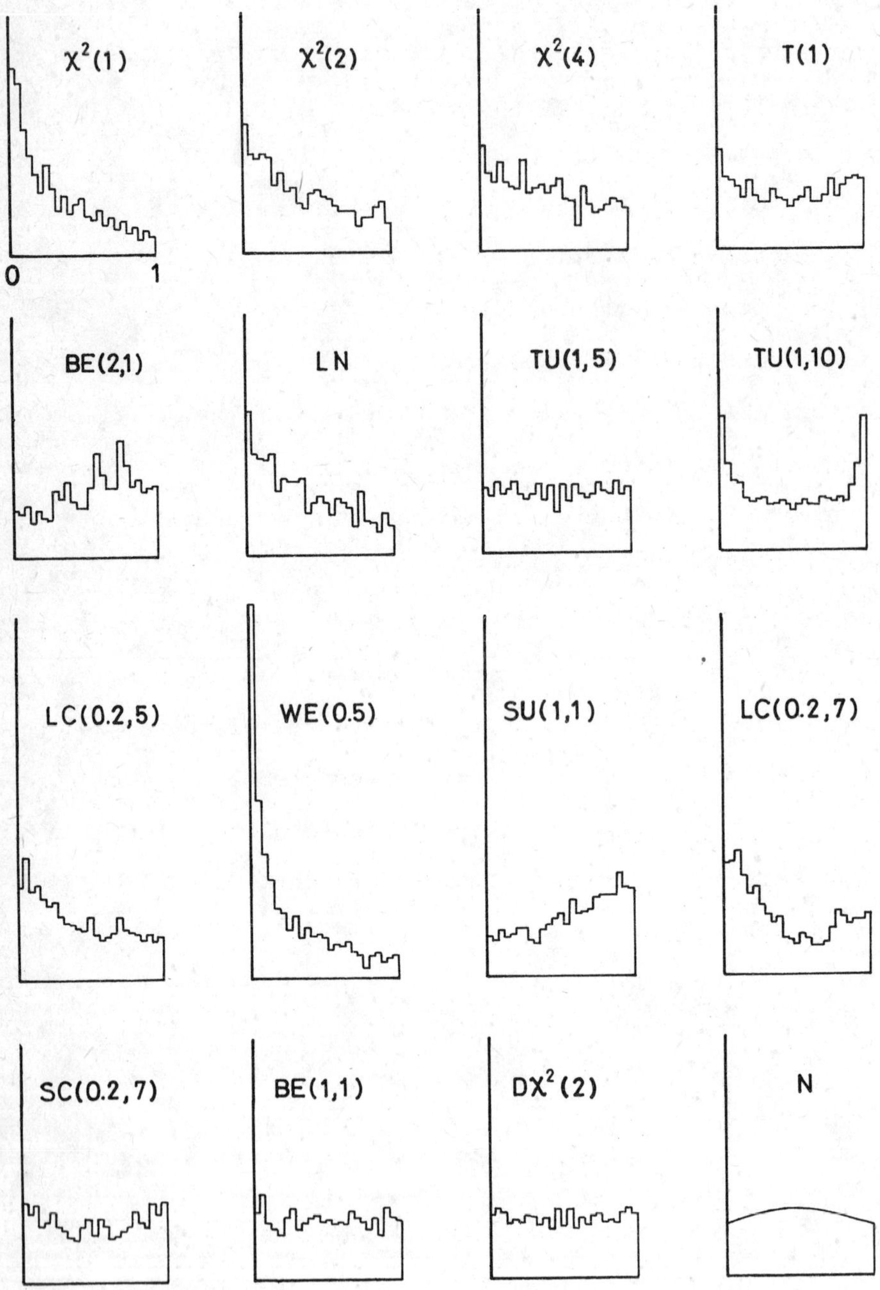

REFERENCES

[1] BONDESSON, L. (1974) Characterization of probability laws through constant regression. Z. Wahrscheinlichkeitstheorie verw. Gebiete 30, 93-115.

[2] DURBIN, J. (1961), Some methods of constructing exact tests. Biometrika 48, 41-55.

[3] KAGAN, A.M.,LINNIK, Yu.V., RAO, C.R. (1973) Characterization problems in mathematical statistics. Wiley, New York.

[4] KOSIK, P., SARKADI, K., (1982) Comparison of multisample tests of normality. In W. Grossmann et al., Probability and Statistical Inference, Reidel, Amsterdam, 183-190.

[5] LAHA, R.G. (1959), On the laws of Cauchy and Gauss. Ann. Math. Statist. 30, 1165-1174.

[6] MAULDON, J.G. (1956) Characteristic properties of statistical distributions. Quarterly J. Math. Oxford, Ser. 2, 7, 155-160.

[7] PEARSON, E.S., D'AGOSTINO, R.B., BOWMAN, K.O. (1977) Tests for departure from normality: comparison of powers. Biometrika 64, 231-246.

[8] PETROV, A.A. (1951) Proverka hipotezi o normalnost raspredelenii po malim viborkam. Dokladi A.N. 76, 355-358.

[9] SARKADI, K. (1967), On testing for normality. Proc. 5th Berkeley Symp. on Math. Stat. Probability I., 373-387.

[10] SHAPIRO, S.S., WILK, M.B., CHEN, N.J. (1968), A comparative study of various tests for normality. J. Amer. Statist. Assoc. 63, 1343-1372.

[11] STÖRMER, H. (1964), Ein Test zum Erkennen von Normalverteilungen. Z. Wahrscheinlichkeitstheorie verw. Gebiete 2, 420-428.

[12] WILK, M.B., SHAPIRO, S.S. (1968), The joint assessment of normality of several independent samples. Technometrics 10, 825-839.

[13] ZINGER, A.A. (1956), Ob odnoi zadache A.N. Kolmogorova. Vestnik Leningrad. Un. 1, 53-56.

Proc. of the 3rd Pannonian Symp. on Math. Stat., Visegrád, Hungary 1982
J. Mogyoródi, I. Vincze, W. Wertz, eds

ON THE CLOSENESS OF DISTRIBUTIONS OF SUMS OF INDEPENDENT INTEGER VALUED RANDOM VARIABLES

A. KOVÁTS

Department of Probability Theory, Eötvös Loránd University, Budapest, Hungary

In [1], Zolotarev uses the pseudomoments for estimating the distance between distributions of sums of independent random variables.

<u>Definition 1.</u>

Let X and Y be random variables with distribution F and G. The quantity

$$v(r) = \int_{-\infty}^{\infty} |x|^r |d(F-G)|$$

is called the $r^{\underline{th}}$ pseudomoment of X and Y.

In this paper we give some estimations for integer valued random variables using the notion of factorial pseudomoments.

<u>Definition 2.</u>

Let X and Y be nonnegative integer valued random variables. Then

$$\beta(r) = \sum_{k=0}^{\infty} k(k-1)\dots(k-r+1)|P(X=k)-P(Y=k)|$$

is called the $r^{\underline{th}}$ factorial pseudomoment of X and Y.

In our first theorem we formulate a Bergström-type inequality, which contains, as special cases, an estimation of Zolotarev [1] and a result of Franken [2]. With the help of this inequality in the second theorem we prove an estimation connected with asymptotic expansions.

Let $\{X_j\}$ and $\{Y_j\}$, $j=1,2,\dots,n$ be independent, nonnegative, integer valued random variables with distribution functions $\{F_j\}$ and $\{G_j\}$, $j=1,2,\dots,n$. Introduce the following notations

$$\alpha_j(r)=\sum_{k=0}^{\infty} k(k-1)\dots(k-r+1)[P(X_j=k)-P(Y_j=k)],$$

$$A(r) = \sum_{j=1}^{n} |\alpha_j(r)|,$$

$$\beta_j(r)=\sum_{k=0}^{\infty} k(k-1)\dots(k-r+1)|P(X_j=k)-P(Y_j=k)|,$$

$$B(r) = \sum_{j=1}^{n} \beta_j(r)$$

$$r=1,2,\dots;\ j=1,2,\dots,n.$$

Theorem 1.

If $A(1) = A(2)=\dots=A(m)=0$ and $B(m+1)$ is finite, then

$$\Big| \prod_{j=1}^{n}{}^{*} F_j - \prod_{j=1}^{n}{}^{*} G_j - \sum_{v=1}^{s} \sum_{1\le\ell_1\le\dots\le\ell_v\le n} \Big(\prod_{k=1}^{v}{}^{*}(F_{\ell_k}-G_{\ell_k}) * \prod_{\substack{1\le i\le n\\ i\ne\ell_1,\ell_2,\dots,\ell_v}}{}^{*} G_i\Big)\Big| \le$$

$$\le K(m,s)\, B^{s+1}(m+1),$$

where the sign $*$ denotes the convolution and $K(m,s)$ is constant depending only on m and s.

In order to show the point of this theorem we formulate some special cases of it.

Corollary 1.1.

If $s=0$, then Theorem 1 gives the following estimation

$$\Big| \prod_{j=1}^{n}{}^{*} F_j - \prod_{j=1}^{n}{}^{*} G_j\Big| \le K(m)B(m+1) .$$

That is, in case of nonnegative, integer valued random variables our estimation is more precise than Zolotarev's in the sense, that factorial pseudomoments substitute the usual pseudomoments.

Corollary 1.2.

Let $\{X_j\}$ and $\{Y_j\}$, $j=1,2,\ldots,n$, respectively be identically distributed. In this case Theorem 1 gives the following estimation

$$|F_1^{*n} - G_1^{*n} - \sum_{j=0}^{s} \binom{n}{j} (F_1-G_1)^{*j} * G_1^{*(n-j)}| \le K(m,s)(n\,\beta_1(m+1))^{s+1}.$$

Using an estimation of this type Bergström [3] proved the classical Edgeworth-expansion. By the help of this inequality one can prove asymptotic expansions for such limit theorems, where the limit distributions are infinitely divisible.

The proof of Theorem 1 is based on the following two lemmas

Lemma 1.

Let $\{a_j\}$ and $\{b_j\}$, $j=1,2,\ldots,n$ be complex numbers, such that $|a_j+b_j|\le 1$ and $|b_j|\le 1$. Then

$$|\prod_{j=1}^{n}(a_j+b_j) - \sum_{v=0}^{s} \sum_{1\le \ell_1\le\ldots\le\ell_v\le n} \left(\prod_{k=1}^{v} a_{\ell_k} \prod_{\substack{1\le i\le n\\ i\ne \ell_1,\ell_2,\ldots,\ell_v}} b_i\right)| \le$$

$$\le \sum_{1\le\ell_1\le\ldots\le\ell_{s+1}\le n} \prod_{k=1}^{s+1} |a_{\ell_k}| .$$

One can prove this lemma by induction.

Let P and Q be signed measures of bounded variation on the nonnegative integers and let f and g be their Fourier-transforms. Introduce $F(x)=\sum_{k<x} P(k)$, $G(x)=\sum_{k<x} Q(k)$.

Lemma 2. (Caregradsky [4])

$$\sup_x |F(x) - G(x)| \le \frac{1}{4\pi} \int_{-\pi}^{\pi} \left|\frac{f(t)-g(t)}{\sin\frac{t}{2}}\right| dt .$$

Proof of Theorem 1.

Let $\{f_j\}$ and $\{g_j\}$, $j=1,2,\ldots,n$ be the characteristic functions of distributions $\{F_j\}$ and $\{G_j\}$, $j=1,2,\ldots,n$. On the basis of our assumptions and Lemma 1 the following estimations are true

$$|f_j(t)-g_j(t)|=\left|\sum_{k=0}^{\infty}(e^{itk}-\sum_{j=1}^{m}\binom{k}{j}(e^{it}-1)^j)(P(X_j=k)-P(Y_j=k))\right|\le$$

$$\le\sum_{k=m+1}^{\infty}\binom{k}{m+1}\left|e^{it}-1\right|^{m+1}\left|P(X_j=k)-P(Y_j=k)\right|\le$$

$$\le\frac{|2\sin\frac{t}{2}|^{m+1}}{(m+1)!}\beta_j(m+1),$$

$$\left|\prod_{j=1}^{n}f_j(t)-\prod_{j=1}^{n}g_j(t)-\sum_{v=1}^{s}\sum_{1\le\ell_1\le\ldots\le\ell_v\le n}(\prod_{k=1}^{v}(f_{\ell_k}-g_{\ell_k})\prod_{\substack{1\le i\le n\\ i\ne\ell_1,\ell_2,\ldots,\ell_v}}g_i)\right|\le$$

$$\le\sum_{1\le\ell_1\le\ldots\le\ell_{s+1}\le n}\prod_{k=1}^{s+1}|f_{\ell_k}-g_{\ell_k}|\le(\sum_{j=1}^{n}|f_j-g_j|)^{s+1}\le$$

$$\le\left(\frac{|2\sin\frac{t}{2}|^{m+1}}{(m+1)!}B(m+1)\right)^{s+1}.$$

Applying Lemma 2 we can get our assertion with

$$K(m,s)=\frac{2^{ms+s+m-1}}{\pi((m+1)!)^{s+1}}\int_{-\pi}^{\pi}\left|\sin\frac{t}{2}\right|^{ms+m+s}dt.$$

In Theorem 2 we prove such an estimation, which gives an opportunity to prove asymptotic expansions that concern arrays of random variables and use factorial pseudomoments.

With the notations of Theorem 1 let $F_j=F$, $G_j=G$, $\alpha_j(r)=\alpha(r)$, $\beta_j(r)=\beta(r)$, $f_j=f$, $g_j=g$ and $m=1$.

Theorem 2.

If $\beta(s+2)$ is finite, then

$$\left|F^{*n}-G^{*n}-\sum_{v=1}^{s}\sum_{j=1}^{v}\binom{n}{j}\Delta^{v+j}\,G^{*(n-j)}\sum_{\substack{i_1+i_2+\ldots+i_j=v+j\\ i_\ell\geq 2}}\prod_{\ell=1}^{j}\frac{\alpha(i_\ell)}{i_\ell!}\right|\leq$$

$$\leq K(1,s)(n\beta(2))^{s+1}+$$

$$+\sum_{v=1}^{s}\binom{n}{v}\left\{\sum_{j=s+v+1}^{v(s+1)}C_jL_{j,v}+\sum_{k=2(v-1)}^{(s+2)(v-1)}C_{k+s+2}\frac{\beta(s+2)}{(s+2)!}\sum_{\ell=0}^{v-1}\sum_{i+j=k}L_{i,\ell}L_{j,v-\ell-1}\right\},$$

where

$$\Delta G(x)=G(x)-G(x-1)\ ;$$

$$C_j=\frac{1}{4\pi}\int_{-\pi}^{\pi}\left|2\sin\frac{t}{2}\right|^{j-1}dt\ ;$$

$$L_{j,v}=\sum_{\substack{i_1+i_2+\ldots+i_v=j\\ i_\ell\geq 2}}\prod_{\ell=1}^{v}\frac{\beta(i_\ell)}{i_\ell!}$$

Corollary 2.1.

Let $\{X_{j,n}\}$ and $\{Y_{j,n}\}$, $j=1,2,\ldots,n;\ n=1,2,\ldots$ be independent, identically distributed, nonnegative, integer valued random variables with distribution functions F_n and G_n, respectively. Let us denote

$$\beta(n,r)=\sum_{k=0}^{\infty}k(k-1)\ldots(k-r+1)\,|P(X_{j,n}=k)-P(Y_{j,n}=k)|,$$

$$\alpha(n,r)=\sum_{k=0}^{\infty}k(k-1)\ldots(k-r+1)\,|P(X_{j,n}=k)-P(Y_{j,n}=k)|,$$

$$\alpha(n,1)=0.$$

If $\beta(n,s+2)$ are finite and there exist numbers $\beta^*(r), r=1,2,\ldots,s+2$ such that $\beta(n,r)=\frac{\beta^*(r)}{n^r}$, then the following assertion is true as a consequence of Theorem 2

$$F_n^{*n}-G_n^{*n}-\sum_{v=1}^{s}\sum_{j=1}^{v}\binom{n}{j}\Delta^{v+j}\,G_n^{*(n-j)}\sum_{\substack{i_1+\ldots+i_j=v+j\\ i_\ell\geq 2}}\prod_{\ell=1}^{j}\frac{\alpha(n,i_\ell)}{i_\ell!}=$$

$$=O\left(\frac{1}{n^{s+1}}\right).$$

Proof of Theorem 2.

The proof is based again on Lemma 2, therefore we estimate the absolute value of the corresponding Fourier-transform. Since we can use the inequality of Theorem 1, only the following expression should be estimated

$$D = \left| \sum_{v=1}^{s} \left[\binom{n}{v} (f-g)^v g^{n-v} - \sum_{j=1}^{v} \binom{n}{j} g^{n-j} (z-1)^{v+j} \sum_{\substack{i_1+\ldots+i_j=v+j \\ i_\ell \geq 2}} \prod_{\ell=1}^{j} \frac{\alpha(i_\ell)}{i_\ell !} \right] \right|,$$

where $z = e^{it}$.

Since

$$f - g = \sum_{j=2}^{s+1} \frac{\alpha(j)}{j!} (z-1)^j + \frac{\gamma(s+2)}{(s+2)!} (z-1)^{s+2} \tag{1}$$

where

$$|\gamma(s+2)| \leq \beta(s+2), \tag{2}$$

and

$$|\alpha(r)| \leq \beta(r) \ , \ r=1,2,\ldots,s+1 \ , \tag{3}$$

therefore D can be estimated as follows. Let us put (1) into D, do the possible reductions, use the estimations (2), (3) and we get

$$D \leq \sum_{v=1}^{s} \binom{n}{v} \left\{ \sum_{j=s+v+1}^{v(s+1)} |z-1|^j L_{j,v} + \right.$$

$$\left. + \sum_{k=2(v-1)}^{(s+2)(v-1)} |z-1|^{k+s+2} \frac{\beta(s+2)}{(s+2)!} \sum_{\ell=0}^{v-1} \sum_{i+j=k} L_{i,\ell} L_{j,v-\ell-1} \right\} .$$

Then using Lemma 2 and Theorem 1 we can obtain the statement of Theorem 2.

REFERENCES

1 V.M. ZOLOTAREV: On the closeness of distributions of two sums of independent random variables. The Theory of Probability and Its Applications, 1965, X,3 (in Russian).

2 P. FRANKEN: Approximation der Verteilungen von Summen unabhängiger nichtnegativer ganzzahliger Zufallsgrössen durch Poissonsche Verteilungen. Mathematische Nachrichten, 1963, 27, 5-6 (in German).

3 H. BERGSTRÖM: On Asymptotic Expansions of Probability Functions. Skandinavisk Aktuarietidskrift 1951, H. 1-2.

4 I.P. CAREGRADSKY: On uniform approximation of the binomial distribution by infinitely divisible laws. The Theory of Probability and Its Applications, 1958, III,4 (in Russian).

5 J. MILASAVICIUS: On the convergence of Grigelionis-Franken asymptotic expansions. Applications of Probability Theory and Mathematical Statistics, Vilnius, 1980, 3 (in Russian).

6 A. KOVÁTS: On Bergström's identity. Applications of Probability Theory and Mathematical Statistics, Vilnius, 1980, 3 (in Russian).

Proc. of the 3rd Pannonian Symp. on Math. Stat., Visegrád, Hungary 1982
J. Mogyoródi, I. Vincze, W. Wertz, eds

HOW TO PROVE THE CLT FOR THE LORENTZ PROCESS BY USING PERTURBATION THEORY?

A. KRÁMLI*, D. SZÁSZ**

*Computer and Automation Institute, Hungarian Academy of Sciences, Budapest

**Mathematical Institute, Hungarian Academy of Sciences, Budapest, Hungary

INTRODUCTION

The planar Lorentz process $\{(q(t),p(t);t \geq 0\}$ where $q(t),p(t) \in \mathbb{R}^2$ $|p(t)|=1$ is the motion of a particle among periodically situated smooth, convex scatterers (see Figure). The motion is uniform with elastic collisions at the scatterers. Suppose that the distribution of the initial point $(q(0),p(0))$ is absolutely continuous in the three-dimensional phase-space.

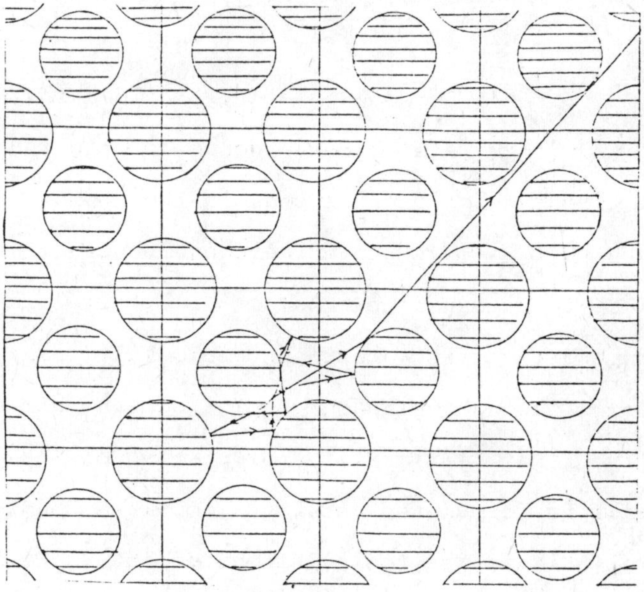

Figure

In [4] we have started to work out the probability theory of random walks with internal degrees of freedom in order to study various problems concerning the Lorentz process and the related Lorentz-gas. Now as a first step in realizing this strategy, we give a new proof for the famous theorem of Bunimovich and Sinai [1] stating that

$$\frac{q(At)}{\sqrt{A}} \xrightarrow{C[0\infty)} W_{\Sigma}(t)$$

where $W_{\Sigma}(t)$ is a Wiener process on the plane with a nonsingular covariance-matrix Σ.

In fact, for the simplicity of exposition, this talk concentrates upon the principal difficulty in obtaining this result by our approach. Namely in [4] we expanded in a power series the leading eigenvalue of the Fourier transform matrix of a random walk with internal degrees of freedom and showed how the CLT is implied by this expansion. Since the Lorentz process can be approximated by a sequence of random walks with internal degrees of freedom, (cf.[1]), this approximation also yields the CLT for the Lorentz process, provided the CLT can be proven with a uniform rate of convergence for the random walks with internal degrees of freedom used in the approximation.

We note that we only prove the convergence of the one-dimensional distribution since that of the finite dimensional distributions and tightness questions are mainly technical.

To avoid loading the reader by introducing him into the mistery of the Sinai billiard, the properties of its Markov partition will be formulated here as axiom-like conditions.

Several notions and tools introduced in [4] will be used here without special reference, nonetheless the following statement of a famous lemma of H. Wielandt will be fundamental in the proofs presented here and it is worth recalling it immediately (cf. [2]):

Let $A=\{a_{jk}\}$ be a matrix with non-negative elements and let $C=\{c_{jk}\}$ be an arbitrary complex matrix. If $|c_{jk}|\leq a_{jk}$ for every j and k, then the absolute value of the largest eigenvalue of C is less than or equal to that of A.

UNIFORM ANALYCITY OF THE RESOLVENTS

Next we recall the necessary information on perturbation theory, which can be found in standard textbooks (cf. e.g. [3], II.§ 1.).

For our purposes it is sufficient to investigate linear operators acting in finite dimensional Euclidean spaces $\mathbb{R}^n$; n can tend to infinity.

The set of all eigenvalues of an operator T will be denoted by spec (T).

Let T(K) be a holomorphic function of such operators depending on a complex vector-valued parameter $K=(K_1,\dots,K_d)$- this means that T(K) is the sum of an absolute convergent power series

$$\sum_{\substack{n=0 \\ j_1+\dots+j_d=n}}^{\infty} A_{j_1,\dots,j_d} K_1^{j_1} \dots K_d^{j_d} \qquad \text{if}$$

$||K||$ is sufficiently small.

Set $A_{0,\dots,0}=T$ and

$$A(K)= \sum_{\substack{n=1 \\ j_1+\dots+j_d=n}}^{\infty} A_{j_1,\dots,j_d} K_1^{j_1}\dots K_d^{j_d}$$

The main object of (analytic) perturbation theory is the investigation of the dependence of spec $T(K)$ on K.

Statement A. If $\lambda\in$ spec (T), then the projection P_λ onto the maximal invariant subspace of T belonging to λ (i.e. $P_\lambda P_\lambda = P_\lambda$, $P_\lambda T = TP_\lambda$, $P_\lambda T$ has the unique eigenvalue λ and $\dim(P_\lambda \mathbb{R}^n) \geq \dim(P\, \mathbb{R}^n)$ if P satisfies the preceding three conditions) can be represented in the following integral form

$$P_\lambda = \frac{1}{2\pi i} \int_\Gamma (T-\xi I)^{-1} d\xi \tag{1}$$

where I is the identity operator, Γ is a closed contour surrounding λ and containing no other eigenvalues of T.

$R_T(\xi) = (T-\xi I)^{-1}$ is called the resolvent operator of T. $R_T(\xi)$ exists for every $\xi \notin \mathrm{spec}(T)$.

Definition B.

The radius $\rho(T)$ of the set spec(T) is called the spectral radius of the operator T, i.e.

$$\rho(T) = \max_{\lambda\in \mathrm{spec}(T)} |\lambda| .$$

Statement C. $\rho(T) = \limsup_{n\to\infty} \|T^n\|^{\frac{1}{n}}$

moreover, if $A_1A_2 = A_2A_1$ then

$$\rho(A_1+A_2) \leq \rho(A_1) + \rho(A_2)$$

$$\rho(A_1A_2) \leq \rho(A_1)\rho(A_2)$$

The proof of the following standard statement will be recalled in a form more convenient for our purposes.

Statement D. $R_{T(K)}(\xi)$ is holomorphic in ξ and K in the domain $\xi \notin \mathrm{spec}(T(K))$.

Proof. Let e.g. $\xi_0 \notin \mathrm{spec}(T)$, then $R_T(\xi_0)$ exists. Further

$$T(K) - \xi I = T - \xi_0 I - (\xi-\xi_0)I + A(K) = (I - [(\xi-\xi_0)I - A(K)]R(\xi_0))(T-\xi_0 I).$$

So

$$R_{T(K)}(\xi)=R_T(\xi_0)\{I-[(\xi-\xi_0)I-A(K)]R(\xi_0)\}^{-1}, \qquad (2)$$

i.e. $R_{T(K)}(\xi)$ can be defined when the Neumann series for the inverse operator on the right hand side of (2) converges. E.g. the condition

$$(|\xi-\xi_0|+\rho(A(K)R_T(\xi_0))<1 \qquad (3)$$

is sufficient for it.

<u>Corollary E.</u> Let $\{T^{(\nu)}(K),\ \nu\in\mathbb{Z}\}$ be a family of holomorphic operator functions with the following two properties:

(i) for every $\nu\in\mathbb{Z}$, ξ_0 is a simple eigenvalue of $T^{(\nu)}$ and there exists a $\delta_1>0$ such that $T^{(\nu)}$ has no other eigenvalues in the neighbourhood $|\xi-\xi_0|<\delta_1$

(ii) there exists a $\delta_2>0$ such that, for every $\frac{\delta_1}{4}<|\eta|<3\cdot\frac{\delta_1}{4}$, $\nu\in\mathbb{Z}$ and $||K||<\delta_2$,

$$\rho(A^{(\nu)}(K)R_{T^{(\nu)}}(\xi_0+\eta))<1$$

Then

(iii) for every $\nu\in\mathbb{Z}$, $R_{T^{(\nu)}(K)}(\xi)$ is holomorphic in the domain

$$||K||<\delta_2,\ \frac{\delta_1}{4}<|\xi-\xi_0|<\frac{3\delta_1}{4}.$$

<u>Corollary F.</u> Let $\delta_1<1$. Under the conditions of Corollary E, for every $\nu\in\mathbb{Z}$ and $||K||<\delta_2$, the set spec $(T^{(\nu)}(K))\cap\{\xi,|\xi-\xi_0|<\frac{\delta_1}{2}\}$ consists of a unique simple eigenvalue $\xi^{(\nu)}(K)$; moreover, in the neighbourhood $|\xi-\xi_0|<\frac{\delta_1}{2}$, $P_{\xi^{(\nu)}(K)}$ and consequently $\xi^{(\nu)}(K)$, too is a holomorphic function of K.

Corollaries E and F can be proved by standard methods based on the integral formula (1). The only crucial point is replacing the usual $||.||$ by $\rho(.)$ in conditions of type (3).

AN INFINITE FAMILY OF RANDOM WALKS WITH INTERNAL DEGREES OF FREEDOM

The objects which we intend to apply the results of the preceding paragraph to were introduced in [4]. Here we briefly repeat their definitions.

Definition 1. Let $\zeta_t^{(\nu)}$ be a Markov chain with the following (finite) state space and transition probabilities

$$\{x\in \mathbb{Z}^d \otimes (1,2,...,\nu)\}$$

$$\text{Prob } (\zeta_t^{(\nu)}=(x,k)\,|\,\zeta_{t-1}^{(\nu)}=(y,j))=q_{jk}(x-y).$$

$\zeta_t^{(\nu)}$ is a random walk with ν internal degrees of freedom.

Set $\{q_{jk}^{(\nu)}(x)\}_{(j,k=1,...,\nu)}=Q^{(\nu)}(x)$

Condition 2. (finite horizon) There exists a constant D such that $Q^{(\nu)}(x)=0$ for all $||x||>D$.

Condition 3. (ergodicity) The stochastic dxd matrix $Q^{(\nu)}=\Sigma_{x\in\mathbb{Z}^d} Q^{(\nu)}(x)$ is irreducible and aperiodic. So there exists a unique stationary vector $\mu^{(\nu)}: Q^*\mu^{(\nu)}=\mu^{(\nu)}$.

Condition 4. (spatial symmetry)

$$\sum_{x\in\mathbb{Z}^d} x(Q^{(\nu)}(x)\mathbb{1}^{(\nu)},\mu^{(\nu)})=0 \quad \text{where} \quad \mathbb{1}^{(\nu)}=\underbrace{(1,...,1)}_{\nu}$$

Set $\alpha^{(\nu)}(K)=\sum_{x\in\mathbb{Z}^d} Q^{(\nu)}(x)\, e^{i(x,K)}$ the Fourier-transform matrix of $Q^{(\nu)}(x)$.

Condition 5. (uniform ergodicity, i.e. Doeblin condition) There exists a $\delta_1>0$ such that for every ν $Q^{(\nu)}$ has the only eigenvalue 1 outside the region $|\xi|<1-\delta_1$.

Statement 6. The Fourier-transforms $\alpha^{(\nu)}(K)$ are uniformly holomorphic, i.e., in $||K||<\delta_2$, $\alpha^{(\nu)}(K)$ can be expanded in a power series up to any fixed order with coefficients (including that of the remainder term) whose spectral radii are uniformly bounded in ν.

Proof. Statement 6 follows from Condition 2 and Statement C. The n-th momentum matrices of $\zeta_t^{(\nu)}$ (the coefficients of the n-th order terms in the Taylor expansion of $\alpha^{(\nu)}(K)$) can be majorized in the sense of the famous Wielandt lemma by $D^n Q$. (Analogous arguments are described in detail in [4].)

Condition 5 together with Statements C and 6 provides that the family $\alpha^{(\nu)}(K)$ satisfies the conditions of corrollaries E and F.

Statement 7. Set $\lambda^{(\nu)}(K)$ the largest eigenvalue of $\alpha^{(\nu)}(K)$ for $||K||<\delta_2$. The Taylor expansion of $\lambda^{(\nu)}(K)$ looks like

$$\lambda^{(\nu)}(K)=1+\frac{1}{2}(\sigma^{(\nu)}K,K)+\dots$$

Statement 7 is a consequence of Condition 4 and Corollaries E and F. A detailed proof for a single random walk with internal degrees of freedom can be found in [4].

In this talk we assume that the dxd matrix $\sigma^{(\nu)}$ does not depend on ν $(\sigma^{(\nu)}=\sigma)$ and it is positive definite.

Statement 8. There exist positive numbers $\rho<1$ and D_1 such that for every $\nu\in\mathbb{Z}$ and $||K||<\delta_2$

$$||((I-P_{\lambda^{(\nu)}(K)})\alpha^{(\nu)}(K))^n||<D_1\nu\rho^n \tag{4}$$

Proof. It follows from the Wielandt lemma that for an $\nu\times\nu$ matrix $A=\{a_{jk}\}$ majorized by const Q $\max|a_{jk}|<\text{const}\,\rho(Q)$. On the other hand $||A||\leq\sqrt{\Sigma|a_{jk}|^2}<\nu\rho(A)$. So (4) follows from Corollary F.

Statement 9. There exists a positive number D_2 such that for every ν and $||K||<\delta_2$

$$||P_{\lambda^{(\nu)}(K)}-P_{\lambda_0^{(\nu)}}||<D_2\nu||K||$$

The proof is similar to the preceding one. We can choose D_2 as the spectral radius of the first momentum matrix of $\zeta_t^{(\nu)}$. Here Corollary E is to be used.

MAIN THEOREM

If besides Conditions 2-5 $\nu^2=\sigma(n)$ holds too and $\text{Prob}\,(\zeta_0^{(\nu)}=(0,j))=\mu_j^{(\nu)}$ then, denoting $\zeta_t^{(\nu)}=(\eta_t^{(\nu)},\varepsilon_t^{(\nu)})$ the distribution of $n^{-1/2}\eta_n^{(\nu)}$ tends weakly to the d-dimensional Gaussian distribution $N(0,\sigma)$.

Proof. The modification of the standard proof for CLT to our case (cf. Proof of pre-theorem 2.1 in [4]) says that we only need the convergence

$$\int_{||K||<\delta} |((\alpha^{(\nu)})^n (\tfrac{s}{\sqrt{n}})\mathbb{1}^{(\nu)},\mu^{(\nu)})-e^{-1/2(\sigma K,K)}| \to 0$$

uniformly in ν. This last relation follows from Statements 6-9.

REFERENCES

1 BUNIMOVICH L. A. - SINAI YA. G. (1981).
Statistical properties of Lorentz gas with periodic configuration of scatterers.
Commun. Math. Physics. 78, 479-497.

2 GANTMAKHER F. R. (1967). Theory of Matrices. 3rd edition. Nauka, Moscow (in Russian).

3 KATO T. (1980). Perturbation Theory for Linear Operators 2nd edition. Springer, Berlin-Heidelberg-New York.

4 KRÁMLI A. - SZÁSZ D. (1981). Random walks with internal degrees of freedom. I. Local limit theorems. Z. für Wahrscheinlichkeitstheorie (submitted).

Proc. of the 3rd Pannonian Symp. on Math. Stat., Visegrád, Hungary 1982
J. Mogyoródi, I. Vincze, W. Wertz, eds

GROUP ALGEBRAS-VALUED RANDOM VARIABLES

M. KRUTINA

Charles University, Prague, Czechoslovakia

ABSTRACT.

Let G be a locally compact Abelian group, $\hat{G}$ its dual group and let $L^1(G)$, $L^2(G)$ be defined with respect to the Haar measure. $L^2(G)$-valued random variables realizable in $L^1(G) \cap L^2(G)$ with the convolution product establish a commutative linear algebra if the group G is σ-compact. It is proved that the probability distribution of every such random variable ξ is uniquely determined by the probability distribution of its random process $(\tilde{\xi}(\alpha), \alpha \in \hat{G})$, and vice versa.

Denote $\mathbb{N}$ the set of all natural numbers, and $\mathbb{C}$ the set of all complex numbers. Let $(\Omega, \mathcal{S}, P)$ be a probability space, and let B be a complex Banach space. A mapping X from Ω to B is weakly measurable if, for any bounded linear functional $f \in B^*$, the $\mathbb{C}$-valued function $f(X(*))$ is $\mathcal{S}$-measurable. Two weakly measurable mappings X,Y from Ω to B are equivalent if, for any $f \in B^*$, $P\{\omega : f(X(\omega)) = f(Y(\omega))\} = 1$. In this case we write $X \sim Y$. We shall call a class of mutually equivalent weakly measurable mappings from Ω to B as a B-valued random variable. If $H \subset B$ then a B-valued random variable ξ is said to be realizable in H if there exists $X \in \xi$ such that, for any $\omega \in \Omega$, $X(\omega) \in H$. The probability distribution of a B-valued random variable ξ is defined as the probability measure PX^{-1} on $\mathcal{F}(B) = \sigma\{f^{-1}(M) : f \in B^*, M \in \mathcal{B}_{\mathbb{C}}\}$ where $X \in \xi$, $\sigma\mathcal{M}$ denotes the minimal σ-algebra generated by a class of sets $\mathcal{M}$, and $\mathcal{B}_{\mathbb{C}}$ denotes the class of Borel sets in $\mathbb{C}$. Obviously, if X and Y are equivalent weakly measurable mappings then $PX^{-1}=PY^{-1}$ on $\mathcal{F}(B)$. So the definition is justified.

In what follows, G is assumed to be a locally compact Abelian group. Let λ be an invariant Haar measure on the class of Borel sets in G. As usual, the symbols $L^1(G)$ and $L^2(G)$ will designate all $\mathbb{C}$-valued Borel measurable functions $x=x(t)$ on G such that $\int_G |x(t)|d\lambda(t)<\infty$, and $\int_G |x(t)|^2 d\lambda(t)<\infty$, respectively. Functions which are equal almost everywhere, will be considered as identical. Let $\hat{G}$ denote the group of characters of the group G. As known, the Abelian group $\hat{G}$ is locally compact in its weak topology.

The Fourier transform of a function $x \in L^1(G) \cap L^2(G)$ is a function $\hat{x}$ on $\hat{G}$ of the form $\hat{x}(\alpha)=\int_G x(t)\alpha(t)d\lambda(t), \alpha\in\hat{G}$. The function $\hat{x}$ is continuous, and there is an invariant Haar measure $\hat{\lambda}$ on the class of Borel sets in $\hat{G}$ such that the Fourier transform has a unique extension to an isometric mapping φ from $L^2(G)$ into $L^2(\hat{G})$.

We shall denote by L the set of all continuous $\mathbb{C}$-valued functions on $\hat{G}$ with compact supports. In the following we shall often write $L^1, L^2, \hat{L}^2$ instead of $L^1(G), L^2(G), L^2(\hat{G})$, respectively. If X is a mapping from Ω to $L^1(G)\cap L^2(G)$, we write $\hat{X}(\alpha,\omega), \hat{X}(\omega)$, and $X(t,\omega)$ instead of $\widehat{X(\omega)}(\alpha), \widehat{X(\omega)}$, and $X(\omega)(t)$, respectively. If $v\in L^2, \alpha\in\hat{G}$, and K is a compact subset of G, then we denote by f_v a functional on L^2 of the form $f_v(x)=\int_G x(t)v(t)d\lambda(t), x\in L^2$. Symbols f_α and $f_{\alpha,K}$ will designate functionals on $L^1\cap L^2$ of the form $f_\alpha(x)=\hat{x}(\alpha)$, and $f_{\alpha,K}(x)=\int_K x(t)\alpha(t)d\lambda(t)$, $x\in L^1\cap L^2$, respectively.

LEMMA 1 Let G be σ-compact. If $z\in L$, then there is a sequence (f_n) of functionals on $L^1\cap L^2$ such that, for any $n\in\mathbb{N}$,

$$f_n = \sum_{j=1}^{j_n} c_{nj} f_{\alpha_{nj}},$$

$j_n \in \mathbb{N}$, $c_{nj} \in \mathbb{C}$, $\alpha_{nj} \in \hat{G}$, $j \in \{1,\ldots,j_n\}$, and, for any $x \in L^1 \cap L^2$,

$$f_{\varphi^{-1}z}(x) = \lim_{n\to\infty} f_n(x).$$

If $\alpha \in \hat{G}$, then, for any $x \in L^1 \cap L^2$,

$$f_\alpha(x) = \lim_{n\to\infty} f_{\alpha,K_n}(x)$$

where (K_n) is a nondecreasing sequence of compact subsets of G such that $\bigcup_{n=1}^{\infty} K_n = G$.

PROOF. We shall make use of the fact that there is a countable base (V_n) of neighbourhoods at the unit element, say $\tilde{\alpha}$, of $\hat{G}$; e.g., define $V_n = \bigcap_{t\in K_n} \{\alpha \in \hat{G} : |\alpha(t) - \tilde{\alpha}(t)| < \frac{1}{n}\}$, $n \in \mathbb{N}$. So we can assume that $V_n \subset V_m$ for $n>m$. Construct, for any $n \in \mathbb{N}$, a neighbourhood W_n of $\tilde{\alpha}$ such that $W_n^{-1} W_n \subset V_n$. Let $n \in \mathbb{N}$, and Q be a compact support of z. The collection $\{\alpha W_n : \alpha \in Q\}$ is an open cover of Q. It contains a finite subcover $\{\alpha_{n1} W_n, \ldots, \alpha_{nj_n} W_n\}$. Hence we obtain a decomposition $\{A_{n1}, \ldots, A_{nj_n}\}$ of Q where A_{nj} is a Borel subset of $\alpha_{nj} W_n$, $j \in \{1,\ldots,j_n\}$. We can assume that, for any $j \in \{1,\ldots,j_n\}$, $A_{nj} \neq \emptyset$, and choose $\alpha'_{nj} \in A_{nj}$. Define a functional f_n on $L^1 \cap L^2$ by

$$f_n(x) = \sum_{j=1}^{j_n} z(\alpha'_{nj}) \hat{\lambda}(A_{nj}) f_{\alpha'_{nj}}(x), \quad x \in L^1 \cap L^2.$$

From the isometry and linearity of the Fourier transform it follows that, for any $x,y \in L^2$, $\int_G x(t)y(t)\,d\lambda(t) = \int_{\hat{G}} (\varphi x)(\alpha)(\varphi y)(\alpha)\,d\hat{\lambda}(\alpha)$. Choose $x \in L^1 \cap L^2$, and $\varepsilon > 0$. Clearly $|f_{\varphi^{-1}z}(x) - f_n(x)| =$

$$= \left| \int_{\hat{G}} \hat{x}(\alpha) z(\alpha)\,d\hat{\lambda}(\alpha) - \sum_{j=1}^{j_n} \hat{x}(\alpha'_{nj}) z(\alpha'_{nj}) \hat{\lambda}(A_{nj}) \right| \leq \sum_{j=1}^{j_n} \int_{A_{nj}} |\hat{x}(\alpha) z(\alpha) - \hat{x}(\alpha'_{nj}) z(\alpha'_{nj})|\,d\hat{\lambda}(\alpha).$$

As the function $\hat{x}z$ is uniformly continuous on $\hat{G}$ with the support Q, we can find an $n_0 \in \mathbb{N}$ such that for any $n > n_0$, $j \in \{1,\ldots j_n\}$, $\alpha \in A_{nj}$, we have $|\hat{x}(\alpha)z(\alpha)-\hat{x}(\alpha'_{nj})z(\alpha'_{nj})|<\varepsilon$ because of $\alpha_{nj}^{-1}\alpha \in W_n^{-1}W_n$. Now $|f_{\varphi^{-1}z}(x)-f_n(x)|<\varepsilon\hat{\lambda}(Q)$. Hence $f_{\varphi^{-1}z}(x)=\lim_{n\to\infty} f_n(x)$. The second part is evident.

LEMMA 2 If G is σ-compact, and if X is a mapping from Ω to L^2 such that, for any $\omega \in \Omega$, $X(\omega) \in L^1 \cap L^2$, then a necessary and sufficient condition for X to be weakly measurable is that, for any character $\alpha \in \hat{G}$, $\hat{X}(\alpha,*)$ be a $\mathbb{C}$-valued $\mathfrak{J}$-measurable function.

PROOF. The necessity of the condition follows immediately from Lemma 1. Conversely, the set L is dense in $\hat{L}^2$ so that $\varphi^{-1}L$ is dense in L^2. Hence it is sufficient to show that, for any $z \in L$, $f_{\varphi^{-1}z}(X(*))$ is $\mathfrak{J}$-measurable. This we also obtain using Lemma 1.

We define the convolution product $*$ on $L^1 \cap L^2$ by the formula $x*y(s)=\int_G x(t)y(t^{-1}s)d\lambda(t)$, $x,y \in L^1 \cap L^2$, $s \in G$. It is known that $*$ is commutative, and $x*y \in L^1 \cap L^2$.

Now let X, X', Y, Y' be weakly measurable mappings from Ω to L^2 such that $X \sim X'$, $Y \sim Y'$, and, for any $\omega \in \Omega$, $X(\omega)$, $X'(\omega)$, $Y(\omega)$, $Y'(\omega)$ belong to $L^1 \cap L^2$. Write $X*Y(\omega)$ in place of $X(\omega)*Y(\omega)$. If G is σ-compact, then the weakly measurable mappings $X*Y$ and $X'*Y'$ are equivalent, Indeed, from Lemma 2, and from relation $\widehat{x*y}(\alpha)=\hat{x}(\alpha)\hat{y}(\alpha)$, $x,y \in L^1 \cap L^2$, $\alpha \in \hat{G}$, it follows that the mappings $X*Y$ and $X'*Y'$ are weakly measurable. In addition, for any $\alpha \in \hat{G}$, and for any compact subset K of G, $f_{\alpha,K}(X)=f_{\alpha,K}(X')$ a.s., $f_{\alpha,K}(Y)=f_{\alpha,K}(Y')$ a.s. By Lemma 1, we have that $f_\alpha(X)=f_\alpha(X')$ a.s., $f_\alpha(Y)=f_\alpha(Y')$ a.s. From this, it holds that $f_\alpha(X*Y)=f_\alpha(X'*Y')$ a.s. Using an auxiliary sequence of functionals (f_n) defined in Lemma 1 we obtain that, for any $z \in L$, $f_{\varphi^{-1}z}(X*Y)=f_{\varphi^{-1}z}(X'*Y')$ a.s. Hence we deduce that $X*Y \sim X'*Y'$.

DEFINITION Let G be a locally compact and σ-compact Abelian group. Let ξ,η be $L^2(G)$-random variables realizable in $L^1(G)\cap L^1(G)$. Suppose $X\in\xi$, $Y\in\eta$ such that, for any $\omega\in\Omega$, $X(\omega)$, $Y(\omega)$ belong to $L^1(G)\cap L^2(G)$, and let $c\in\mathbb{C}$, $\alpha\in\hat{G}$. Denote $\xi*\eta$, $\xi+\eta$, and $c\xi$ the $L^2(G)$-random variables which are represented by $X*Y$, $X+Y$, and cX, respectively. The symbol $\tilde{\xi}(\alpha)$ will denote the $\mathbb{C}$-valued random variable represented by $\hat{X}(\alpha,*)$. We shall call $\xi*\eta$ as a convolution product of ξ an η.

The definition is justified, because it does not depend upon the choice of the representatives of random variables in hand. However, if G were not σ-compact, the definition would be false. Let U be a symmetric neighbourhood of the unit e of G with a compact closure $\overline{U}$. Thus $G_0 = \bigcup_{n=1}^{\infty} \overline{U}^n$ is a subgroup of G which is both open and closed. There is a decomposition $\{t_d G_0 : d\in D\}$ of G where D is a noncountable index set, and $t_d\in G$, $d\in D$. Notice that $0<\lambda(U)<\infty$. Assume that $\Omega=D$, and that $\mathcal{S}$ is the σ-algebra of all subsets of D which are countable or have a countable complement. Let $D=D_1\cup D_2$ where D_1 and D_2 are disjoint and noncountable. Define mappings X,Y from D to L^2 as follows: $X(t,d)=1$, $t\in t_d U$, $d\in D_1$, $Y(t,d)=1$, $t\in t_d^{-1}U^2$, $d\in D_1$, and, otherwise, $X(t,d)=0$, $Y(t,d)=0$. Both mappings are weakly measurable. Of course, for any $f\in(L^2)^*$, the sets $\{d\in D: f(X(d))\neq 0\}$, $\{d\in D: f(Y(d))\neq 0\}$ are countable. In addition, for any $d\in D$, $X(d)$ and $Y(d)$ belong to $L^1\cap L^2$. However, $\hat{X}(\tilde{\alpha},d)=\lambda(U)$, $d\in D_1$, $\hat{Y}(\tilde{\alpha},d)=\lambda(U^2)$, $d\in D_1$, and, otherwise, $\hat{X}(\tilde{\alpha},d)=0$, $\hat{Y}(\tilde{\alpha},d)=0$, where $\tilde{\alpha}$ denotes the unit of G. Thus $\hat{X}(\tilde{\alpha},*)$ and $\hat{Y}(\tilde{\alpha},*)$ are not $\mathcal{S}$-measurable functions. Moreover, $X*Y$ is not a weakly measurable mapping. Indeed, let $v\in L^2$ such that $v(t)=1$, $t\in U$, and $v(t)=0$ else. If $d\in D_1$, then $f_v(X*Y(d))=\int_U X*Y(s,d)\,d\lambda(s)=$

$\int\int_{UG} X(t,d)Y(t^{-1}s,d)d\lambda(t)d\lambda(s)=\int_G \int_{t^{-1}U} X(t,d)Y(s,d)d\lambda(s)d\lambda(t)=$

$\int_{t_dU} \int_{t^{-1}U} Y(s,d)d\lambda(s)d\lambda(t)$. For $t\in t_dU$ it is $t^{-1}U\subset t_d^{-1}U^2$. Hence the latter integral is equal to $\int_{t_dU} \int_{t^{-1}U} d\lambda(s)d\lambda(t)=(\lambda(U))^2$. If $d\in D_2$ then $f_v(X*Y(d))=0$.

THEOREM 1 Let G be a locally compact and σ-compact Abelian group. The set of all $L^2(G)$-random variables realizable in $L^1(G)\cap L^2(G)$ is a commutative linear algebra with respect to the operations of convolution product, addition, and scalar multiplication.

The proof follows immediately from the properties of the convolution product on $L^1\cap L^2$.

In what follows, let $\mathcal{M}\cap H$ be the class $\{E\cap H:E\in\mathcal{M}\}$ where $\mathcal{M}$ is a class of sets and H is a set. Denote by $\mathcal{K}$ the σ-algebra $\sigma\{\{z\in \mathbb{C}^{\hat{G}}:z(\alpha)\in M\}:\alpha\in\hat{G},M\in \mathcal{B}_{\mathbb{C}}\}$. Further, let us write $\mathcal{F}$ instead of $\mathcal{F}(G)$.

LEMMA 3 If G is σ-compact, then

$\mathcal{F}\cap L^1=\sigma\{f_\alpha^{-1}(M):\alpha\in\hat{G},M\in\mathcal{B}_{\mathbb{C}}\}\cap L^1=\varphi^{-1}(\mathcal{K}\cap\hat{L}^2)\cap L^1$.

PROOF. Using Lemma 1, we easily obtain that

$\sigma\{f_{\varphi^{-1}z}^{-1}(M):z\in L,M\in\mathcal{B}_{\mathbb{C}}\}\cap L^1=\sigma\{f_\alpha^{-1}(M):\alpha\in\hat{G},M\in\mathcal{B}_{\mathbb{C}}\}\cap L^1$. The first equality follows now from the relation $\mathcal{F}=\sigma\{f_{\varphi^{-1}z}^{-1}(M):z\in L,M\in\mathcal{B}_{\mathbb{C}}\}$. The second equality is evident.

Suppose that G is σ-compact. Let ξ be a L^2-random variable realizable in $L^1\cap L^2$. Define its random process as the family $(\tilde{\xi}(\alpha),\alpha\in\hat{G})$. Let $X\in\xi$ such that, for any $\omega\in\Omega$, $X(\omega)\in L^1\cap L^2$. The probability distribution of the random process $(\tilde{\xi}(\alpha),\alpha\in\hat{G})$ is a measure ν on the measurable space $(\mathbb{C}^{\hat{G}},\mathcal{K})$ such that, for any $n\in\mathbb{N}$, $\alpha_1,\dots,\alpha_n\in\hat{G}$, and for any Borel set M in $\mathbb{C}^n$, it holds that

$\nu\{z\in \mathbb{C}^{\hat{G}}:(z(\alpha_1),\dots,z(\alpha_n))\in M)=P\{\omega:(\hat{X}(\alpha_1,\omega),\dots,\hat{X}(\alpha_n,\omega))\in M\}$. If $E_1,E_2\in\mathcal{K}$ such that $E_1\cap\varphi(L^1\cap L^2)=E_2\cap\varphi(L^1\cap L^2)$, then $\nu(E_1)=P\hat{X}^{-1}(E_1\cap\varphi(L^1\cap L^2))=P\hat{X}^{-1}(E_2\cap\varphi(L^1\cap L^2))=\nu(E_2)$. We may define a probability measure ν_0 on the measurable space $(\mathbb{C}^{\hat{G}}\cap\varphi(L^1\cap L^2),$ $\mathcal{K}\cap\varphi(L^1\cap L^2))$ as follows: $\nu_0(E\cap\varphi(L^1\cap L^2))=\nu(E),E\in\mathcal{K}$. Let μ denote the probability distribution of ξ. Clearly, for any $F_1,F_2\in\mathcal{F}$ with the property $F_1\cap L^1=F_2\cap L^1$, we obtain $\mu(F_1)=PX^{-1}(F_1)=$ $=PX^{-1}(F_1\cap L^1)=PX^{-1}(F_2\cap L^1)=PX^{-1}(F_2)=\mu(F_2)$. Define a probability measure μ_0 on the measurable space $(L^1\cap L^2,\ \mathcal{F}\cap L^1)$ as follows: $\mu_0(F\cap L^1)=\mu(F),F\in\mathcal{F}$.

THEOREM 2 Let G be a locally compact and σ-compact Abelian group. Let μ be the probability distribution of a $L^2(G)$-random variable ξ realizable in $L^1(G)\cap L^2(G)$, and let ν be the probability distribution of the random process $(\tilde{\xi}(\alpha),\alpha\in\hat{G})$. Then, for any $E\in\mathcal{K}$, and for any $F\in\mathcal{F}$, we have

$$\nu(E)=\mu_0(\varphi^{-1}(E\cap L^2(\hat{G}))\cap L^1(G)),E\in\mathcal{K},\quad\text{and}$$

$$\mu(F)=\nu_0(\varphi(F\cap L^1(G))),F\in\mathcal{F}.$$

PROOF. Let $X\in\xi$ be such that, for any $\omega\in\Omega$, $X(\omega)\in L^1\cap L^2$. Assume first that $E\in\mathcal{K}$. By Lemma 3 we have $\varphi^{-1}(E\cap\hat{L}^2)\cap L^1\in\mathcal{F}\cap L^1$. Thus $\mu_0(\varphi^{-1}(E\cap\hat{L}^2)\cap L^1)=PX^{-1}(\varphi^{-1}(E\cap\hat{L}^2)\cap L^1)=P\hat{X}^{-1}(E\cap\varphi(L^1\cap L^2))=\nu(E)$. Applying Lemma 3, we obtain $\varphi(F\cap L^1)\in\mathcal{K}\cap\varphi(L^1\cap L^2),F\in\mathcal{F}$. In addition, $\nu_0(\varphi(F\cap L^1))=P\hat{X}^{-1}(\varphi(F\cap L^1))=PX^{-1}(F\cap L^1)=\mu(F)$, Q.E.D.

REFERENCES

1 P.R. HALMOS, Measure Theory, Princeton - New Jersey, D. Van Nostrand Co., Inc. 1950

2 M.A. NAJMARK, Normed Rings (in Russian), Moscow, GITTL 1956

3 K. Winkelbauer JR., Group Algebras - Valued Random Variables (in Czech), Bratislava, Doctor Thesis PF UK 1975

Proc. of the 3rd Pannonian Symp. on Math. Stat., Visegrád, Hungary 1982
J. Mogyoródi, I. Vincze, W. Wertz, eds

ON INHOMOGENEOUS MARKOV CHAIN DESCRIBING RANDOM WALK IN A BAND WITH HOMOGENEOUS SECOND COMPONENT

L. LAKATOS

Computer Applications and Service Company,
Budapest, Hungary

1. Let us consider an inhomogeneous Markov chain $\xi(t)$, $t\geq 0$ with state space $\{0,1,\dots,N-1,N\}$ and transition probabilities for $[t,t+\Delta](\Delta\to 0)$:

$$P\{k\xrightarrow{[t,t+\Delta]}k+r\}=\delta_{or}+a_r(t)\Delta+o(\Delta),\ k\in[d,N-c],-d\leq r\leq c;$$

$$P\{k\xrightarrow{[t,t+\Delta]}r\}=\delta_{kr}+u_{kr}(t)\Delta+o(\Delta),\ 0\leq k\leq d-1,\ 0\leq r\leq N; \qquad (1)$$

$$P\{k\xrightarrow{[t,t+\Delta]}r\}=\delta_{kr}+v_{kr}(t)\Delta+o(\Delta),\ N-c+1\leq k\leq N,\ 0\leq r\leq N,$$

where $a_o(t)<0$, $u_{kk}(t)<0$, $v_{kk}(t)<0$; $a_r(t)/r\neq 0/$, $u_{kr}(t)$, $v_{kr}(t)$ $(k\neq r)$ nonnegative continuous functions for which the relations

$$\sum_{r=-d}^{c} a_r(t)=0;\quad \sum_{r=0}^{N} u_{kr}(t)=\sum_{r=0}^{N} v_{kr}(t)=0,\ 0\leq k\leq N,$$

hold.

This chain describes a one-dimensional random walk in $[0,N]$ with bounded positive and negative jumps.

We associate with $\xi(t)$ a homogeneous second component $\eta(t)\in R^d$, and create the process $\zeta(t)=\{\xi(t),\eta(t)\}$. So we come to a Markov process with homogeneous second component introduced and investigated by Ezhov and Skorohod in [2] and [4]. We determine $\eta(t)$ depending on the value of $\xi(t)$ as follows:

$$\begin{array}{lll} 1. \text{ if } \xi(t-0)=k,\ \xi(t+0)=k,\ d\le k\le N-c, & \text{then } \eta(t)=\varepsilon(t); \\ 2. \text{ if } \xi(t-0)=k,\ \xi(t+0)=r,\ d\le k\le N-c, & \text{then } \eta(t)=\varepsilon_{r-k}(t); \\ 3. \text{ if } \xi(t-0)=k,\ \xi(t+0)=k,\ k\in[0,d-1] & \text{or } k\in[N-c+1,N], \\ & \text{then } \eta(t)=\varepsilon_{kk}(t); \\ 4. \text{ if } \xi(t-0)=k,\ \xi(t+0)=r,\ k\in[0,d-1] & \text{or } k\in[N-c+1,N], \\ & \text{then } \eta(t)=\varepsilon_{kr}(t) \end{array} \qquad (2)$$

where $\varepsilon(t)$, $\varepsilon_j(t)$, $\varepsilon_{kr}(t)\in R^d$ are homogeneous processes. $\eta(t)$ coincides with one of them depending on the actual jump of the Markov chain.

Moreover, let

$$M(e^{i(\lambda,\varepsilon(s)-\varepsilon(t))}|\xi(x)=\text{const},x\in[t,s])=\exp\{\int_t^s f(x,\lambda)dx\};$$

$$M(e^{i(\lambda,\varepsilon_r(t+0)-\varepsilon_r(t-0))}|\xi(t+0)-\xi(t-0)=r)=\varphi_r(t,\lambda);$$

$$M(e^{i(\lambda,\varepsilon_{kk}(s)-\varepsilon_{kk}(t))}|\xi(x)=\text{const},x\in[t,s])=\exp\{\int_t^s f_k(x,\lambda)dx\};$$

$$M(e^{i(\lambda,\varepsilon_{kr}(t+0)-\varepsilon_{kr}(t-0))}|\xi(t-0)=k,\xi(t+0)=r)=\varphi_{kr}(t,\lambda),$$

where $f(t,\lambda)$ and $f_k(t,\lambda)$ are cumulants of processes with independent increments in R^d, $\varphi_r(t,\lambda)$ and $\varphi_{kr}(t,\lambda)$ are characteristic functions for certain distributions in R^d.

2. The described Markov chain with homogeneous second component is regular. We derive the direct system of Kolmogorov differential equations for the transition probabilities $(s>t)$

$$P_{\ell k}(t,s,A)=P\{\xi(s)=k,\ \eta(s)-\eta(t)\in A|\xi(t)=\ell\}.$$

Here A is the set of possible values of the second component.

Let

$$P_{\ell k}(t,s,\lambda)=\int_{R^d} e^{i(\lambda,x)}P_{\ell k}(t,s,dx).$$

We have

$$P_{\ell k}(t,s+\Delta,\lambda)=\sum_{i=0}^{d-1} P_{\ell i}(t,s,\lambda)[\delta_{ik}+u_{ik}(s,\lambda)\Delta+o(\Delta)]+$$

$$+\sum_{i=d\vee(k-c)}^{(N-c)\wedge(k+d)} P_{\ell i}(t,s,\lambda)[\delta_{0,k-i}+a_{k-i}(s,\lambda)\Delta+o(\Delta)]+ \qquad (3)$$

$$+\sum_{i=N-c+1}^{N} P_{\ell i}(t,s,\lambda)[\delta_{ik}+v_{ik}(s,\lambda)\Delta+o(\Delta)],$$

where $a\wedge b$ and $a\vee b$ are correspondingly the minimum and maximum of values a and b. The limits of summation in the second summand of right part ensure for the process at moment s to be in the interval $[d,N-c]$ and the value of jump for $[s,s+\Delta]$ may change between $-d$ and c.

To establish (3) e.g. for $k\in[d,N-c]$ it is enough to present it in the form

$$P_{\ell k}(t,s+\Delta,A)=$$

$$=\sum_{i=0}^{d-1}\int_{R^d} P_{\ell i}(t,s,dx)[\delta_{ik}+u_{ik}(s)\Delta+o(\Delta)]P\{\varepsilon_{ik}(s+\Delta)-\varepsilon_{ik}(s)+x\in A\}+$$

$$+\sum_{\substack{i=d\vee(k-c)\\ \ell\neq i}}^{(N-c)\wedge(k+d)}\int_{R^d} P_{\ell i}(t,s,dx)[a_{k-i}(s)\Delta+o(\Delta)]P\{\varepsilon_{k-i}(s+\Delta)-\varepsilon_{k-i}(s)+x\in A\}+$$

$$+\int_{R^d} P_{\ell k}(t,s,dx)[1+a_0(s)\Delta+o(\Delta)]P\{\varepsilon(s+\Delta)-\varepsilon(s)+x\in A\}+$$

$$+\sum_{i=N-c+1}^{N}\int_{R^d} P_{\ell i}(t,s,dx)[\delta_{ik}+v_{ik}(s)\Delta+o(\Delta)]P\{\varepsilon_{ik}(s+\Delta)-\varepsilon_{ik}(s)+x\in A\}$$

and to apply the Fourier transform. E.g. the local transition probabilities $a_r(t,\lambda)\Delta+o(\Delta)$ and $1+a_0(t,\lambda)\Delta+o(\Delta)$ are got by the following way:

$$[a_r(t)\Delta+o(\Delta)]Me^{i(\lambda,\varepsilon_r(t+\Delta)-\varepsilon_r(t))}=[a_r(t)\Delta+o(\Delta)]\varphi_r(t,\lambda)=$$

$$=a_r(t,\lambda)\Delta+o(\Delta);$$

$$[1+a_o(t)\Delta+o(\Delta)]Me^{i(\lambda,\varepsilon(t+\Delta)-\varepsilon(t))}=[1+a_o(t)\Delta+o(\Delta)].$$

$$\cdot\exp\{\int_t^{t+\Delta} f(x,\lambda)dx\}=1+[a_o(t)+f(t,\lambda)]\Delta+o(\Delta)=1+a_o(t,\lambda)\Delta+o(\Delta).$$

From (3) follows

$$\frac{\partial P_{\ell k}(t,s,\lambda)}{\partial s}=\sum_{i=0}^{d-1} P_{\ell i}(t,s,\lambda)u_{ik}(s,\lambda)+\sum_{i=N-c+1}^{N} P_{\ell i}(t,s,\lambda)v_{ik}(s,\lambda)+$$

$$+\sum_{i=d\vee(k-c)}^{(N-c)\wedge(k+d)} P_{\ell i}(t,s,\lambda)a_{k-i}(s,\lambda),\ 0\leq k\leq N. \qquad (4)$$

So we proved

Theorem 1. If

$$P_{\ell k}(t,s,\lambda)=\int_{R^d} e^{i(\lambda,z)}P\{\xi(s)=k,\eta(s)-\eta(t)\in dz|\xi(t)=\ell\},$$

where $P\{\xi(s)=k|\xi(t)=\ell\}$ are the transition probabilities for inhomogeneous Markov chain $\xi(t)$ with local characteristics (1), the second component $\eta(t)$ is determined by (2), then $P_{\ell k}(t,s,\lambda)$ are the solution of the system of equations (4) for any fixed $0\leq\ell\leq N$.

Introduce the generating functions

$$P_\ell(t,s,\lambda,\theta)=\sum_{k=0}^{N} P_{\ell k}(t,s,\lambda)\theta^k;\quad a(t,\lambda,\theta)=\sum_{k=-d}^{c} a_k(t,\lambda)\theta^k;$$

$$w_i(t,\lambda,\theta)=\begin{cases}\sum_{k=0}^{N} u_{ik}(t,\lambda)\theta^k, & \text{if } i\in[0,d-1],\\ \sum_{k=0}^{N} v_{ik}(t,\lambda)\theta^k, & \text{if } i\in[N-c+1,N].\end{cases}$$

Then, as it can be easily seen,

$$\frac{\partial P_\ell(t,s,\lambda,\Theta)}{\partial s}=P_\ell(t,s,\lambda,\Theta)a(s,\lambda,\Theta)-$$

$$-\sum_{i\in K_o} P_{\ell i}(t,s,\lambda)\Theta^i a(s,\lambda,\Theta)+\sum_{i\in k_o} P_{\ell i}(t,s,\lambda)w_i(s,\lambda,\Theta)= \tag{5}$$

$$=P_\ell(t,s,\lambda,\Theta)a(s,\lambda,\Theta)+\sum_{i\in K_o} P_{\ell i}(t,s,\lambda)[w_i(s,\lambda,\Theta)-\Theta^i a(s,\lambda,\Theta)],$$

where K_o is the union of sets $[0,d-1]$ and $[N-c+1,N]$.

3. (5) is a linear differential equation for $P_\ell(t,s,\lambda,\Theta)$, so according to [5] its solution is

$$P_\ell(t,s,\lambda,\Theta)=\exp\{\int_t^s a(x,\lambda,\Theta)dx\}\{\Theta^\ell+\int_t^s \exp\{-\int_t^y a(x,\lambda,\Theta)dx\}\cdot$$

$$\cdot\sum_{i\in K_o} P_{\ell i}(t,y,\lambda)[w_i(y,\lambda,\Theta)-\Theta^i a(y,\lambda,\Theta)]dy\}. \tag{6}$$

We introduce the auxiliary process with independent increments $\zeta^*(t)=\{\xi^*(t),\eta(t)\}$ with state space $\{0,\pm1\pm2,\dots\}xR^d$ and transition probabilities of $\xi^*(t)$ for $[t,t+\Delta]$

$$P\{k\xrightarrow{[t,t+\Delta]}k+r\}=\delta_{or}+a_r(t)\Delta+o(\Delta).$$

Then

$$M\Theta^{\xi^*(s)-\xi^*(t)}e^{i(\lambda,\eta(s)-\eta(t))}=\exp\{\int_t^s a(x,\lambda,\Theta)dx\}$$

and let

$$\rho_k(t,s,\lambda)=\frac{1}{2\Pi i}\oint_{|\Theta|=1}\exp\{\int_t^s a(x,\lambda,\Theta)dx\}\frac{d\Theta}{\Theta^{k+1}} \tag{7}$$

Now comparing in (6) the coefficients at Θ^k we obtain

<u>Theorem 2.</u> $P_{\ell k}(t,s,\lambda)$ is the solution of equation

$$P_{\ell k}(t,s,\lambda)=\rho_{k-\ell}(t,s,\lambda)+\sum_{i\in K_o}\int_t^s P_{\ell i}(t,y,\lambda)\cdot$$

$$\cdot[\sum_{j=0}^N w_{ij}(y,\lambda)\rho_{k-j}(y,s,\lambda)-\sum_j a_j(y,\lambda)\rho_{k-j-i}(y,s,\lambda)]dy, \tag{8}$$

where $\rho_k(t,s,\lambda)$ are determined by (7).

4. In order to compute $P_{\ell k}(t,s,\lambda)(k=0,1,\ldots,N)$ it is necessary to determine the unknown $P_{\ell i}(t,s,\lambda)$, $i\in K_0$. Let in (8) k take on values $0,1,\ldots,d-1$, $N-c+1,\ldots,N$ and introduce the notations

$$\vec{P}_\ell(t,s,\lambda)=\{P_{\ell 0}(t,s,\lambda),P_{\ell 1}(t,s,\lambda),\ldots,P_{\ell,d-1}(t,s,\lambda),$$
$$P_{\ell,N-c+1}(t,s,\lambda),\ldots,P_{\ell N}(t,s,\lambda)\};$$
$$\vec{\rho}_\ell(t,s,\lambda)=\{\rho_{-\ell}(t,s,\lambda),\rho_{1-\ell}(t,s,\lambda),\ldots,\rho_{d-1-\ell}(t,s,\lambda),\tag{9}$$
$$\rho_{N-c+1-\ell}(t,s,\lambda),\ldots,\rho_{N-\ell}(t,s,\lambda)\};$$
$$f_{ik}(t,s,\lambda)=\sum_{j=0}^{N} w_{ij}(t,\lambda)\rho_{k-j}(t,s,\lambda)-\sum_j a_j(t,\lambda)\rho_{k-j-i}(t,s,\lambda);$$
$$F(t,s,\lambda)=||f_{ik}(t,s,\lambda)||_{i,k\in K_0}.$$

Now from (8) we obtain the system of equations

$$\vec{P}_\ell(t,s,\lambda)=\vec{\rho}_\ell(t,s,\lambda)+\int_t^s\vec{P}_\ell(t,y,\lambda)F(y,s,\lambda)dy.$$

We fix certain T, then according to the theory of Volterra integral equations [11] in $0\le t\le s\le T$

$$\vec{P}_\ell(t,s,\lambda)=\vec{\rho}_\ell(t,s,\lambda)+\int_t^s\vec{\rho}_\ell(t,y,\lambda)G(y,s,\lambda)dy,$$

where $G(t,s,\lambda)$ is the resolvent of matrix $F(t,s,\lambda)$. Let

$$\vec{f}_k(t,s,\lambda)=\begin{pmatrix} f_{0k}(t,s,\lambda)\\ f_{1k}(t,s,\lambda)\\ \vdots\\ f_{d-1,k}(t,s,\lambda)\\ f_{N-c+1,k}(t,s,\lambda)\\ \vdots\\ f_{Nk}(t,s,\lambda)\end{pmatrix}\tag{10}$$

then

$$P_{\ell k}(t,s,\lambda)=\rho_{k-\ell}(t,s,\lambda)+\int_t^s \vec{P}_{\ell}(t,y,\lambda)\vec{f}_k(y,s,\lambda)dy=$$

$$=\rho_{k-\ell}(t,s,\lambda)+\int_t^s \vec{\rho}_{\ell}(t,y,\lambda)\vec{f}_k(y,s,\lambda)dy+ \tag{11}$$

$$+\int\int_{t\le x\le y\le s} \vec{\rho}_{\ell}(t,x,\lambda)G(x,y,\lambda)\vec{f}_k(y,s,\lambda)dxdy,$$

$$d\le k\le N-c.$$

We have proved

Theorem 3. The transition probabilities $P_{\ell k}(t,s,\lambda)$ of $\zeta(t)$ are determined by (11), where $\rho_k(t,s,\lambda)$ are given by (7), $\vec{f}_k(t,s,\lambda)$ by (9) and (10); $G(t,s,\lambda)$ is the resolvent of matrix $F(t,s,\lambda)$.

Similar processes with countable state space were investigated in [1], [3] and [6-10].

The author thanks Professor I.I.Ezhov (Kiev) for the attention to his work during his stay in Kiev.

REFERENCES

1 АННАЕВ Т. - ЛАКАТОШ Л.: О вероятностях состояний неоднородной системы с неординарным поступлением и парным обслуживанием требований, Acta Math. Acad. Sci. Hungar., 37, 1981 (4), pp. 307-311.

2 ЕЖОВ И.И.: Исследования по теории случайных процессов с дискретной компонентой, докторская диссертация, Киев, 1967.

3 ЕЖОВ И.И. - ЛАКАТОШ Л.: Про ймовирности станив системи масового обслуговувания, що описуеться неодноридним ланцюгом маркова, Доповиди АН УРСР, сер. А, 1980 № 6, ст. 13-16.

4 ЕЖОВ И.И. - СКОРОХОД А.В.: Марковские процессы, однородные по второй компоненте, ч. I., II., Теория вероятностей и ее применения, 1969, № 1. стр. 3-14, № 4. стр. 679-692.

5 КАМКЕ Э.: Справочник по обыкновенным дифференциальным уравнениям, изд. Наука, М., 1971.

6 ЛАКАТОШ Л.: О вероятностях состояний системы массового обслуживания, описываемой неоднородной цепью Маркова, препринт 80.24 "Марковские процессы и аддитивные функционалы от их траекторий", стр. 41-46, Институт Математики АН УССР, Киев, 1980.

7 ЛАКАТОШ Л.: Цепи Маркова с ограниченными снизу скачками и однородной второй компонентой, Publ. Math. Debrecen, 28, 1981 (3-4), pp. 327-334.

8 ЛАКАТОШ Л.: Цепи Маркова с ограниченными скачками и их применение, кандидатская диссертация, Институт математики АН УССР, Киев, 1981.

9 ОЛИЙНЫК И.Д.: Исследование одного класса стохастических систем о доходами, кандидатская диссертация, КГУ, Киев, 1975.

10 ТАИРОВ М.: Исследование систем массового обслуживания, локальные характеристики которых зависят от времени, кандидатская диссертация, КГУ, Киев, 1978.

11 ТРИКОМИ Ф.ДЖ.: Интегральные уравнения, Изд. иностр. лит., М., 1960.

Proc. of the 3rd Pannonian Symp. on Math. Stat., Visegrád, Hungary 1982
J. Mogyoródi, I. Vincze, W. Wertz, eds

ON OPTIMAL REPLACEMENT POLICY

A. LEŠANOVSKY

Mathematical Institute, Czechoslovak Academy of Sciences, Prague, Czechoslovakia

ABSTRACT

A system with a single activated unit which can be in k+1 states 0,1,...,k is considered. Inspections of the system are carried out at discrete time instants n=0,1,2,... At these moments it is possible to replace the unit by another one whose state is better than that of the inspected unit. The algorithm suggested in the paper for finding the replacement strategy which is optimal in the sense of average income per unit time among all the $\frac{k(k+1)}{2}$ possible ones requests to investigate not more than 2k of them.

The present paper deals with a system with a single activated unit. We do not assume (as is usually done) that the unit is completely effective until it fails. We suppose that the unit can be in k+1 states denoted by 0,1,...,k ($k \geq 2$ and finite) at any time. The state i, $i \in \{0,1,...,k\}$ can be interpreted as a level of the wear of the unit. The states 0 and k correspond respectively to the full operative ability of the unit, and to the failure of the unit. Let us put $K=\{0;1;...;k\}$.

Let us suppose that inspections of the system are carried out at discrete time instants t=0,1,2,... and that we have the possibility to replace the unit inspected at t by another one which is at t in a better state than the unit inspected at t, for every $t=0,1,2,\ldots$. Concerning the changes of states of a unit we assume:

1) The probability that the unit used in the system during $(t; t+1]$, $t=0,1,2,\ldots$, is in state j at $t+1$ under the condition that it is in state i at t depends only on i and j, i.e. this probability depends neither on t nor on the changes of states of the units used in the system before t, nor on the particular unit inspected at t. Let us denote this probability by p_{ij}.
2) The relations

   $p_{ij}=0$ for all $i\in K-\{k\}$, $j\in K-\{i;\ i+1;\ k\}$,

   $p_{i,i+1}\neq 0$ for all $i\in K-\{k\}$

   hold.

If the unit used during the interval $(t-1;\ t]$, $t=1,2,\ldots$, between two successive inspections of the system fails during $(t-1;t]$, we must replace it at t. On the other hand, if it does not fail during $(t-1;t]$ and is in state i at t, then one of several possible actions - not to replace or to replace by a unit which is in some of the states $0,1,\ldots,\ i-1$ - can be taken at t. We are interested in stationary replacement strategies which are determined by couples $(n;N)$ from the set

$$Q=\{(n;N);\ n\in K-\{0\},\ N\in K-\{k\},\ n>N\}, \qquad (1)$$

and by the following rule: The unit inspected at t is to be replaced at t if and only if its state at t is $\geq n$, and this unit is to be replaced by another one, which is in state N at t.

Let $R(N)$, where $R(N)>0$ for every $N\in K-\{k\}$, be the costs for replacement of a unit by another one, which is in state N, i.e. the price of a unit in state N, and let m_{ij}, for $i\in K-\{k\}$, $j\in\{i;i+1;k\}$, be the income of the system achieved during the interval (say (t;t+1]) between two successive inspections of the system under the condition that the states of the unit used during this interval are i at t and j at t+1. We assume that the sequence $\left\{p_{ii}\, m_{ii}+p_{i,i+1}\, m_{i,i+1}+p_{ik}\, m_{ik}\right\}_{i=0}^{k-1}$ is decreasing.

The aim of the present paper is to find the replacement strategy whose average income per unit time is maximal among all the stationary replacement strategies described above. More precisely, if we put

$$C=\max\{C_n^N;(n;N)\in Q\}, \tag{2}$$

where C_n^N is the average income per unit time of the system controlled by the replacement strategy determined by the couple (n;N), then we will find the couple (z;Z) with the property

$$(z,Z)=\min_{\prec}\{(n;N);(n;N)\in Q, C_n^N=C\}, \tag{3}$$

where $\min_{\prec}$ denotes the minimum value in the sense of the lexicographical ordering on Q

$$(n;N)\prec(n';N') \text{ if } n<n' \text{ or } n=n' \text{ and } N<N'.$$

The papers [1] and [2] deal with the same model. The former assumes that by every replacement we activate a unit in the best possible state (state 0), i.e. only the replacement strategies determined by the couples (n;0), $n\in K-\{0\}$, are studied in [1]. It shows some properties of the sequence $\{C_n^o\}_{n=1}^k$ and introduces algorithms for finding the replacement strategy optimal in the sense of the average income per unit time among those considered without calculating all the values of $C_1^o, C_2^o, \ldots, C_k^o$. The present paper is hence a direct generalization of [1] and, on the other hand, it uses the results of [1] in a high degree. The paper [2] is devoted to some relations between the total expected income of the system achieved during a period of finite length and the state of the unit inspected at the initial instant of this period.

1. ANALYSIS OF THE SYSTEM

Let us denote by D_n^N and R_n^N, respectively, the expected length of the period between two successive replacements of units and the expected income of the system achieved during this period when the system is controlled by the replacement strategy determined by the couple $[n;N]$. It is easy to see that $D_n^N = D_n[N]$ and $R_n^N = R_n[N]$, where $D_n[N]$ and $R_n[N]$ are defined in [1] if we put $R = R[N]$ in the expression for $R_n[N]$. Thus we have

$$D_n^N = \sum_{j=N}^{n-1} P_j^N \quad \text{for all } (n;N) \in Q, \tag{4}$$

$$R_n^N = -R(N) + \sum_{j=N}^{n-1} m(j) P_j^N \quad \text{for all } (n;N) \in Q, \tag{5}$$

where

$$P_j^N = \frac{1}{1-p_{jj}} \cdot \prod_{i=N}^{j-1} \frac{p_{i,i+1}}{1-p_{ii}} \quad \text{for all } j,\ N \in K-\{k\}, j \geq N, \tag{6}$$

$$m(j) = m_{jj} p_{jj} + m_{j,j+1} p_{j,j+1} + m_{jk} p_{jk} \quad \text{for all } j \in K-\{k\}. \tag{7}$$

By similar reasoning as in [1] we obtain

$$C_n^N = \frac{R_n^N}{D_n^N} = \frac{-R(N) + \sum_{j=N}^{n-1} m(j) P_j^N}{\sum_{j=N}^{n-1} P_j^N} \quad \text{for all } (n;N) \in Q, \tag{8}$$

and

$$C^N_{n'} = \frac{C^N_n D^N_n + \sum_{j=n}^{n'-1} m(j)\ P^N_j}{D^N_n + \sum_{j=n}^{n'-1} P^N_j} \quad \text{if } (n;N)\in Q, (n';N)\in Q,\ n'>n. \qquad (9)$$

Let us have $N\in K-\{k\}$ fixed for the moment and let us put

$\tilde{K}=\{i;i\in K,i\leq k-N\}$,

$\tilde{k}=k-N$,

$\tilde{R}(0)=R(N)$,

$\tilde{m}_{ij}=m_{i+N,j+N}$ for all $i,j\in\tilde{K},i<\tilde{k}$,

$\tilde{p}_{ij}=p_{i+N,j+N}$ for all $i,j\in\tilde{K},i<\tilde{k}$.

The sequence $\left\{\tilde{m}(j)\right\}_{j=0}^{\tilde{k}-1}$ is obviously decreasing and

$$C^N_n = \tilde{C}^o_{n-N} \quad \text{for every } (n;N)\in Q, \qquad (10)$$

where $\tilde{C}^o_{n-N}$ has the same meaning as C^o_{n-N} but, however, in the model with the state-space $\tilde{K}$, transition probabilities $\tilde{p}_{ij}$, costs $\tilde{R}(0)$ for replacement of a unit by a unit in state 0, and incomes achieved by the system during a unit time interval $\tilde{m}_{ij}$. In the model "with wave" Theorem 2 of [1] is true and, with respect to (10), we can transform its assertions to the sequence $\left\{C^N_n\right\}_{n=N+1}^{k}$ for every $N\in K-\{k\}$.

<u>Theorem 1:</u> Let us put for every $N\in K-\{k\}$

$$z_N=\min\{n;n\in K,n>N,C^N_n=\max\{C^N_i,i\in K,i>N\}\}. \qquad (11)$$

Then for every $N\in K-\{k\}$

$$\left\{C^N_n\right\}_{n=N+1}^{z_N} \text{ is increasing,} \qquad (12)$$

$$\left\{C_n^N\right\}_{n=z_N+1}^{k} \text{ is decreasing,} \tag{13}$$

and in the case that $z_N \neq k$ the following implications are true

a) if $C_{z_N}^N \neq m(z_N)$ then $C_{z_N}^N > C_{z_N+1}^N$;

b) if $C_{z_N}^N = m(z_N)$ then $C_{z_N}^N = C_{z_N+1}^N$.

Further, we can transform also the assertions concerning relations between $\tilde{C}_n^o$ and $\tilde{m}(n)$ obtained during the proof of Theorem 2 in [1].

<u>Theorem 2:</u> Let $N \in K-\{k\}$ then

$$C_n^N < m(n) \text{ for all } n \in K, n < z_N, n > N, \tag{14}$$

$$C_n^N > m(n) \text{ for all } n \in K-\{k\}, n > z_N, \tag{15}$$

and if $z_N \neq k$ then

$$C_{z_N}^N \geq m(z_N). \tag{16}$$

From (2), (3) and (11) it is obvious that

$$C = \max\left\{C_{z_N}^N ; N \in K-\{k\}\right\}, \tag{17}$$

and that the optimal couple $(z;Z)$ fulfils

$$z = z_Z. \tag{18}$$

Hence, we find that for the solution of our problem it is sufficient to know the values z_N for all $N \in K-\{k\}$, and to find the maximum (17) and the minimum (in the sense of the lexicographical ordering) of the couples $(z_N;N)$ which fulfil $C_{z_N}^N = C$. If we would look for the values of z_N for every $N \in K-\{k\}$ separately then we should not reach, however, the requested simplicity of the solution because the number of those C_n^N which are to calculate is still very large. Our aim is to construct an algorithm for finding the optimal couple $(z;Z)$ which operates much more effectively. We shall base it on the following four theorems.

<u>Theorem 3:</u> The first component z of the optimal couple $(z;Z)$ has the property

$$z=\min\{z_N; N\in K-\{k\}\}. \tag{19}$$

<u>Proof:</u> Let (19) be not true. Then there exists $N\in K-\{k\}$ such that $z_N<z$, i.e. $(z_N;N)\prec(z;Z)$, so that (see (3))

$$C^N_{z_N} < C. \tag{20}$$

If $Z<z_N$ then from (14) and (18) we obtain

$$C^Z_{z_N} < m(z_N)$$

and according to (9), (16), and (20) we have

$$C^Z_z \leqq \frac{C^Z_{z_N} D^Z_{z_N} + \sum_{j=z_N}^{z-1} m(z_N) P^Z_j}{D^Z_{z_N} + \sum_{j=z_N}^{z-1} P^Z_j} \leqq$$

$$\leqq \max\left\{C^Z_{z_N}; m(z_N)\right\} = m(z_N) \leqq C^N_{z_N} < C.$$

Further, if $Z\geqq z_N$ then (8), (16), and (20) imply

$$C^Z_z \; \frac{\sum_{j=Z}^{z-1} m(Z) P^Z_j}{\sum_{j=Z}^{z-1} P^Z_j} = m(Z) \leqq m(z_N) \leqq C^N_{z_N} < C.$$

The last two relations contradict (17) so that (19) is verified.

<u>Theorem 4:</u> Let $a\in K$, and $N\in K-\{k\}$ be such that $a>N+1$. Then $z_N<a$ if and only if

$$C_a^N \geq m(a-1). \tag{21}$$

Proof: According to Theorem 1 we know that $z_N < a$ if and only if

$$C_{a-1}^N \geq C_a^N. \tag{22}$$

We shall verify that (21) and (22) are equivalent. If (21) is true but (22) is not, then, using (9), we obtain the following impossible relation

$$C_a^N < \frac{C_a^N D_{a-1}^N + C_a^N P_{a-1}^N}{D_{a-1}^N + P_{a-1}^N} = C_a^N.$$

Similarly, if (22) holds but (21) does not, then

$$C_a^N > \frac{C_a^N D_{a-1}^N + C_a^N P_{a-1}^N}{D_{a-1}^N + P_{a-1}^N} = C_a^N.$$

Theorem 5: Let $n_o \in K-\{0\}$, $N_o \in K-\{k\}$ and $N_o < n_o$ such that

$$z_{N_o} = n_o, \tag{23}$$

$$z_N > z_{N_o} \quad \text{for all } N \in K,\ N < N_o. \tag{24}$$

Let us denote

$$U_n^{N_o} = \{N; N \in K, N_o < N < n_o - 1, C_{n_o}^N \geq m(n_o - 1)\}, \tag{25}$$

$$V_{n_o}^{N_o} = \begin{cases} \{N; N \in K, N_o < N \leq n_o - 1, C_{n_o}^N \geq m(n_o)\} \cup \{N_o\}, \text{if } n_o \neq k, \\ \{N; N \in K, N_o \leq N \leq k-1\} \text{ if } n_o = k, \end{cases} \tag{26}$$

$$C(V_{n_o}^{N_o}) = \max\left\{C_{n_o}^N; N \in V_{n_o}^{N_o}\right\}. \tag{27}$$

If

$$U_{n_o}^{N_o} = \emptyset, \tag{28}$$

then

$$z = n_o, \tag{29}$$

$$Z = \min\{N; N\in V^{N_o}_{n_o}, C^{N}_{n_o} = C(V^{N_o}_{n_o})\}. \qquad (30)$$

Proof: From (23), (24), (28) and Theorem 4 we obtain

$$z_N \geqq n_o \text{ for all } N\in K,\ N<n_o-1. \qquad (31)$$

Further, $(z_N; N)\in Q$ for every $N\in K-\{k\}$ so that

$$z_N \geqq N+1 \geqq n_o \text{ for all } N\in K-\{k\},\ N\geqq n_o-1 \qquad (32)$$

evidently holds. The relations (31), (32) and Theorem 3 prove (29). According to (3), (18), and (29) we have

$$C = \max\left\{C^{N}_{n_o}; N\in K, N\leqq n_o-1\right\},$$

$$Z\in\{N; N\in K, N\leqq n_o-1, z_N=n_o\},$$

and with respect to (17), (23), and (24) we obtain

$$C = \left\{\max C^{N}_{n_o}; N\in K, N\leqq n_o-1, z_N=n_o\right\} = \max\left\{C^{N}_{n_o}; N\in W^{N_o}_{n_o}\right\} \qquad (33)$$

$$Z\in W^{N_o}_{n_o}, \qquad (34)$$

where

$$W^{N_o}_{n_o} = \{N; N\in K, N_o<N\leqq n_o-1, z_N=n_o\}\cup\{N_o\}.$$

The relations (23), (31), and in case that $n_o\neq k$, these ones together with Theorem 2, imply that the sets $V^{N_o}_{n_o}$ and $W^{N_o}_{n_o}$ are identical. The value of z is fixed by (29). The second component of the optimal couple (z;Z) is thus equal - according to (3), (27), (29), (33) and (34) - to the minimum on the right-hand side of (30).

Theorem 6: Let $n_o\in K-\{0\}$, $N_o\in K-\{k\}$ and $N_o<n_o$ be such that (23) and (24) are true, $U^{N_o}_{n_o} \neq \emptyset$ and

$$N_1 = \min\left\{N; N\in U^{N_o}_{n_o}\right\}, \qquad (35)$$

where the set $U_{n_o}^{N_o}$ is determined by (25). Then $N_1 \in K-\{k\}$ and

$$z_{N_1} < z_N \text{ for all } N \in K, N < N_1. \tag{36}$$

Proof: From (23) and (35) and Theorem 4 we obtain

$$z_{N_1} < z_{N_o}, \tag{37}$$

and

$$z_N \geqq z_{N_o} \text{ for all } N \in K, N_o < N < N_1. \tag{38}$$

Further, according to (25) and (35) $N_1 \in U_{n_o}^{N_o}$ so that $N_1 < n_o - 1 \leqq k-1$ and (24), (37), and (38) imply (36).

2. ALGORITHM FOR FINDING THE OPTIMAL COUPLE (z;Z)

Let us find first the value of z_o, e.g. by one of the procedures introduced in [1]. If $z_o=1$, then we know from Theorem 3 that $z=1$ and, obviously, $Z=1$. On the other hand, if $z_o \in K-\{0;1\}$ we carry out the following inductive step of the algorithm.

Let us have a couple $(n_o;N_o) \in Q$ such that (23) and (24) hold and let us suppose that the values of $D_{n_o}^{N_o}$ and $R_{n_o}^{N_o}$ are known. We shall calculate $C_{n_o}^{N}$ for $N \in K$ such that $N_o < N \leqq n_o - 1$ according to the rule A: We start with $C_{n_o}^{N_o+1}$ and after calculating $C_{n_o}^{N}$ $(N_o < N < n_o - 1)$ we compare $C_{n_o}^{N}$ with $m(n_o-1)$ and provided that $C_{n_o}^{N} < m(n_o-1)$ we proceed to $C_{n_o}^{N+1}$. With the sequential calculation of $C_{n_o}^{N}$ according to the rule A we can use e.g. the following recurrent relations

$$D_{n_o}^{N+1} = \left[D_{n_o}^{N} - P_N^N\right] \cdot \frac{1-p_{NN}}{p_{N,N+1}} \text{ for } n_o \in K-\{0\},\ N \in K, N < n_o - 1, \tag{39}$$

$$R_{n_o}^{N+1} = \left[R_{n_o}^{N} + R(N) - m(N) P_N^N\right] \cdot \frac{1-p_{NN}}{p_{N,N+1}} - R(N+1) \text{ for } n_o \in K-\{0\},\quad N \in K, N < n_o - 1. \tag{40}$$

There are two possibilities:

1) We had to calculate $C_{n_o}^{N}$ for all N such that $N \in K$ and $N_o < N \leq n_o - 1$ because $C_{n_o}^{N} < m(n_o-1)$ for all $N \in K$ fulfilling $N_o < N < n_o - 1$, i.e. (28) holds. Theorem 5 states that $z = n_o$ and yields the instruction how to find the value of Z. If $n_o \neq k$, then we must compare $C_{n_o}^{N}$ (by now known) with $m(n_o)$ for all $N \in K$ such that $N_o < N < n_o - 1$. Further, after constructing the set $V_{n_o}^{N_o}$ according to (26) we must find the maximum (27) and the minimum (30). The latter is equal to Z.

2) The last value calculated according to the rule A was $C_{n_o}^{N_1}$, where $N_o < N_1 < n_o - 1$, because $C_{n_o}^{N_1} \geq m(n_o-1)$ and $C_{n_o}^{N} < m(n_o-1)$ for all $N \in K$ such that $N_o < N < N_1$, i.e. (35) is true. Theorem 6 implies $z_{N_1} < n_o$. Let us find z_{N_1}. For this purpose we can use the rule B of the following form: We start with calculation of $C_{n_o-1}^{N_1}$. After calculating $C_{n_o-i}^{N_1}$ $(1 \leq i < n_o - N_1 - 1)$ we compare this value with $m(n_o-i-1)$ and we proceed to $C_{n_o-i-1}^{N_1}$ if and only if $C_{n_o-i}^{N_1} \geq m(n_o-i-1)$. Theorem 4 guarantees that the last value $C_{n_o-i_o}^{N_1}$ calculated according to this rule B, where $1 \leq i_o \leq n_o - N_1 - 1$, has the property

$$z_{N_1} = n_o - i_o. \tag{41}$$

With the sequential calculation of $C_{n_o-i}^{N_1}$ according to the rule B we can use the following recurrent relations

$$D_{n-1}^{N_1} = D_n^{N_1} - P_{n-1}^{N_1} \quad \text{for all } N_1 \in K - \{k\}, n \in K, n > N_1 + 1, \tag{42}$$

$$R_{n-1}^{N_1} = R_n^{N_1} - m(n-1) P_{n-1}^{N_1} \quad \text{for all } N_1 \in K - \{k\}, n \in K, n > N_1 + 1. \tag{43}$$

Let us now put

$$N_o' = N_1, \text{ and } n_o' = n_o - i_o. \tag{44}$$

The couple $(n_o'; N_o')$ is obviously an element of Q, and from (41), (44) and Theorem 6 we obtain that (23) and (24) are fulfilled if we substitute n_0' for n_0 and N_0' for N_0. Finally, the values of $D_{n_0'}^{N_0'}$ and $R_{n_0'}^{N_0'}$ are known. In this way, all the initial assumptions of the inductive step are fulfilled and we carry it out once more, now starting with the couple $(n_0';N_0')$, of course.

It is easy to see that $n_0' \leqq n_0-1$, and $N_0' \geqq N_0+1$. Further, the first application of the inductive step of this algorithm starts with the couple $(z_0;0)$ so that if i of its applications are to be carried out, then the i-th one starts with a couple $(n_0;N_0)$ which fulfills

$n_0 \leqq z_0-i+1 \leqq k-i+1,$

$N_0 \geqq i-1,$

$N_0<n_0.$

Hence

$$i < \frac{k}{2} + 1, \qquad (45)$$

so that after less than $\frac{k}{2}+1$ applications of the inductive step of this algorithm we obtain the optimal couple $(z;Z)$.

Note: The assumption that the values of $D_{n_0}^{N_0}$ and $R_{n_0}^{N_0}$ are known is necessary if we use (39), (40), (42), and (43) for the recurrent calculation of C_n^N. If we do it in a different way, it may be sufficient to know only $C_{n_0}^{N_0}$, i.e. the value of their quotient.

<u>Theorem 7:</u> Let F be the number of those C_n^N which are to be calculated if the algorithm introduced above is used. Then

$$F=F_0+z_0-1, \qquad (46)$$

where F_0 is the number of those C_n^N which are to be calculated by the (chosen) procedure used for finding the value of z_0.

<u>Proof:</u> Let J applications of the inductive step be carried out

and let q_j, $j=1,\ldots,J$, be the number of those C_n^N which are to be calculated with its j-th application. Finally, let its j-th application start with the couple $(n_o^{(j)};N_o^{(j)})$. It can easily be seen that

$$n_o^{(1)}-N_o^{(1)}=z_o,$$

$$n_o^{(j+1)}-N_o^{(j+1)}=n_o^{(j)}-N_o^{(j)}-q_j \text{ for all } j=1,..,J-1,$$

$$n_o^{(J)}-N_o^{(J)}=q_J+1,$$

so that

$$\sum_{j=1}^{J} q_j=z_o-1.$$

It is evident that F is the sum of F_o and of $\sum_{j=1}^{J} q_j$ of the numbers of C_n^N calculated with the initial step and with J applications of the inductive step. Thus (46) is proved.

Note: The relation (46) implies that

$$F<2k. \tag{47}$$

If we admit that "in the average" z_o is equal to $\frac{k+1}{2}$ and if we look for z_o by the procedure suggested in Section 2 of [1], then $F_o=z_o$ so that F equals to k "in the average". In other words, by the algorithm considered in the present section the replacement strategy optimal among $\frac{k(k+1)}{2}$ possible ones may be found but for this purpose it is necessary to investigate less than 2k (k "in the average") of them.

The operation of this algorithm is demonstrated on the following example.

Example: Let

$k=9$,

$m(0)=15$, $m(1)=13$, $m(2)=12$, $m(3)=11.9$, $m(4)=9$, $m(5)=7.9$,

$m(6)=5.9$, $m(7)=5.8$, $m(8)=5.1$,

$R(i)=90.\ \exp(-i)$ for all $i\ K-\{0;9\}$,

$R(0)=R(1)+11$,

$p_{ii}=\frac{10-i}{20}$ for all $i\in K-\{9\}$,

$p_{i,i+1}=\frac{9}{20}$ for all $i\in K-\{8;9\}$,

$p_{8,9}=\frac{18}{20}$,

$p_{ik}=\frac{i+1}{20}$ for all $i\in K-\{8;9\}$.

Table 1 contains the values of C_n^N for all $\{n;N\}\in Q$. Those which were calculated by the algorithm in question are underlined and the process of the same is indicated by arrows. The columnal maxima $C_{z_N}^N$, $N\in K-\{9\}$, are denoted by one asterisk with the exception of the largest of them, i.e. the optimum we are looking for, which has two asterisks. In the first step we find that $z_0=6$. The set U_6^0 is not empty because $2\in U_6^0$. More precisely, $N_1=\min U_6^0=2$ because $1\in U_6^0$. Hence the first application of the inductive step abides by the second part of its description given above and finishes with the result $z_2=5$. Now we have $U_5^2=\emptyset$, so that $z=5$. We form the set V_5^2. We see that $V_5^2=\{2;3\}$ and $C=\max\{C_5^2;C_5^3\}=C_5^3=8.992>C_5^2$ so that $Z=3$. Summarily $(z;Z)=(5;3)$.

REFERENCES:

1 KASUMU, R.A., LEŠANOVSKÝ, A.: On optimal replacement policy, to appear in the Journal Aplikace Matematiky.

2 LEŠANOVSKÝ, A.: On dependences of the expected income of a system on its initial state, to appear in the journal IEEE Transactions on Reliability.

Table 1

n \ N	0	1	2	3	4	5	6	7	8
1	-7.055	-	-	-	-	-	-	-	-
2	1.970	-5.210	-	-	-	-	-	-	-
3	4.501	2.166	4.692	-	-	-	-	-	-
4	5.601	4.393	7.641	8.987	-	-	-	-	-
5	5.898	4.984	7.924*	8.992**	7.846	-	-	-	-
6	5.997*	5.192	7.921	8.785	7.866*	7.445*	-	-	-
7	5.995	5.219	7.802	8.506	7.524	6.889	5.722	-	-
8	5.992	5.231*	7.742	8.374	7.378	6.715	5.749*	5.730*	-
9	5.986	5.230	7.703	8.296	7.286	6.595	5.653	5.520	5.073*

Proc. of the 3rd Pannonian Symp. on Math. Stat., Visegrád, Hungary 1982
J. Mogyoródi, I. Vincze, W. Wertz, eds

ON THE STABILITY OF A CHARACTERIZATION OF THE RECTANGULAR DISTRIBUTION

E. LUKACS

The Catholic University of America, Washington, D.C.
USA

1. INTRODUCTION

In an earlier paper [1] the author obtained the following result.

Theorem. Let $X_1, X_2, \ldots, X_n$ be a sample from a population with a continuous distribution function $F(x)$. Suppose that $F(x)$ is non-degenerate and symmetric and that the moment of order 4 of $F(x)$ exists. Let

$$S = \frac{1}{n(n-1)} \sum_{j \neq k} \{2X_j^3 X_k - 4X_j X_k - 6X_j^2 X_k^2\} + \frac{4}{n} \sum_j X_j^2$$

$$T = \frac{1}{n(n-1)} \sum_{j \neq k} \{X_j^4 X_k - 2X_j^2 X_k^3\} + \frac{1}{n} \sum_j (2X_j^3 - X_j)$$

$$\Lambda = \sum_j X_j$$

be three statistics and suppose that S as well as T have constant regression on Λ. Then $F(x)$ is the rectangular (uniform) distribution over the interval $(-1, +1)$.

In the present paper a stability theorem corresponding to this earlier result is derived.

2. DEFINITIONS AND LEMMAS

Definition 1. Let X and Y be two random variables and assume that the conditional expectation $\mathcal{E}(Y|X)$ exists and let $\varepsilon_1>0$. We say that Y has ε_1-regression on X if

$$\rho(\omega) = \mathcal{E}(Y|X) - \mathcal{E}(y) \tag{2.1}$$

is such that

$$|\rho(\omega)| \leq \varepsilon_1 \quad \text{a.e.} \tag{2.1a}$$

Definition 2. If $\mathcal{E}(Y|X) = \mathcal{E}(Y)$, that is if $\rho(\omega)\equiv 0$ then we say that Y has constant regression on X.

Lemma 1. Let X,Y be two random variables and assume that $\mathcal{E}(X)$ exists. Y has constant regression on X if, and only if,

$$\mathcal{E}(Y\ e^{i\,t\,X}) = \mathcal{E}(Y)\,\mathcal{E}(e^{i\,t\,X}) \tag{2.2}$$

for all real t.

For the proof see [2] chapter 6.

Lemma 2. Let X,Y be two random variables such that (Y) exists. If Y has ε_1-regression on X then

$$|\mathcal{E}(Y\ e^{i\,t\,X}) - \mathcal{E}(Y)\ \mathcal{E}(e^{i\,t\,X})| \leq \varepsilon_1 \tag{2.3}$$

Proof of Lemma 2. We multiply (2.1) by $e^{i\,t\,X}$ and take expectations and get

$$\mathcal{E}(Y\ e^{i\,t\,X}) - \mathcal{E}(Y)\mathcal{E}(e^{i\,t\,X}) = \mathcal{E}(\rho\ e^{i\,t\,X}) = \int \rho(\omega)e^{i\,t\,X(\omega)}dP_X(\omega)$$

hence (2.3) holds.

3. THE THEOREM

Let X_1, X_2, X_n be a sample of n from a population with population distribution function F(x) and characteristic function f(t). Suppose that F(x) is non-degenerate and symmetric and that the moment of order 4 of F(x) exists. Let

$$S = \frac{1}{n(n-1)} \sum_{j\neq k} \{2X_j^3 X_k - 4X_j X_k - 6X_j^2 X_k^2\} + \frac{4}{n} \sum_j X_j^2 \qquad (3.1)$$

$$T = \frac{1}{n(n-1)} \sum_{j\neq k} \{X_j^4 X_k - 2X_j^2 X_k^3\} + \frac{1}{n} \sum_j 2X_j^3 - X_j \qquad (3.2)$$

$$\Lambda = \sum_j X_j \qquad (3.3)$$

by three statistics. Let $k>k_1>1$ and let ℓ be such that $f(\ell)= \frac{1}{k_1}$. Assume further that $\mathfrak{n} = [0,a]$ is an interval in which the relation

$$\frac{1}{k_1} > |f(t)| \geq \frac{1}{k} > 0 \qquad (3.4)$$

is satisfied for $t \in \mathfrak{n} \setminus [0,\ell]$.

Let $\varepsilon>0$ and suppose that S has ε_1 regression on Λ while T has constant regression on Λ, $\varepsilon_1 < \frac{\varepsilon}{1+k^{n-2}}$. Assume further that f, f′ and f″ are bounded in $\mathfrak{n}$. Then

$$t\, f'' + 2f' + t\, f = \varepsilon_1(f'' + f) + 2\varepsilon_2 f' \qquad (3.5)$$

where ε_1 and ε_2 can be made arbitrarily small. The coefficients of the equation

$$t\, f'' + 2f' + t\, f = 0 \qquad (3.5a)$$

are close to the coefficients of equation (3.5). It is therefore conjectured that the solution

$$f(t) = \frac{\sin t}{t}$$

of (3.5a) will be close to the solution of (3.5).

<u>Proof</u>. We put in (2.2) $Y = S$, $X = \Lambda$ and write

$$\mathcal{E}(S\, e^{i t \Lambda}) - \mathcal{E}(S)\mathcal{E}(e^{i t \Lambda}) = \eta(t) = \eta \qquad (3.6)$$

where according to Lemma 2

$$|\eta| \leq \varepsilon_1 < \frac{\varepsilon}{1+k^{n-2}} \qquad (3.6a)$$

We put $\mathcal{E}(S) = c$ so that

$$\mathcal{E}(S\, e^{i t \Lambda}) = \mathcal{E}(S)\,\mathcal{E}(e^{i t \Lambda}) + \eta = c[f(t)]^n + \eta$$

In view of the symmetry of F(x) we have $\mathcal{E}(T) \equiv 0$ and from (2.2) that

$$\mathcal{E}(T\, e^{i t \Lambda}) = 0 \tag{3.7}$$

As in the earlier paper [1] we express $\mathcal{E}(S\, e^{i t \Lambda})$ and $(T\, e^{i t \Lambda})$ in terms of the derivatives of f(t) and get the differential equations

$$2f'''f'(f)^{n-2} + 2(f')^2(f)^{n-2} - 3(f'')^2(f)^{n-2} - 4f''(f)^{n-1} = c(f)^n + \eta \tag{3.8a}$$

$$f^{iv}f'(f)^{n-2} - 2f''f'''(f)^{n-2} - 2f'''(f)^{n-1} - f'(f)^{n-1} = 0. \tag{3.8b}$$

Since $f(t) \neq 0$ in $\mathfrak{n}$ we divide (3.8a) and (3.8b) by $[f]^{n-2}$ and obtain

$$2f'''f' + 2(f')^2 - 3(f'')^2 - 4f''f = c(f)^2 + \frac{\eta}{[f]^{n-2}} \tag{3.9a}$$

$$f^{iv}f' - 2f''f''' - 2f'''f - f'f = 0 \tag{3.9b}$$

We rewrite (3.9b) as

$$2f^{iv}f' + 2f'''f'' + 4f''f' - 6f'''f'' - 4f'''f - 4f''f' = 2f'f \tag{3.9c}$$

We integrate (3.9c) and get

$$2f'''f' + 2(f')^2 - 3(f'')^2 - 4f''f = f^2 + c_1 \tag{3.10}$$

Comparing (3.9a) and (3.10) one gets

$$cf^2 + \frac{\eta}{[f]^{n-2}} = f^2 + c_1 \tag{3.11}$$

or

$$(c-1)f^2 = c_1 + \frac{\eta}{[f]^{n-2}}$$

We put t = 0, then

$$c - 1 = c_1 - \eta(0) \tag{3.12}$$

hence

$$[c_1 - \eta(0)]f^2 = c_1 + \frac{\eta}{[f]^{n-2}}$$

so that

$$c_1(f^2 - 1) = \eta(0)f^2 + \frac{\eta}{[f]^{n-2}}$$

We see from (3.6a) and (3.4) that

$$|c_1(f^2 - 1)| < |\eta(0)| + |\eta|k^{n-2}$$

or

$$|c_1 \; 1 - f^2 \;| < \varepsilon_1 + \frac{\varepsilon_1}{1+k^{n-2}} k^{n-2} < 2\varepsilon_1 < 2\varepsilon$$

Since $f^2<1$ in $\mathfrak{n}[0,\ell]$ we see that

$$|c_1| < \frac{2\varepsilon}{1-f^2}$$

In $\mathfrak{n} - [0,\ell]$ one has $\frac{1}{k_i^2} > f^2 > \frac{1}{k^2}$ and since f is decreasing in $[0,a]$ $\operatorname{Min}_{t\in\mathfrak{n}} |f(t)| = f(a)$.

Let $A = \operatorname{Min}_{t\in[\ell,a]} (1-f^2)$ then $0<A<1$ so that

$$|c_1| < \frac{2\varepsilon}{A} \tag{3.13}$$

It follows from (3.12) that

$$c \leq 1 + \frac{2\varepsilon}{A} + \eta(0) \tag{3.14}$$

We have to solve (3.10) when $t\in\mathfrak{n}$.

It follows from (3.10) that

$$2f'(f''' + f') - (f'' + f)^2 - 2f''(f'' + f) = c_1 \tag{3.15}$$

Since

$$(f')^2 \frac{d}{dt}[\frac{f''+f}{f'}] = (f''' + f')f' - f''(f'' + f)$$

it follows from (3.15) that

$$2(f')^2 \frac{d}{dt} [\frac{f''+f}{f'}] = (f'' + f)^2 + c_1 \tag{3.16}$$

where c_1 is small.

Since $c_2 \neq i\alpha_1$, one has $f''(0) + f'(0) \neq 0$. Therefore there exists a neighborhood $\mathcal{N}_1$ of the origin where neither $f(t)$ nor $f''(t) + f'(t)$ vanish. We restrict t to the set $\mathcal{N}_1 \cap \mathcal{N} = \mathcal{N}_2$.

Let

$$B = \min_{t \in \mathcal{N}_2} [f''(t) + f'(t)] \quad .$$

Then $B \neq 0$ and

$$2\left(\frac{f}{f''+f}\right)^2 \frac{d}{dt}\left[\frac{f''+f}{f'}\right] = 1 + \frac{c_1}{(f''+f)^2} \leq 1 + \frac{c_1}{B^2} \tag{3.17}$$

Let

$$h(t) = \frac{f''+f}{f'} \quad . \tag{3.18}$$

Then (3.17) becomes

$$h^{-2}\, h' \leq \frac{1}{2} + \frac{c_1}{2B^2} \quad . \tag{3.19}$$

We can then write

$$h^{-2}\, h' = \frac{1}{2} + \frac{c_1}{2B^2} + \lambda(t) \tag{3.20}$$

where $\lambda(t)$ is small and non-positive, $|\lambda(t)| \leq \delta$ (δ small).

Therefore

$$[h(t)]^{-1} = -\left[\frac{1}{2} + \frac{c_1}{2B^2}\right] t - \int_0^t \lambda(\tau)d\tau \quad . \tag{3.21}$$

We write

$$D = 1 + \frac{c_1}{B^2} \tag{3.22}$$

$$\mu(t) = \int_0^t \lambda(\tau)d\tau \leq 0, \ |\mu(t)| \leq \delta t \tag{3.23}$$

so that we see from (3.16) and (3.18) that

$$\frac{f}{f''+f} = -\frac{D}{2} + \mu(t) \tag{3.24}$$

$$f' = -\frac{D}{2}\, tf'' - \frac{D}{2}\, tf + \mu(f'' + f)$$

$$\frac{2}{D}\, f' = -tf'' - tf + \frac{2}{D}\, \mu(f'' + f)$$

$$tf'' + \frac{2f'}{D} + tf = \frac{2}{D}\,\mu(f'' + f) \tag{3.25}$$

$$\frac{1}{D} = \frac{1}{1+\frac{c_1}{B^2}} = \frac{B^2}{B^2+c_1} = 1 + \frac{B^2}{B^2+c_1} - 1 = 1 + \frac{B^2-B^2-c_1}{B^2+c_1} =$$

$$= 1 - \frac{c_1}{B^2+c_1}$$

$$tf'' + 2f' + tf = 2f' - \frac{2f'}{D} + \frac{2}{D}\,\mu(f'' + f) \tag{3.26}$$

$$= 2f' - \left(1 - \frac{c_1}{B^2+c_1}\right) 2f' + \frac{2}{D}\,\mu(f'' + f)$$

$$tf'' + 2f' + tf = 2f' \frac{c_1}{B^2+c_1} + \frac{2\mu}{D}\,(f'' + f)$$

$$\left(t - \frac{2\mu}{D}\right) f'' + 2f'\left(1 - \frac{c_1}{B^2+c_1}\right) + f\left(t - \frac{2\mu}{D}\right) = 0\ . \tag{3.27}$$

We write

$$\varepsilon_1 = \frac{2\mu}{\delta} \qquad\qquad \varepsilon_2 = \frac{c_1}{B^2+c_1} \tag{3.28}$$

and see that ε_1 and ε_2 are so small that

$$(t - \varepsilon_1)f'' + 2(\ 1 - \varepsilon_2)f' + (t - \varepsilon_1)f = 0 \tag{3.29}$$

$$tf'' + 2f' + tf = \varepsilon_1(f'' + f) + 2\,\varepsilon_2\, f' \tag{3.29a}$$

If one assumes that f, f′, f″ are bounded in the interval under consideration then the right hand side of (3.29a) can be made arbitrarily small and approaches zero as the ε's tend to zero. The equation

$$tf'' + 2f' + tf = 0 \tag{3.30}$$

can be solved by substituting $f = \frac{g}{t}$, it becomes then

$$g'' + g = 0 \tag{3.31}$$

which has the solution

$$g(t) = \lambda_1\, e^{i\,t} + \lambda_2\, e^{-i\,t} \quad (\lambda_1, \lambda_2 \text{ constants of integration})$$

so that

$$f(t) = \frac{\lambda_1\, e^{i\,t} + \lambda_2\, e^{-i\,t}}{t}$$

is a solution of (3.30), λ_1 and λ_2 can be determined from $f(t) = f(-t)$ and $f(0) = 1$ and one obtains the solution of (3.30)

$$f(t) = \frac{\sin t}{t} \tag{3.32}$$

i.e. the characteristic function of the rectangular distribution.

Since the coefficients of the equations (3.30) and (3.29) are close to each other it is conjectured that the solutions of (3.30) and (3.31) will also be close to each other, that is that f(t) will be close to the characteristic function of the rectangular distribution.

REFERENCES

1 EUGENE LUKACS, A Characterization of the Rectangular Distribution, Stochastic Processes and Their Applications 9, 273-279 (1979).

2 E. LUKACS - R.G. LAHA, Applications of Characteristic Functions. Charles Griffin and Co., London, 1964.

Proc. of the 3rd Pannonian Symp. on Math. Stat., Visegrád, Hungary 1982
J. Mogyoródi, I. Vincze, W. Wertz, eds

JOINT DIFFUSION FOR PARTICLES WITH RANDOM COLLISIONS ON THE LINE

P. LUKÁCS

Computer and Automation Institute,
Hungarian Academy of Sciences, Budapest, Hungary

For a one-dimensional system of particles with "random" and elastic collisions the trajectories of distinct particles are considered in the diffusion limit. Depending on the order of the growth of the initial distance between two particles the joint distribution of their rescaled paths converges to different limit.

Key-words: Infinite particle system, limit theorems, Wiener process.

1. INTRODUCTION

We consider a system of randomly positioned particles on the real line. At time zero each of these particles is given a random velocity. These velocities are independent random variables with a common distribution. The particles now embark on a uniform motion on the real line, interrupted only by elastic collision whenever two particles meet. All particles have equal mass and they collide elastically, therefore in case of collision particles exchange their velocities. Consider now the actual motion $y(t)$ $t \geq 0$ of a marked particle on the line. This model was introduced by Harris [1] in 1965 who also proved the existence of the motion and obtained a Gaussian approximation for the position of the marked particle after a long time. The assumption on the initial position of the particles was that they form a Poisson process on the real line. The finiteness of the absolute momenta of the velocity was also supposed. For the same model F. Spitzer proved that under the usual normalization the path of a single marked particle tends to the Wiener process on C[0,1] [2]. It was pointed out that by a result of C. Stone [6] one can automatically obtain weak convergence of the corresponding measures on $D[0,\infty)$.

In 1975 Szatzschneider [3] investigated a modified model where the particles started almost from a lattice of the integers. He obtained that the limit process is a non-Wiener Gaussian one.

A joint generalization of the Spitzer's and Szatzschneider's results was proved by P. Major and D. Szász [4] where the initial position of the particles forms a two-sided renewal process. Here the limit process is in general a non-Wiener Gaussian process. (For a wider class of initial positions a limit process was also obtained.)

For the previous model with additional assumption on the asymptotic behaviour of the initial position process D. Szász proved a limit theorem for the joint distribution of the rescaled trajectories of particles when the initial distance of two particles increases in an appropriate way [5].

The existence of non-Wiener limit process in natural collision model is unexpected. This strange fact is closely related to the invariance of the order of particles in the one-dimensional collision model. Therefore R. Dobrushin proposed to investigate a model in which the particles only collide with some positive probability π if they meet and with probability $1-\pi$ they pass through each other. This is the so called random collision assumption.

As an application of P. Major and D. Szász's result we proved the following proposition [10]. If the initial position of the particles forms a two-sided renewal process and the symmetric distribution of the velocities is concentrated on -1 and +1, then the normalized path of a single particle tends weakly to the Wiener process on $C[0,\infty)$.

Now we use another method to get in the preceding model a Wiener approximation for the joint paths of n marked particle. The result is a modification of the result for Poisson initial position of Kipnis-Lebowitz-Presutti-Spohn [6].

2. RESULTS

We denote by $x_i(0)$, v_i the initial position and initial velocity of the particles $i=0,\pm1,\pm2,\ldots$.

We make the following assumption:

(i) the initial position of the particles forms a two-sided renewal process, where the gap between the particles has (m,σ^2) as the first two momenta

(ii) the velocities of the particles are concentrated on ±1 with equal probability

(iii) when two particles meet they collide elastically with probability π.

1-Theorem:

For the test particle starting from $y(0)$ the rescaled path $A^{-1/2}(y(At)-y(0))$ tends weakly to $w(t)+b(t)$ in $D[0,1)$ as A tends to infinity. The $w(t)$ and $b(t)$ are independent Wiener processes with local dispersions $((\sigma^2+m^2)\cdot m^{-1})^{1/2}$ and $(m(1-\pi)\cdot\pi^{-1})^{1/2}$ respectively.

Without loss of generality we can assume that at time zero we have two tagged particles: one at the origin and the other at the point $f(A)$.

2-Theorem:

For two test particles starting from 0 and $f(A)$ respectively the joint distribution of the rescaled paths of the test particles tends weakly in $D[0,1)$ to the vector of Wiener processes $(\gamma_1(t), \gamma_2(t))$ when the initial position of the particles forms a λ intensity Poisson process. There are three possibilities depending on the order of the growth of $f(A)$:

1. If $\lim_{A\to\infty} A^{-1}f(A)=0$ then $\gamma_i(t)=w(t)+b_i(t)$, $i=1,2$

   where the Wiener processes $b_i(t)$, $i=1,2$ are independent.

2. If $\lim A^{-1}f(A)=\infty$ then $\gamma_i(t)=w_i(t)+b_i(t)$, $i=1,2$

   where the Wiener processes $w_i(t)$, $b_i(t)$ $i=1,2$ are independent.

3. If $\lim A^{-1}f(A)= a$ then

   $\gamma_1(t)=w_L(t)+w_R(t-a)+b_1(t)$

   $\gamma_2(t)=w_L(t-a)+w_R(t)+b_2(t)$

   where the Wiener processes $w_L(t)$ and $w_R(t)$ are independent with local dispersions $(4\lambda)^{-1/2}$ respectively.

Let us remark that the Wiener processes denoted by $w(t)$ and by $b_i(t)$, $i=1,2$ are in all cases independent in the consequence of the 1-Theorem.

In order to comment the 2-Theorem we would remark that in the first case the two test particles meet approximately the same part of the particle system and this is the cause of the common part $w(t)$, while in the second case they see different part of the particle system. In the third case the delay with which the two particles meet the same part of the particle system is apperceivable and it is equal to a.

In this case $w_L(t)$ resp. $w_R(t)$ reflect the effect of the system on the left (resp. right) hand side of the left (resp. right) test particle.

It will not cause additional difficulties to prove an analogous proposition to the preceding 2-Theorem for the joint diffusion of n test particles, starting from $q_i f(A)$ $i=1,2,\dots,n$ respectively.

When the initial position of the particles forms a renewal process the following modified proposition can be proved by the same idea as the proof of the preceding theorem.

2-Proposition:

For two test particles starting from 0 and $\sigma(\sqrt{A})$* the joint distribution of the rescaled paths of the test particles tends weakly to $(w(t)+b_1(t),\ w(t)+b_2(t))$ where the Wiener processes $w(t)$, $b_i(t)$ $i=1,2$ are independent with local dispersion $((m^2+\sigma^2)/2m)^{1/2}$, $(m(1-\pi)/\pi)^{1/2}$, $(m(1-\pi)/\pi)^{1/2}$ respectively.

3. Proof of the 1-Theorem:

Without loss of generality we can assume that our test particle starts from the zero point of the real line.

In order to describe the path of the test particle we must consider only those particles on the right hand side of the test particle which have initial velocity -1 while on the left hand side only those which have initial velocity +1.

---

* $\lim_{A\to\infty} \sigma(A^{1/2})A^{-1/2} = 0$

After this random selection the retained particles form also a two-sided renewal process. Let us denote by X_i the gap between the i-th and (i-1)-th retained particles ($i=\pm 1, \pm 2, \ldots$). Since the selection was made with 1/2 probability $EX_i = 2m$ and $D^2 X_i = (\sigma^2 + m^2)/2$.

Let us introduce the notations

α_i is the relative index of the first particle from the right hand side of the test particle with which the test particles collide after the (i-1)-th collision with particle coming from the right hand side of the test particle

β_i has analogous meaning for particles on the left hand side of the test particle.

From the random collision assumption follows that the α_i's and β_i's are geometrically distributed independent random variables with expectation $1/\pi$ and variance $(1-\pi)/\pi^2$. They are also independent of the X_i's.

Denote

$$R_i := \sum_{j=\alpha_1+\ldots+\alpha_{i-1}+1}^{\alpha_1+\ldots+\alpha_i} \frac{1}{2} X_j \qquad i=1,2,\ldots$$

and

$$L_i := \sum_{j=\beta_1+\ldots+\beta_{i-1}+1}^{\beta_1+\ldots+\beta_i} \frac{1}{2} X_j \qquad i=1,2,\ldots .$$

Thus R_i, (L_i) is the forth (back) part of the motion of the test particle on $\mathbb{R}^1$ between the (i-1)-th and i-th collision with particles coming from the right (left) hand side of the test particle. The $\frac{1}{2}$ multiplication factors come from the geometry of the model.

Denote

$$Y_i = R_i - L_i$$

$$Z_i = R_i + L_i$$

and

$$N_t = \max\{k : Z_1 + Z_2 + \ldots + Z_k \le t\}.$$

Now we have all the notations to describe the rescaled path $A^{-1/2}y(At)$ of the test particle.

$$A^{-1/2}y(At) = A^{-1/2} \sum_{i=1}^{N_{At}} Y_i + \hat{Y}_{N_{At}+1} \tag{1}$$

where $\hat{Y}_{N_{At}+1}$ denotes the residual part of the motion up to t (cf.fig.1)

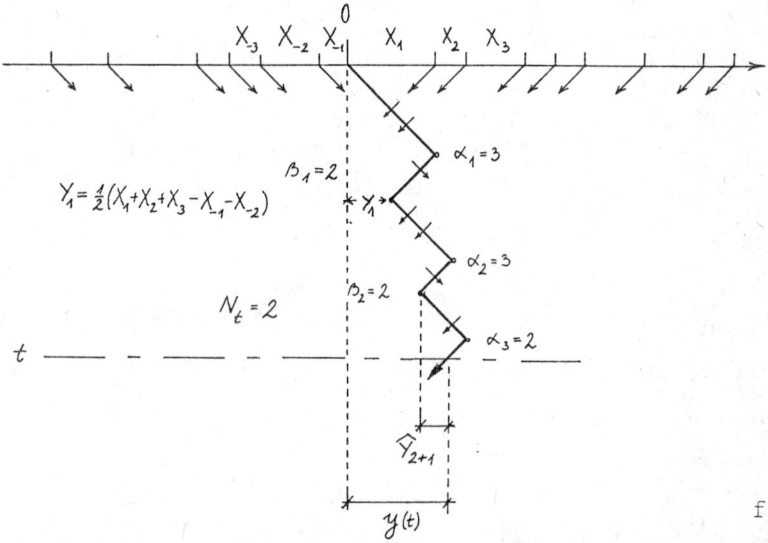

fig. 1.

If we would prove that there exists a positive constant c that $t^{-1}N_t \xrightarrow{P} c$, and that the residual part tends in probability to the 0 element of the space D[0,1], as $A\to\infty$, then we are able to use the generalization of the Anscombe theorem to the weak convergence in D[0,1] (Billingsley [7] p.146) Easy to see that

$$t^{-1}N_t \xrightarrow{a.s.} c \text{ where } c=1/EZ_1. \tag{2}$$

As a consequence of the model assumptions $\lim_{t\to\infty} N_t=\infty$ a.s.

By the definition of N_t

$$Z_1+\ldots+Z_{N_t} \leq t < Z_1+\ldots+Z_{N_t+1}.$$

Dividing this inequality by N_t from the Strong Law of Large Numbers for sum of random number of independent random variables follows (2). A short calculation shows that $EZ_1=2m/\pi$.

It is trivial that

$$|\hat{Y}_{N_{At}+1}| \leq Z_{N_{At}+1}. \tag{3}$$

By (2) we can use the generalized Anscombe theorem and it follows that

$$A^{-1/2} Z_{N_{At}+1} \xrightarrow{\mathcal{D}} 0 \in D[0,1]^{*} \tag{4}$$

Since $A^{-1/2}Z_{N_{At}+1}$ is the difference of two random elements $A^{-1/2}(Z_1+\ldots+Z_{N_{At}+1})$ and $A^{-1/2}(Z_1+\ldots+Z_{N_{At}})$ which have the same weak limit in $D[0,1]$. By (3) and (4) follows

$$\hat{Y}_{N_{At}+1} \longrightarrow 0 \in D[0,1].$$

Hence to prove weak convergence for the rescaled path we must analyze only the first part in (1).

By simple algebraic transformation

$$A^{-1/2} \sum_{i=1}^{N_{At}} Y_i = A^{-1/2} \sum_{i=1}^{\alpha_1+\ldots+\alpha_{N_{At}}} \frac{1}{2} \tilde{X}_i - A^{-1/2} \sum_{i=1}^{\beta_1+\ldots+\beta_{N_{At}}} \frac{1}{2} \tilde{X}_{-i}$$

$$+ \frac{EX_1}{2} A^{-1/2} \sum_{i=1}^{N_{At}} (\tilde{\alpha}_i - \tilde{\beta}_i) \tag{5}$$

where ~ denote the centralized version of the rv's.

$^{*} \xrightarrow{\mathcal{D}}$ denotes the convergence in distribution

In order to use the Anscombe theorem for the three sums in (5) we must prove that the following convergences in probability hold for some constant d

$$\lim_{t\to\infty} \frac{\alpha_1+\ldots+\alpha_{N_t}}{t} = d \tag{6}$$

$$\lim_{t\to\infty} \frac{\beta_1+\ldots+\beta_{N_t}}{t} = d. \tag{7}$$

Since $N_t \xrightarrow{a.s.} \infty$ as $t\to\infty$ and by the Strong Law of Large Numbers the desired convergences hold not only in probability but also a.s., and obviously $d=1/2m$.

Hence by the Anscombe theorem we get that

$$A^{-1/2}y(At) \xrightarrow{\mathcal{D}} w(t)+b(t)$$

where the $w(t)$ and $b(t)$ are Wiener processes with local dispersion $((\sigma^2+m^2)/2m)^{1/2}$ and $(m(1-\pi)/\pi)^{1/2}$, respectively.

So it remains to prove the independence of the two Wiener processes.

For this goal we split the sums in (5) in two parts

$$A^{-1/2}\sum_{i=1}^{\alpha_1+\ldots+\alpha_{N_{At}}} \frac{1}{2}\tilde{X}_i - A^{-1/2}\sum_{i=1}^{\beta_1+\ldots+\beta_{N_{At}}} \frac{1}{2}\tilde{X}_{-i} = \tag{8}$$

$$= A^{-1/2}\sum_{i=1}^{dAt} \frac{1}{2}(\tilde{X}_i - \tilde{X}_{-i}) +$$

$$+A^{-1/2}\sum_{i=1}^{\alpha_1+\ldots+\alpha_{N_{At}}-dAt} \frac{1}{2}\tilde{X}_i - A^{-1/2}\sum_{i=1}^{\beta_1+\ldots+\beta_{N_{At}}-dAt} \frac{1}{2}\tilde{X}_{-i}$$

$$\frac{EX}{2}A^{-1/2}\sum_{i=1}^{N_{At}}(\tilde{\alpha}_i-\tilde{\beta}_i) = \frac{EX}{2}A^{-1/2}\left\{\sum_{i=1}^{cAt}(\tilde{\alpha}_i-\tilde{\beta}_i) + \sum_{i=1}^{N_{At}-cAt}(\tilde{\alpha}_i-\tilde{\beta}_i)\right\} \tag{9}$$

The first sums in both decompositions with constant upper limit are trivially independent by the initial assumptions. If we prove that the residual sums in the right hand sides of (8) and (9) tend in probability to the zero function of the space D[0,1] then we can prove the desired independence using the following fact:

if $U_n = U'_n + U''_n$, $V_n = V'_n + V''_n$ $n=1,2,\ldots$ are random elements of a metric space and if $U_n \xrightarrow{\mathcal{D}} U$, $V_n \xrightarrow{\mathcal{D}} V (n\to\infty)$ where the U'_n and V'_n are independent, while the U''_n and $V''_n \xrightarrow{P} 0$ then U and V are independent.

The residual sums tend in probability to 0 for all t. Their tightness follows from the tightness of the left hand sides in (8) and (9) and from the tightness of the first sums in the right hand sides in (8) and (9), since tightness is additive.

4. PROOF OF THE 2-THEOREM:

In order to calculate the covariance of the rescaled paths of two different test particles, we introduce some additional notations.

Denote X_i^+ (resp. X_i^-) the gap between the initial positions of the i-th and (i-1)-th particles with initial velocity +1 (resp.-1) for $i=\pm1,\pm2,\ldots$.

Denote

$$M_A^+ = \{\max\{k: X_1^+ + \ldots + X_k^+ \leq f(A)\}+1$$

and

$$M_A^- = \max\{k: X_1^- + \ldots + X_k^- \leq f(A)\}.$$

Let us denote the path of the test particle starting from zero (resp. f(A)) by $y_1(t)$ (resp. $y_2(t)$).

The $\{\alpha_i'\},\{\beta_i'\},N_t'$ (resp. $\{\alpha_i''\},\{\beta_i''\},N_t''$) have similar meaning as the $\{\alpha_i\},\{\beta_i\},N_t$ in the proof of the 1-Theorem.

It is obvious that

$$A^{-1/2}(y_2(At)-f(A))= \tag{10}$$

$$=A^{-1/2}\left\{\sum_{i=M_A^-+1}^{M_A^-+\alpha_1''+\dots+\alpha''_{N_{At}}}\frac{1}{2}\tilde{X}_i^- - \sum_{i=-M_A^++1}^{-M_A^++\beta_1''+\dots+\beta''_{N_{At}}}\frac{1}{2}\tilde{X}_i^+ + \frac{EX_1}{2}\sum_{i=1}^{N''_{At}}(\tilde{\alpha}_i''-\tilde{\beta}_i'')\right\}$$

In the expression of the $cov_A^{1,2}(s,t)=cov(A^{-1/2}y_1(As),A^{-1/2}y_2(At))$ only those $\tilde{X}_i^+$ (resp. $\tilde{X}_i^-$) play role for which the i-th particle with velocity +1 (resp. -1) is on the left (resp. right) hand side of the left (resp. right) test particle.

By the Prékopa theorem for Poisson process the variables X_i^+ and X_i^- $i\leq\max\{M_A^+,M_A^-\}$ are independent [9]. From (5) and (10) is evident that

$$cov_A^{1,2}(s,t)=\frac{1}{4A}\sum_{i=M_A^-+1}^{E\min\{\alpha_1'+\dots+\alpha'_{N'_{As}},\ M_A^-+\alpha_1''+\dots+\alpha''_{N''_{At}}\}}D^2(X_i^-)$$

$$+\frac{1}{4A}\sum_{i=1}^{E\min\{\beta_1'+\dots+\beta'_{N'_{As}},\ -M_A^++\beta_1''+\dots+\beta''_{N''_{At}}\}}D^2(X_i^+)$$

Let us denote $cov_{1,2}(s,t)=\lim_{A\to\infty} cov_A^{1,2}(s,t)$.

Similarly to the proof of (2) we can conclude that

$$\lim_{A\to\infty}(f(A))^{-1}M_A^+=\lim_{A\to\infty}(f(A))^{-1}M_A^-=C=(2m)^{-1} \quad \text{a.s.} \tag{11}$$

By (2), (6), (7), (11) follows that the covariance $cov_{1,2}(s,t)$ has the following form in the three cases respectively.

1. for $\lim_{A\to\infty} f(A)\cdot A^{-1}=0$ $\qquad cov_{1,2}(s,t)=\frac{\sigma^2+m^2}{2m}\min(s,t)$

2. for $\lim_{A\to\infty} f(A)\cdot A^{-1}=\infty$ $\qquad cov_{1,2}(s,t)=0$

3. for $\lim_{A\to\infty} f(A)A^{-1}=a$ $\quad cov_{1,2}(s,t)=\frac{\sigma\ +m^2}{2m}(\min(s,[t-a]^+)+\min([s-a]^+,t))$

where $s,t\geq 0$, $\sigma^2=m^2=\lambda^{-2}$.

From the form of the covariance, and the form of the local dispersion of the limit process for one test particle follow the desired decompositions of the processes $\gamma_1(t)$ and $\gamma_2(t)$ in the different cases. Since for two Gaussian processes $\xi(t)$ and $\eta(t)$ from $\xi(0)=\eta(0)$ and from $cov\ (\xi(t),\eta(t))/(D(\xi(t))\cdot D(\eta(t)))=1$ for all t follows that $\xi(t)=\eta(t)$.

5. PROOF OF THE 2-PROPOSITION

When the initial position of the particles forms a renewal process the independence of the variables $(X_i^+)_{i\leq M_A^+}$ and $(X_i^-)_{i\leq M_A^-}$ does not follow but their covariance can be majorized by $D(X_i^-)\cdot D(X_j^+)$ for $i\leq M_A^-$, $j\leq M_A^+$ so their contribution to the covariance of the rescaled paths is at least $const\cdot D(X_1^-)D(X_1^+)\ \sigma^2(\sqrt{A})\cdot A^{-1}$, hence it vanishes when A tends to ∞.

ACKNOWLEDGMENT

We are very indebted to Prof. D. Szász for the valuable remarks and aids.

REFERENCES

1 HARRIS, T.E.: Diffusion with "Collisions" between Particles, J. Appl. Prob. 2. (1965)

2 SPITZER, F.: Uniform Motion with Elastic Collision of an Infinite Particle System, J. Math. Mech. Vol. 18. No. 10 (1969)

3 SZATZSCHNEIDER, E.: A version of the Harris-Spitzer "random constant velocity" model for infinite system of particles, Studia Math. 63. (1978) (Springer's LNM 472 (1975)).

4 MAJOR, P. - SZÁSZ, D.: On the Effect of Collisions on the Motion of an Atom in R^1, Ann. Prob. Vol 8. No. 6. (1980).

5 SZÁSZ, D.: Joint Diffusion on the Line, J. Stat. Phys. Vol. 23. No. 2. (1980).

6 KIPNIS, C. - LEBOWITZ, J.L. - PRESUTTI, E. - SPOHN, H.: Self-Diffusion for Particles with Stochastic Collisions in One Dimension (private communication).

7 BILLINGSLEY, P.: Convergence of Probability Measures. New York, Wiley (1968).

8 STONE, C.: Weak convergence of stochastic processes defined on semi-infinite time intervals, Proc. Amer. Math. Soc. 14 (1963)

9 PRÉKOPA, A.: On Secondary Processes generated by a Random Point Distribution of Poisson Type, Ann. Univ. Sci. Budapest, Eötvös Sect. Math. Hung. 1 (1958)

10 LUKÁCS, P.: On the Inhomogeneous and Random Collisions on the Line, Proc. of the Conference on Limit Theorems in Probability and Statistics, (1982) (to appear).

Proc. of the 3rd Pannonian Symp. on Math. Stat., Visegrád, Hungary 1982
J. Mogyoródi, I. Vincze, W. Wertz, eds

MINIMAL SUFFICIENCY. THE STRUCTURE OF STATISTICAL SPACES

GY. MICHALETZKY

Department of Probability Theory,
Eötvös Loránd University, Budapest, Hungary

SUMMARY

Let $(\Omega, \mathfrak{A}, \mathfrak{P})$ be a statistical space. The following conditions are known to be equivalent: The statistical space is (i) compact (ii) coherence (iii) weakly dominated (iv) uniform complete. These conditions assure the existence of minimal sufficient σ-field. In this paper we prove the equivalence of the following properties: (i) for every measure class $\mathfrak{P}'$ got from $\mathfrak{P}$ by changing some measure with equivalent ones there exists a minimal sufficient σ-field and (ii) the statistical space - roughly speaking - can be divided into two parts; one of these parts is "almost" uniform complete, and the other one is so "rich" that the only sufficient σ-field is the σ-field containing all events.

INTRODUCTION

In 1949 P. R. Halmos and L.J. Savage have shown that if the measure class is dominated - i.e. there exists a probability measure such that the elements of $\mathfrak{P}$ are absolutely continuous with respect to this measure - then there exists a minimal sufficient σ-field. This is the classic result. How can the assumption of dominatedness be weakened?

In 1965 T.S.Pitcher has introduced the notion of compact statistical space. This means the compactness of the unit ball of a topological vector space defined by means of the measure class $\mathfrak{P}$.

M. Hasegawa and M.D.Perlman (1974, 1975) defined the notion of coherent statistical space.

In 1972 D. Mussmann defined the notion of weakly dominated statistical space. This is a direct generalization of the notion of dominatedness, namely this means that there exists a localizable measure λ such that the elements of $\mathcal{P}$ are absolutely continuous with respect to λ and moreover the corresponding Radon-Nikodym derivates exist.

In 1980 F. Göndőcs and G. Michaletzky constructed the notion of uniform complete statistical space. Every measure $P\in\mathcal{P}$ determines a semimetric on the equivalence classes of events. These semimetrics define a uniform structure, and the uniform completeness of the statistical space means that this uniform structure is complete.

These conditions (compactness, coherence, weakly domination, uniform completeness) guarantee the existence of a minimal sufficient σ-field. Moreover - what is somehow surprising - these conditions are equivalent. Although they use absolutely different tools, and approach from absolutely different sides, the same condition is derived. The question arises: What is the common in these approaches?

If we examine carefully these conditions we can observe that they do not use the concrete values of measures P, only the structure of events having measure zero is essential.

At the same time there are statistical spaces having minimal sufficient σ-fields which are not uniform complete but the concrete values of the elements of measure class $\mathcal{P}$ are occasionally such that this assures the existence of minimal sufficient σ-field. Let us consider an example. Let $\Omega=[0,2)$, $\mathcal{A}$ be the Borel subsets of this interval and for every $0\leq x<1$

define a measure $P_x=\frac{1}{2}(\varepsilon_x+\varepsilon_{(1+x)})$, where ε_x denotes the measure concentrated to the point x, and finally for every $0\leq c\leq 1$ let $Q_c=c\lambda_1+(1-c)\lambda_2$ where λ_1 and λ_2 are the Lebesgue measures on $[0,1)$ and $[1,2)$, respectively. Now fix a number c and let $\mathfrak{P}_c=\{P_x \mid 0\leq x<1\}\cup\{Q_c\}$. Define a sub-σ-field $\mathfrak{F}$ of $\mathfrak{A}$ as follows:

$$\mathfrak{F}=\{A\in\mathfrak{A} \mid 1+A\cap[0,1)=A\cap[1,2)\}.$$

It is easy to show that for any c the σ-field $\mathfrak{F}$ can be the minimal sufficient σ-field if there exists any. (Observe that these statistical spaces are not uniform complete but they have the same null-sets.) If $c=\frac{1}{2}$ then $\mathfrak{F}$ is minimal sufficient for $\mathfrak{P}_c$, but if $c\neq\frac{1}{2}$ - i.e. we change the measure class slightly - the σ-field $\mathfrak{F}$ does not remain a sufficient one, so in this case there does not exist minimal sufficient σ-field.

These suggest that if for every measure class $\mathfrak{P}'$ obtained from $\mathfrak{P}$ by changing some measure to equivalent ones there exists a minimal sufficient σ-field then - perhaps - the statistical space has the properties above, namely it is compact, uniform complete, etc.

But this is not the case. Nevertheless a similar assertion holds. In order to formulate this first we must say what it means precisely -"changing some measures to equivalent ones".

First we enlarge measure class $\mathfrak{P}$:

$$\mathfrak{P}_o=\{\Sigma a_i P_i \mid (P_i)\subset\mathfrak{P},\ a_i\geq 0,\ \Sigma a_i=1\}$$

(the convergence of the infinite sum is taken in total variation norm of measures). The measure class $\mathfrak{P}_o$ has obviously the same null sets as the measure class $\mathfrak{P}$, and this enlargement does not have any influence to sufficiency of a given σ-field. Finally let

$\mathcal{P}^x = \{Q \mid \text{there exists a measure } P \in \mathcal{P}_0 \text{ such that } Q << P\}$.

For any measure class $\mathcal{Q}$ let $\mathcal{N}(\mathcal{Q})$

$\mathcal{N}(\mathcal{Q}) = \{A \in \mathcal{A} \mid Q(A) = 0 \text{ for every } Q \in \mathcal{Q}\}$.

We have $\mathcal{N}(\mathcal{P}^x) = \mathcal{N}(\mathcal{P})$. Let

$\mathcal{M}(\mathcal{P}) = \{\mathcal{Q} \subset \mathcal{P}^x \mid \mathcal{N}(\mathcal{Q}) = \mathcal{N}(\mathcal{P})\}$.

The precise formulation of our assumption is the following: for every measure class $\mathcal{Q} \in \mathcal{M}(\mathcal{P})$ there exists a minimal sufficient σ-field. (I)

Before continuing our considerations let us examine two further examples. These are the typical ones. In fact the first example is not concrete, it is rather a remark. If $(\Omega, \mathcal{A}, \mathcal{P})$ is such that for every $\mathcal{Q} \in \mathcal{M}(\mathcal{P})$ the only sufficient σ-field is the whole σ-field $\mathcal{A}$, then obviously it is the minimal one. The second example is the following: let $\Omega = [0,1)$, the σ-field $\mathcal{A}$ is generated by the one-point sets of Ω, and let P_x be concentrated to the point x, $0 \le x < 1$, finally let λ be the Lebesgue measure restricted to $\mathcal{A}$. Define $\mathcal{P}$ as follows

$\mathcal{P} = \{P_x \mid 0 \le x < 1\} \cup \{\lambda\}$

In this case $\mathcal{P}^x$ is the same as the measure class $\mathcal{P}_0$ and it contains those measures which can be written in the form $c\lambda + \Sigma a_i P_{x_i}$ where $(x_i) \subset [0,1)$ and $c \ge 0$, $a_i \ge 0$, $c + \Sigma a_i = 1$. For every measure we can define its support A_Q : if $Q = c\lambda + \Sigma a_i P_{x_i}$ then A_Q contains the points x_i for which $a_i > 0$. Denote the class of these sets by $\mathcal{C}$:

$\mathcal{C} = \{A_Q \mid Q \in \mathcal{P}^x\}$.

It is obvious that the union of any denumerable subset of $\mathcal{C}$ is an event and for any event $A \in \mathcal{A}$ either A or its complementer is denumerable. On the other hand the measure λ takes the values

0 or 1 - on the denumerable subsets its value is 0, on the complementer of a denumerable subset its value is 1.

It can be easily proved that for any measure class $\mathcal{Q} \subset \mathcal{P}^x$ for which $\cup A_Q = [0,1)$ (i.e. $\mathcal{N}(\mathcal{Q}) = \mathcal{N}(\mathcal{P})$) there exists a minimal sufficient σ-field.

In the sequel we shall see that these examples are the typical ones. Every statistical space which we are dealing with (i.e. which has property (I)) can be built up as - roughly speaking - the union of two spaces of these types.

<u>Notations</u> Denote by $\mathfrak{A}$ the factor algebra of $\mathfrak{F}$ with respect to $\mathcal{N}(\mathcal{P})$. $\mathfrak{A} = \mathfrak{F}/\mathcal{N}(\mathcal{P})$. $\mathfrak{A}$ is a Boole-algebra. The lattice-algebraic operations will be denoted by $<, \wedge, \vee$ (the corresponding set-theoretic operations are $\subset, \cap, \cup$). Denote by $\tilde{A}$ the equivalence class of an event A.

Consider an event $A \in \mathfrak{F}$ and a probability measure P defined on $(\Omega, \mathfrak{F})$. Define the measure P^A as follows

$$P^A(B) = \begin{cases} P(B|A) & \text{if } P(A)>0, \\ 0 & \text{if } P(A)=0. \end{cases}$$

For any $A \in \mathfrak{F}$ denote $\mathfrak{F}|_A = \{B \subset A \mid B \in \mathfrak{F}\}$ and $\mathfrak{A}|_{\tilde{A}} = \{\tilde{B} \in \mathfrak{A} \mid \tilde{B} < \tilde{A}\}$. This latter is called the σ-ideal generated by $\tilde{A}$.

CONSTRUCTION OF "$\mathfrak{C}$" IN AN ARBITRARY STATISTICAL SPACE

A probability measure Q is said to be strictly positive on $\tilde{A}$ (or an A) if for every equivalence class $\tilde{B}$ such that $\tilde{\emptyset} \neq \tilde{B} < \tilde{A}$ we have $Q(B)>0$. In words this means that every element of $\mathcal{P}$ (and of $\mathcal{P}^x$) is absolutely continuous with respect to Q on the event A. I.e. on this event the measure Q dominates the measure class $\mathcal{P}$ (and $\mathcal{P}^x$).

Lemma 1 Let $(\Omega,\mathfrak{A},\mathfrak{P})$ be a statistical space. Take a measure $Q\in\mathfrak{P}^x$. Then there exists a greatest equivalence class $\tilde{A}_Q\in\mathfrak{U}$ such that Q is strictly positive on $\tilde{A}_Q$.

Proof Define a subset $\mathfrak{C}_Q$ of $\mathfrak{U}$ as follows:

$$\mathfrak{C}_Q=\{\tilde{A}\in\mathfrak{U}|Q \text{ is strictly positive on } \tilde{A}\}.$$

Since $\mathfrak{A}$ is a σ-field, $\mathfrak{C}_Q$ is a σ-ideal. Let

$$c=\sup\{Q(A)|\tilde{A}\in\mathfrak{C}_Q\}.$$

For every $n\in\mathbb{N}$ there exists an equivalence class $\tilde{A}_n\in\mathfrak{C}_Q$ for which

$$Q(\tilde{A}_n)\geq c-\frac{1}{n}.$$

In this case $Q(\bigvee_{n\in\mathbb{N}}\tilde{A}_n)=c$. Thus we can choose $\bigvee_{n\in\mathbb{N}}\tilde{A}_n$ as $\tilde{A}_Q$.

The equivalence class $\tilde{A}_Q$ is referred to as the waistbelt of Q.

Definition 1 Denote

$$\mathfrak{C}=\{\tilde{A}_Q|Q\in\mathfrak{P}^x\},$$

$$\mathfrak{C}^x=\{\tilde{A}\in\mathfrak{U}|\ \tilde{A}=\bigvee_{\tilde{B}\in\mathfrak{C}}(\tilde{A}\wedge\tilde{B})\},$$

$$\mathfrak{D}=\{\tilde{A}\in\mathfrak{U}\ \ \tilde{A}\wedge\tilde{B}=\emptyset \text{ for every } \tilde{B}\in\mathfrak{C}^x\}.$$

Remark $\mathfrak{C}$ contains the "small parts of Ω", on these parts the statistical space is dominated. (Consequently $\mathfrak{U}|\tilde{A}_Q$ is complete.) On the other hand the statistical space is not dominated on any subsets of elements belonging to $\mathfrak{D}$. At first glance these are the "irregular" parts of Ω.

Now we consider those parts of Ω which can be built up as the union of relatively few elements in $\mathfrak{C}$. On these events the measure class $\mathfrak{P}$ is furthermore not dominated but it is uniform complete. To this end we need some lemmas.

Lemma 2 Suppose that $(\Omega,\mathfrak{A},\mathfrak{P})$ satisfies (I). Take an event $A\in\mathfrak{A}$, and consider two maximal subsystems of $\mathfrak{C}$ containing pairwise disjoint elements $(\tilde{A}_i)_{i\in I}$, $(\tilde{B}_j)_{j\in J}$ such that $\tilde{A}_i<\tilde{A}$, $\tilde{B}_j<\widetilde{\Omega\setminus A}$. Suppose that $|J|\geq|I|$ (where $|\cdot|$ denotes the cardinality of the corresponding set). In this case every subsystem of $\mathfrak{C}$ whose cardinality is not greater than $|I|$ has a supremum in the Boole-algebra $\mathfrak{A}$.

The idea of proof: Let $(\tilde{C}_k)_{k\in K}\subset\mathfrak{C}$ and suppose that $|K|\leq|I|$. First we divide every $\tilde{C}_k$ using $(\tilde{A}_i)_{i\in I}$ into disjoint parts. Namely let

$$\tilde{D}_k=\bigwedge_{i\in I}(\tilde{C}_k\setminus\tilde{A}_i),\quad \tilde{E}_i=\bigvee_{k\in K}(\tilde{A}_i\wedge\tilde{C}_k).$$

Every $\tilde{D}_k$ is disjoint with every $\tilde{A}_i$, and on the other hand $\tilde{E}_i<\tilde{A}_i$ i.e. it is disjoint with every $(\tilde{B}_j)_{j\in J}$.

Observe that it suffices to prove that $\vee\tilde{D}_k$ and $\vee\tilde{E}_i$ exist. Since $(\tilde{D}_k)_{k\in K}$ and $(\tilde{A}_i)_{i\in I}$ play the same role as $(\tilde{E}_i)_{i\in I}$ and $(\tilde{B}_j)_{j\in J}$ we shall prove that $\vee\tilde{D}_k$ exists.

Since $|I|\geq|K|$ there exists an injection $\rho:K\to I$. Let $\tilde{F}_k=\tilde{D}_k\vee\tilde{A}_{\rho(k)}$. $\tilde{F}_k$ is an element of $\mathfrak{C}$ so there exists a measure P_k such that $P_k(F_k)=1$ and P_k is strictly positive on F_k. Now we shall construct a measure class $\bar{\mathfrak{P}}$ which is on one hand sufficiently rich in the sense that any sufficient σ-field in $(\Omega,\mathfrak{A},\bar{\mathfrak{P}})$ contains every event which is disjoint with F_k'-s - this can be achieved if we take in $\bar{\mathfrak{P}}$ every measure from $\mathfrak{P}^x$ which vanishes on $F_k, k\in K$ - and on the other hand it is very poor in F_k'-s, namely there exists only one measure which does not vanish on F_k. Precisely let

$$\mathfrak{P}_o=\{P\in\mathfrak{P}^x \mid P(F_k)=0 \text{ for every } k\in K\}$$

and

$$\bar{\mathcal{P}}=\{P_k\}_{k\in K}\cup\mathcal{P}_o.$$

According to our assumption (I) there exists a minimal sufficient σ-field. Denote this by $\mathcal{F}$. What is it like? It can be proved that since $\mathcal{P}_o$ is "rich enough" $\mathcal{F}$ contains every B for which $\tilde{B}\wedge\tilde{F}_k=\tilde{\emptyset}$ for every $k\in K$. On the other hand since the structure of $\bar{\mathcal{P}}$ on F_k is rather simple - $P(F_k)=0$ for every $P\in\bar{\mathcal{P}}$ which differs from P_k - if $B\in\mathcal{F}$ and $\tilde{\emptyset}\neq\tilde{B}<\tilde{F}_k$ for a $k\in K$ then $\tilde{B}=\tilde{F}_k$. Thus

$$\mathcal{F}=\{B\in\mathcal{A}|\ \tilde{B}\wedge\tilde{F}_k=\tilde{\emptyset}\ \text{or}\ \tilde{B}\wedge\tilde{F}_k=\tilde{F}_k\ \text{for every}\ k\in K\}.$$

How can the conditional probability $\bar{\mathcal{P}}(A|\mathcal{F})$ be written? Out of the events F_k, $k\in K$, the σ-field $\mathcal{F}$ coincides with the σ-field $\mathcal{A}$ i.e. on these events $\bar{\mathcal{P}}(A|\mathcal{F})$ takes on the values 0 or 1. But on the event F_k we have $\bar{\mathcal{P}}(A|\mathcal{F})=P_k(A|\mathcal{F})=P_k(A_{\rho(k)})$ since $A\cap F_k=A_{\rho(k)}$. P_k is strictly positive on F_k, so $0<P_k(A_{\rho(k)})<1$. Observe that

$$\vee\{\tilde{F}_k,\ k\in K\}\cup\{\tilde{B}\in\mathfrak{A}|\tilde{B}\wedge\tilde{F}_k=\tilde{\emptyset}\ \text{for every}\ k\in K\}=\tilde{\Omega}.$$

Thus knowing the value $\bar{\mathcal{P}}(A|\mathcal{F})$ on these equivalence classes we know it on the whole Ω. This particularly implies that

$$\widetilde{\{0<\bar{\mathcal{P}}(A|\mathcal{F})<1\}}=\bigvee_{k\in K}\tilde{F}_k.$$

Thus $\vee\tilde{F}_k$ exists and $\vee\tilde{D}_k=\vee\tilde{F}_k\setminus\tilde{A}$.

<u>Remark</u> The index set I satisfying the assumptions of Lemma 2 is of course not unique. But if it is not finite then $|I|$ is already unique. If $(\tilde{A}_i)_{i\in I}$ and $(\tilde{A}'_k)_{k\in K}$ are two maximal subsystems of $\mathfrak{C}$ containing pairwise disjoint elements such that $\vee\tilde{A}_i=\tilde{A}$ and $\tilde{A}'_k=\tilde{A}$, then $|I|=|K|$ since any equivalence class $\tilde{A}_i$ or $\tilde{A}'_k$ can be decomposed to at most denumerable pairwise disjoint parts.

Corollary 1 There exists a cardinality α_0 (which is not denumerable) such that

/i/ every subsystem of $\mathfrak{C}$ whose cardinality is less than α_0 has a supremum, and

/ii/ for any $A \in \mathfrak{A}$ either $\mathfrak{A}|_{\tilde{A}} \cap \mathfrak{C}$ or $\mathfrak{A}|_{\widetilde{\Omega \setminus A}} \cap \mathfrak{C}$ has the property that the cardinality of every subsystem of it containing pairwise disjoint elements is less than α_0. (II)

Proof: Let α_0 be the smallest cardinal for which there exists a subsystem of $\mathfrak{C}$ with this cardinality containing pairwise disjoint elements which does not have supremum if there exists such a system at all and let $\alpha_0 = |\mathfrak{C}|+1$ otherwise.

If $\alpha_0 = |\mathfrak{C}|+1$ then $\bigvee_{\tilde{A} \in \mathfrak{C}} \tilde{A}$ exists. Thus Ω can be divided into two parts: $\Omega_{\mathfrak{C}^x}$ and $\Omega_{\mathfrak{D}}$. If $\tilde{A} \in \mathfrak{C}^x$ then $\tilde{A} < \tilde{\Omega}_{\mathfrak{C}^x}$, if $\tilde{A} \in \mathfrak{D}$ then $\tilde{A} < \tilde{\Omega}_{\mathfrak{D}}$, and the Boole-algebra $\mathfrak{A}|_{\widetilde{\Omega}_{\mathfrak{C}^x}}$ is complete.

In any case define

$$\mathfrak{C}^x(\alpha_0) = \{\tilde{A} \in \mathfrak{C}^x \mid \text{ there exists a system } (\tilde{A}_i)_{i \in I} \subset \mathfrak{C} \text{ for which } |I| < \alpha_0 \text{ and } \tilde{A} = \bigvee_{i \in I} \tilde{A}_i\},$$

and

$$\mathfrak{A}(\mathfrak{C}^x(\alpha_0)) = \{\tilde{A} \in \mathfrak{A} \mid \text{ there exists a } \tilde{C} \in \mathfrak{C}^x(\alpha_0) \text{ and a } \tilde{D} \in \mathfrak{D} \text{ for which } \tilde{C} \vee \tilde{D} = \tilde{A}\}.$$

Observe that $\mathfrak{D} \subset \mathfrak{A}(\mathfrak{C}^x(\alpha_0))$ and property (II) implies that for every $A \in \mathfrak{A}$ either $\tilde{A} \in \mathfrak{A}(\mathfrak{C}^x(\alpha_0))$ or $\widetilde{\Omega \setminus A} \in \mathfrak{A}(\mathfrak{C}^x(\alpha_0))$.

On the elements of $\mathfrak{C}^x(\alpha_0)$, since the σ-ideal generated by these equivalence classes is complete as a lattice, the statistical space $\mathfrak{A}$ is uniform complete.

THE ELEMENTS OF $\mathfrak{P}$

Up to now we have analysed the structure of the Boole-algebra $\mathfrak{A}$ under assumption (I). Now we examine the elements of $\mathfrak{P}$.

We already know that any measure $Q\in\mathfrak{P}^x$ dominates the measure class $\mathfrak{P}^x$ on its waist-belt $\tilde{A}_Q$. But which values can it be taken on those parts of Ω which are disjoint to A_Q. Although we cannot say anything about its value on the elements of $\mathfrak{D}$, its value can be characterized entirely on the elements of $\mathfrak{C}^x$ which are disjoint to $\tilde{A}_Q$.

<u>Lemma 3</u> Suppose that $(\Omega,\mathfrak{A},\mathfrak{P})$ has property (I). Let $P\in\mathfrak{P}^x$ be a measure and $\tilde{A}\in\mathfrak{C}^x$ be an equivalence class for which $\tilde{A}\wedge\tilde{A}_P=\tilde{\emptyset}$. If $\tilde{A}\in\mathfrak{C}^x(\alpha_0)$ then $P(A)=0$.

<u>Sketch of proof</u> One has to distinguish three cases. The first case is $\tilde{A}\in\mathfrak{C}$, the second one is $\alpha_0=|\mathfrak{C}|+1$, the third one is $\alpha_0<|\mathfrak{C}|+1$.

In the first case there exists a measure $Q\in\mathfrak{P}^x$ such that $\tilde{A}=\tilde{A}_Q$, and if $\tilde{A}\neq\tilde{\emptyset}$ holds then P would be strictly positive on $\widetilde{\{\frac{dP}{dQ}>0\}}\wedge\tilde{A}_Q$ which is disjoint to $\tilde{A}_P$. This cannot hold, thus $P(A)=0$.

In the second case every subsystem of $\mathfrak{C}$ has a supremum thus $\tilde{\Omega}_{\mathfrak{C}}x$ exists. Consider a maximal pairwise disjoint subsystem of $(\mathfrak{C}\cap\mathfrak{A}|_{\tilde{\Omega}_{\mathfrak{C}}x\setminus\tilde{A}_P})$. If it contains finitely many elements then due to the remark above $P(\Omega_{\mathfrak{C}}x\setminus A_P)=0$, so $P(A)=0$. If it contains infinitely many elements then divide it into two parts $(\tilde{B}_i)_{i\in I}$, $(\tilde{C}_i)_{i\in I}$, and let $\tilde{E}_i=\tilde{B}_i\vee\tilde{C}_i$, $\tilde{B}=\vee\tilde{B}_i$, $\tilde{C}=\vee\tilde{C}_i$ and $\tilde{E}=\vee\tilde{E}_i(=\tilde{\Omega}_{\mathfrak{C}}x\setminus\tilde{A}_P)$.

In the third case we can construct $\tilde{B}$, $\tilde{C}$ similarly. Namely let us choose a maximal pairwise disjoint subsystem of $\mathfrak{C}\cap\mathfrak{A}|_{\tilde{A}}$ and of $\mathfrak{C}\cap\mathfrak{A}|_{\tilde{\Omega}\setminus(\tilde{A}\vee\tilde{A}_P)}$. Denote these by $(\tilde{B}_i)_{i\in I}$, $(\tilde{C}_j)_{j\in J}$. Since $\tilde{A}\in\mathfrak{C}^x(\alpha_0)$ it follows that $|I|<\alpha_0$ and consequently $|J|\geq\alpha_0$. Thus there exists an injection $\rho:I\to J$. Let $\tilde{E}_i=\tilde{B}_i\vee\tilde{C}_{\rho(i)}$. There exist the supremums $\tilde{B}=\vee\tilde{B}_i$, $\tilde{C}=\vee\tilde{C}_{\rho(i)}$, $\tilde{E}=\vee\tilde{E}_i$. Now we can continue our proof in the second and in the third case together. The idea is the same as in the proof of Lemma 2, namely we construct a measure class $\bar{\mathcal{P}}\in\mathcal{M}(\mathcal{P})$ and we use that there exists a minimal sufficient σ-field. Suppose that $P(E)>0$, so there exists a number $0<c\leq 1$ for which

$$c\neq\frac{P(B)}{P(E)}.$$

At the same time $\tilde{E}_i\in\mathcal{C}$, thus there exists a measure $P_i\in\mathcal{P}^X$ such that

$$P_i(E_i)=1,\ \tilde{A}_{P_i}=\tilde{E}_i,\ P_i(B_i)=c.$$

Observe that the measures $(P_i)_{i\in I}$ are mutually singular, the measure P_i is concentrated to E_i. Let

$$\mathcal{P}_o=\{Q\in\mathcal{P}^X \mid Q(E_i)=0 \text{ for every } i\in I\}$$

and finally let

$$\bar{\mathcal{P}}=\{P_i \mid i\in I\}\cup\mathcal{P}_o.$$

Obviously $\bar{\mathcal{P}}\in\mathcal{M}(\mathcal{P})$. (Observe that $P\in\mathcal{P}_o$.) According to assumption (I) there exists a minimal sufficient σ-field in $(\Omega,\mathcal{A},\bar{\mathcal{P}})$. Denote it by $\mathcal{F}$. What is this σ-field like? Since $\mathcal{P}_o$ is "rich", every $D\in\mathcal{A}$ for which $\tilde{D}\wedge\tilde{E}_i=\tilde{\emptyset}$ $i\in I$ belongs to $\mathcal{F}$. On the other hand the measure class $\bar{\mathcal{P}}$ is poor on E_i since for every $i\in I$ there exists only one measure in $\bar{\mathcal{P}}$ such that E_i has positive measure with respect to it - namely this is P_i. Thus $\mathcal{F}$ is the trivial σ-field on E_i. Consequently

$$\mathcal{F}=\{D\in\mathcal{A} \mid \tilde{D}\wedge\tilde{E}_i=\tilde{\emptyset} \text{ or } \tilde{D}\wedge\tilde{E}_i=\tilde{E}_i \text{ for every } i\in I\}.$$

Now let us try to determine the value of $\bar{\mathcal{P}}(B|\mathcal{F})$. Since $E_i\in\mathcal{F}$ and $\tilde{E}=\vee\tilde{E}_i$ exist we have $E\in\mathcal{F}$. At the same time $B\subset E$ thus

$$\bar{\mathcal{P}}(B|\mathcal{F})\chi_{\Omega\setminus E}=0.$$

On the other hand

$$\bar{\mathcal{P}}(B|\mathcal{F})\chi_{E_i}=P_i(B|\mathcal{F})\chi_{E_i}=c\chi_{E_i}.$$

So

$$\bar{\mathfrak{P}}(B|\mathfrak{F})=c\chi_E.$$

But

$$\int_E \chi_B dP=P(B)\neq cP(E)=\int_E \bar{\mathfrak{P}}(B|\mathfrak{F})dP.$$

This would mean that $\mathfrak{F}$ is not sufficient. We have obtained a contradiction thus P(E) must be zero. From this it follows that P(A)=0.

In words P vanishes not only on the elements of $\mathfrak{C}$ which are disjoint to $\tilde{A}_P$ but on the elements of $\mathfrak{C}^x(\alpha_0)$. Observe that if $\tilde{A},\tilde{B}\in\mathfrak{C}^x\setminus\mathfrak{C}^x(\alpha_0)$ then $\tilde{A}\setminus\tilde{B}$ and $\tilde{B}\setminus\tilde{A}$ belong to $\mathfrak{C}^x(\alpha_0)$.

Thus the measure P takes the same value on every element of $\mathfrak{C}\setminus\mathfrak{C}^x(\alpha_0)$. From this can be easily proved

<u>Lemma 4</u> Consider a measure $P\in\mathfrak{P}$ which vanishes on $\mathfrak{D}$. Then there exists a number c_P such that if $\tilde{A}$ is an equivalence class which belongs to $\mathfrak{A}\setminus\mathfrak{A}(\mathfrak{C}^x(\alpha_0))$ and $\tilde{A}\wedge\tilde{A}_P=\tilde{\emptyset}$ then $P(A)=c_P$.

<u>Remark</u> This is rather surprising. This means that if P is a probability measure which vanishes on $\mathfrak{D}$ and for which $\tilde{A}_P=\tilde{\emptyset}$, then it is a zero-one measure. Does there exist such a measure on an arbitrary statistical space?

<u>Lemma 5</u> Suppose that $(\Omega,\mathfrak{A},\mathfrak{P})$ has property (II). Then there exists a measure P_0 for which

$$P_0(A)=0 \text{ if } \tilde{A}\in\mathfrak{A}(\mathfrak{C}^x(\alpha_0))$$

and

$$P_0(A)=1 \text{ if } \tilde{A}\in\mathfrak{A}\setminus\mathfrak{A}(\mathfrak{C}^x(\alpha_0)).$$

<u>Proof</u> Observe that P_0 is not identically zero iff $\mathfrak{A}\setminus\mathfrak{A}(\mathfrak{C}^x(\alpha_0))$ is not empty. ($\mathfrak{A}(\mathfrak{C}^x(\alpha_0))=\mathfrak{A}$ means that $\tilde{\Omega}_{\mathfrak{C}}x$ exists and $\mathfrak{A}|\tilde{\Omega}_{\mathfrak{C}}x$ is a complete Boole-algebra.) .

Define P_0 as above. Since for every $A \in \mathfrak{A}$ either $A \in \mathfrak{U}(\mathfrak{C}^x(\alpha_0))$ or $\widetilde{\Omega \setminus A} \in \mathfrak{U}(\mathfrak{C}^x(\alpha_0))$ thus P_0 can take the value 1 only on one of two disjoint events. So P_0 is finitely additive. Since α_0 is not denumerable, P_0 is a measure.

Remark The measure P_0 vanishes on the elements of $\mathfrak{D} \cup \mathfrak{C}^x(\alpha_0)$.

Corollary 2 Suppose that $(\Omega, \mathfrak{A}, \mathfrak{P})$ has property (I). Then for every measure $P \in \mathfrak{P}$ there exists a number d_P such that $P(A) = P(A \cap A_P) + P(A \cap B_P) + d_P P_0(A)$ for every $A \in \mathfrak{A}$ where $\tilde{B}_P$ is an element of $\mathfrak{D}$. (III)

Sketch of proof The event B_P can be constructed similarly as A_P, namely it is an event for which the relation

$$P(B_P) = \sup\{P(B) \mid \tilde{B} \in \mathfrak{D}\}$$

holds. Consequently the measure $A \to P(A) - P(A \cap A_P) - P(A \cap B_P)$ vanishes on $\mathfrak{D} \cup \mathfrak{U}|_{\tilde{A}_P}$. Thus it can be written in the form $d_P \cdot P_0$.

THE STRUCTURE OF SUFFICIENT σ-FIELDS

The crucial observation of this paragraph is that if $(\Omega, \mathfrak{A}, \mathfrak{Q})$ is a statistical space and $\mathfrak{F}$ is a sufficient σ-field and $P, Q \in \mathfrak{Q}$ then considering the generalized Radon-Nikodym derivate and Lebesgue decomposition of P with respect to Q

$$P(A) = \int_A \frac{dP}{dQ} dQ + P(A \cap N) \text{ where } Q(N) = 0$$

we can choose the R-N derivate $\frac{dP}{dQ}$ and the singularity set N to be $\mathfrak{F}$-measurable. If we fix the measure Q and P varies over $\mathfrak{Q}$, then the singularity set also varies but it can be proved easily that its supremum is just $\widetilde{\Omega \setminus A_Q}$. From this it follows that $A_Q \in \mathfrak{F}$. Now let us concentrate to waistbelt $\tilde{A}_Q$. The measure Q dominates $\mathfrak{Q}$ on it, thus the R-N derivates are determined mod $\mathcal{N}(\mathfrak{Q})$ uniquely on this event. Consequently for every version of $\frac{dP}{dQ}$ the random variable $\frac{dP}{dQ}\chi(A_Q)$ is $\mathfrak{F}$-measurable. For every $Q\in\mathfrak{Q}$ define a σ-field

$$\mathfrak{F}_Q=\sigma(\frac{dP}{dQ}\chi(A_Q)\,|\,P\in\mathfrak{Q}\}.$$

So far we have obtained that $A_Q\in\mathfrak{F}$ and $\mathfrak{F}_Q\subset\mathfrak{F}$ for every $Q\in\mathfrak{Q}$ and for every sufficient σ-field $\mathfrak{F}$. The following lemma examines the behavior of $\mathfrak{F}$ on $\mathfrak{D}$.

Lemma 6 Suppose that $(\Omega,\mathfrak{A},\mathfrak{P})$ has property (I). Then

for every measure class $\mathfrak{Q}\in\mathfrak{M}(\mathfrak{P})$ if $\mathfrak{F}\subset\mathfrak{A}$ is a sufficient σ-field with respect to $\mathfrak{Q}$, then $\mathfrak{D}\subset\mathfrak{F}/\mathfrak{N}(\mathfrak{P})$. (IV)

The proof of this lemma is similar to that of Lemma 3 but it is a little bit more complicated, it requires a more tedious examination. After this we can formulate the following

Lemma 7 Suppose that $(\Omega,\mathfrak{A},\mathfrak{P})$ has properties (II)-(III)-(IV). Consider a measure class $\mathfrak{Q}\in\mathfrak{M}(\mathfrak{P})$ and let

$$\mathfrak{G}=\{A\in\mathfrak{A}\,|\quad \tilde{A}\wedge\tilde{A}_Q\in\mathfrak{F}_Q/\mathfrak{N}(\mathfrak{P})\quad \text{for every } Q\in\mathfrak{Q}\}.$$

In this case $\mathfrak{G}$ is the minimal sufficient σ-field in $(\Omega,\mathfrak{A},\mathfrak{Q})$.

The idea of proof Applying our previous considerations it can be shown that $\mathfrak{G}$ is contained in every sufficient σ-field. To prove the minimality of $\mathfrak{G}$ we have to show that $\mathfrak{G}$ is sufficient.

Every measure Q in $\mathfrak{Q}$ can be written as a linear combination of three measures: the first one is concentrated to A_Q, the second one is concentrated to a set belonging to $\mathfrak{D}$, the third one is P_o. Take an event $A\in\mathfrak{A}$. Our aim is to construct $\mathfrak{Q}(A\,|\,\mathfrak{G})$. We can suppose that $A\in\mathfrak{A}(\mathfrak{C}^x(\alpha_o))$ since $Q(A\,|\,\mathfrak{G})+Q(\Omega\setminus A\,|\,\mathfrak{G})=1$ for every $Q\in\mathfrak{Q}$. Since P_o vanishes on $\mathfrak{A}(\mathfrak{C}^x(\alpha_o))$ it can be shown that we can omit the third component from every measure $Q\in\mathfrak{Q}$. Furthermore if $A\in\mathfrak{A}(\mathfrak{C}^x(\alpha_o))$ then it can be divided into two parts $\tilde{A}_{\mathfrak{C}^x}\in\mathfrak{C}^x(\alpha_o)$, $\tilde{A}_{\mathfrak{D}}\in\mathfrak{D}$.

Since $\mathfrak{D} \subset \mathcal{G}/\mathcal{N}(\mathfrak{Q})$ we have $\mathfrak{Q}(A_{\mathfrak{D}}|\mathcal{G}) = \chi_{A_{\mathfrak{D}}}$. On the other hand on $\tilde{A}_{\mathcal{C}}x$ the statistical space is uniform complete, so invoking F. Göndőcs et al. Theorem $\mathfrak{Q}(A_{\mathcal{C}}x|\mathcal{G})$ exists. (Without referring to this assertion we can visualise this existence as follows. If we restrict ourselves to $\tilde{A}_Q$, then we obtain a dominated statistical space and so we can apply Halmos-Savage theorem - the restriction of $\mathcal{G}$ to A_Q gives just the minimal sufficient σ-field in the dominated case. The values of $\mathfrak{Q}(A_{\mathcal{C}}x|\mathcal{G})$ have to be built up from this restriction. If we know $\mathfrak{Q}(A_{\mathcal{C}}x|\mathcal{G})\chi(A_Q)$ for every $Q \in \mathfrak{Q}$ then we know its level sets and we can form the supremum of the equivalence classes of these level sets since the Boole-algebra $\mathfrak{A}|_{\tilde{A}_{\mathcal{C}}x}$ is complete. Thus we have already known the level sets of the required random variable $\mathfrak{Q}(A_{\mathcal{C}}x|\mathcal{G})$. From this we get this function itself.) Consequently $\mathcal{G}$ is sufficient, so it is the minimal one.

Now we can characterize the statistical spaces having property (I).

<u>THEOREM</u> Let $(\Omega, \mathfrak{A}, \mathfrak{P})$ be a statistical space. The following two assertions are equivalent:

/i/ It has property (I)

/ii/ a, the Boole-algebra $\mathfrak{A}$ has property (II),

b, for every measure $P \in \mathfrak{P}$ there exists a number d_P such that

$$P(A) = P(A \cap A_P) + P(A \cap B_P) + d_P P_o(A),$$

c, for every $\mathfrak{Q} \in \mathcal{M}(\mathfrak{P})$ if $\mathfrak{F}$ is a sufficient σ-field with respect to $\mathfrak{Q}$, then $\mathfrak{D} \subset \mathfrak{F}/\mathcal{N}(\mathfrak{P})$.

<u>Proof</u> /i/ $\Rightarrow$ /ii/ This is just the assertion of Corollary 1, Corollary 2 and Lemma 6.

/ii/ $\Rightarrow$ /i/ This is the assertion of Lemma 7.

ACKNOWLEDGEMENT

The author would like to thank Professor F. Göndőcs for his helpful suggestions.

REFERENCES

F. GÖNDŐCS and G. MICHALETZKY (1980): Construction of minimal sufficient and pairwise sufficient σ-fields. (To appear.)

P. R. HALMOS and L. J. SAVAGE (1949): Application of the Radon-Nikodym theorem to the theory of sufficient statistics, Ann. Math. Stat., 20, 225-241.

M. HASEGAWA and M. D. PERLMAN (1974,1975): On the existence of a minimal sufficient subfield, Ann. Stat., 2, 1049-1055 and Ann. Stat. 3, 1371-1372.

D. MUSSMANN (1972): Vergleich von Experimenten im schwach dominierten Fall, Z. für Wahrscheinlichkeitstheorie verw. Gebiete 24, 295-308.

T. S. PITCHER (1965): A more general property than domination for sets of probability measures, Pac. J. Math., 15, 597-611.

*Proc. of the 3rd Pannonian Symp.
on Math. Stat., Visegrád, Hungary 1982
J. Mogyoródi, I. Vincze, W. Wertz, eds*

ON SOME PROBLEMS FOR PREDICTABLE RANDOM VARIABLES

J. MOGYORÓDI

Department of Probability Theory,
Eötvös Loránd University, Budapest, Hungary

1. Let

$$z_1, z_2, \ldots$$

be nonnegative random variables. Suppose that

$$c = \left\| \sum_{i=1}^{\infty} z_i \right\|_{\infty} < +\infty .$$

Let further $F_1, F_2, \ldots$ be an arbitrary increasing sequence of σ-fields of events. Then for arbitrary $t \in (0, c^{-1})$ we have

$$E(\exp(tA_\infty)) \leq (1-tc)^{-1}$$

where

$$A_\infty = \lim_{n \to +\infty} A_n \ ; \quad A_n = \sum_{i=1}^{n} E(z_i | F_i), \quad n \geq 1.$$

This result is known (cf. e.g. P. A. Meyer [6], Theorem 46.).

Let $\Phi(x)$, $x \geq 0$, be an arbitrary Young function and denote by $\Psi(x)$ its conjugate Young function. For the definition and properties of these we refer to [2], [7]. E.g. the function $\Phi(x) = x \log^+ x$, $x \geq 0$, or

$$\Phi(x) = (x+1)\log(x+1) - x, \quad x \geq 0$$

are Young functions and the conjugate of the last one is

$$\Psi(x) = e^x - x - 1.$$

We say that the Young function $\Phi(x)$ satisfies the growth condition if there exist constants $a > 1$ and $A > 0$ such that for every $x>0$ we have

$$\Phi(ax)\le A\Phi(x).$$

In connection with this we mention that the quantity

$$p=\sup_{x>0}\frac{x\varphi(x)}{\Phi(x)}$$

is finite if and only if Φ satisfies the growth condition. p is called the power of Φ. Here $\varphi(x)$ denotes the right hand side derivative of the convex function $\Phi(x)$.

If a random variable is in L log L then it belongs to the Hardy space $\mathcal{H}_1$, too ([1]). Gundy's converse result states that every nonnegative and predictable (for the definition see section 3.) random variable (which, of course, belongs to $\mathcal{H}_1$) is in turn in L log L (cf. [1]).

The main purpose of the present note is to study in detail the behaviour of the nonnegative predictable random variables. Namely, we shall show that such a random variable belongs to all the Orlicz spaces L^Φ for which

$$x\varphi(x)-\Phi(x)=O(x),\quad x\to+\infty$$

and this condition is also necessary.

2. Lemma 1. Let $\Psi(x)$ be a Young function and suppose that

$$\sum_{i=1}^{\infty} z_i\le c\ ,$$

where $c>0$ is a constant. If for some $t\in(0,c^{-1})$ we have

$$I(t)=\int_0^{+\infty} e^{-t\lambda}d\Psi(\lambda)<+\infty\ ,$$

then

$$E(\Psi(A_\infty))\le(1-tc)^{-1}I(t).$$

Proof. Trivially,

$$E(\Psi(A_\infty))=\int_0^{+\infty} P(A_\infty\ge\lambda)d\Psi(\lambda).$$

According to the Markov inequality and Meyer's above result

$$P(A_\infty\geq\lambda)=P(e^{tA_\infty}\geq e^{t\lambda})\leq e^{-t\lambda}E(e^{tA_\infty})\leq(1-tc)^{-1}e^{-t\lambda}.$$

Consequently,

$$E(\Psi(A_\infty))\leq(1-tc)^{-1}\int_0^{+\infty} e^{-t\lambda}d\Psi(\lambda)=(1-tc)^{-1}I(t)$$

and this proves the assertion.

Now we are in the position to prove

<u>Theorem 1.</u> Let (X_n, F_n), $n\geq 1$ be a nonnegative submartingale. If for some $t\in(0,1)$ the Laplace transform

$$I(t)=\int_0^{+\infty} e^{-t\lambda}d\Psi(\lambda)$$

exists and if $g(x)$ is a nonnegative and increasing convex function tending to infinity as $x\to\infty$, then

$$E(g(X_n^*))\leq E(\Phi(g(X_n)))+(1-t)^{-1}I(t)$$

provided that

$$E(\Phi(g(X_n)))<+\infty.$$

Here

$$X_n^*=\max_{1\leq k\leq n} X_k .$$

<u>Proof.</u> By our supposition there exists an $x_0>0$ such that for $x\geq x_0$ we have $\Phi(g(x))\geq g(x)$. Consequently, from

$$E(\Phi(g(X_n)))<+\infty$$

it follows that $g(X_k)$, $k\leq n$, is of finite expectation and thus $(g(X_k), F_k)$ is a nonnegative submartingale. Consider the events

$$A_1=\{X_1=X_n^*\},\ A_k=\{X_{k-1}^*<X_n^*,\ X_k=X_n^*\},\ k=2,\dots,n.$$

In this case the random variables

$$z_k=\chi_{A_k},\quad k=1,2,\dots,n$$

satisfy

$$\sum_{k=1}^{n} z_k=1.$$

Consequently, the preceding lemma can be applied. For this purpose remark that

$$E(g(X_n^*))=\sum_{k=1}^{n} E(g(X_k)\chi_{A_k})\leq \sum_{k=1}^{n} E(g(E(X_n|F_k))\chi_{A_k})$$

since $(g(X_k), F_k)$ is a nonnegative submartingale.

From this by the Jensen inequality we get

$$E(g(X_n^*))\leq \sum_{k=1}^{n} E(E(g(X_n)|F_k)\chi_{A_k})=\sum_{k=1}^{n} E(g(X_n)E(\chi_{A_k}|F_k))=$$

$$=E(g(X_n)\sum_{k=1}^{n} E(\chi_{A_k}|F_k)).$$

Using the Young inequality on the right hand side we obtain

$$E(g(X_n^*))\leq E(\Phi(g(X_n)))+E(\Psi(\sum_{k=1}^{n} E(\chi_{A_k}|F_k))).$$

The second member can be estimated by the preceding lemma as follows

$$E(\Psi)\sum_{k=1}^{n} E(\chi_{A_k}|F_k)))\leq(1-t)^{-1}I(t).$$

Comparing the so obtained inequalities we finally get

$$E(g(X_n^*))\leq E(\Phi(g(X_n)))+(1-t)^{-1}I(t).$$

This was to be proved.

Remarks. Let $g(x)$ be a Young function and let

$$X_n=E(X|F_n),\quad n\geq 0,$$

be a martingale such that $X_0=0$ a.e. We say that X belongs to the Hardy space $\mathcal{H}_g$ if

$$S=\left(\sum_{i=1}^{\infty} d_i^2\right)^{1/2} \in L^g .$$

Here d_1, d_2, ... denotes the martingale difference sequence corresponding to (X_n, F_n).

Now consider the nonnegative submartingale

$$X_n=|E(X|F_n)|, \quad n\geq 1,$$

where $F_1, F_2, \ldots$ is an increasing sequence of σ-fields. If $g(x)$ is a Young function having finite power that the assertion of Theorem 1 says that $X \in \mathcal{H}_g$, whenever $E(\Phi(g(X_n)))$ for every n (cf. Burkholder, [8]).

Let $g(x)=x$, $x\geq 0$ and consider the Young function

$$\Phi(x)=d((x+1)\log(x+1)-x),$$

where $d>0$ is a constant. Using the result of Theorem 1 it is easy to verify that with appropriately chosen d and $t>0$ the Laplace transform $I(t)$ exists and we have

$$E(X_n^*)\leq d(E((X_n+1)\log(X_n+1))+1)$$

which is the classical inequality of Doob. This implies that $X \in \mathcal{H}_1$ whenever $X \in L\log L$.

In the case when $g(x)$ is a Young function, a necessary and sufficient condition for the validity of the maximal inequality

$$E(g(X_n^*))\leq E(g(cX_n)), \qquad c>0 \text{ constant},$$

is that the conjugate Young function h have finite power. In this case c can be chosen to be the power of h. This is the result of [3]. Theorem 1 above gives a sufficient condition for $E(g(X_n^*))$ to be finite. This is estimated by $E(\Phi(g(X_n)))$. Note that $\Phi(g(x))$ increases more rapidly than $g(x)$.

3. Let $X \in L_1$ and consider the martingale $X_n = E(X | F_n)$ where

$F_1 = (\emptyset, \Omega), F_2, \ldots$

is an increasing sequence of σ-fields of events.

Definition. We say that X is predictable in L^Φ if there exists a sequence

$0 \le \lambda_1 \le \lambda_2 \le \ldots$

of adapted random variables such that for every $n \ge 2$ the inequality

$|X_n| \le \lambda_{n-1}$ and $|X_1| \le \lambda_1$

holds a.e. and

$$\lambda_\infty = \lim_{n \to +\infty} \lambda_n$$

belongs to L^Φ. We define similarly the predictability in L_1.

Theorem 2. If the nonnegative random variable X is predictable in L_1 then with some constant $K_\Phi > 1$ we have

$$E(\Phi(X_n)) \le K_\Phi [E(\lambda_n) + 1]$$

if and only if

$$\xi(x) = x\varphi(x) - \Phi(x) = 0(x)$$

as $x \to +\infty$.

Proof. Garsia has shown that for nonnegative predictable martingales the inequality

$$E(X_n \chi(\lambda_n \ge x) \le X_1 \chi(\lambda_1 \ge x) + xE(\chi(\lambda_n \ge x))$$

holds (cf. [1], Theorem III. 3. 4.). Here $x > 0$ is an arbitrary constant and $\chi(A)$ denotes the indicator of the event A and

$0 \le \lambda_1 \le \lambda_2 \le \ldots$

is a predicting sequence of X in L_1. Integrate the above inequality on $(0, +\infty)$ with respect to the measure generated by the nondecreasing and right continuous function $\varphi(x)$. Using the Fubini theorem we get

$$E(X_n\varphi(\lambda_n))\le X_1\varphi(\lambda_1)+E(\xi(\lambda_n))$$

since

$$\int_0^z x\,d\varphi(x)=\xi(x).$$

Note that $\varphi(x)$ increases and $\lambda_n\ge X_n$. Consequently,

$$E(X_n\varphi(\lambda_n))\ge E(X_n\varphi(X_n))\ge E(\Phi(X_n)).$$

From this

$$E(\Phi(X_n))\le\lambda_1\varphi(\lambda_1)+E(\xi(\lambda_n)).$$

Now if $\xi(x)=0(x)$, we can choose $x_0\ge 0$ such that the inequality

$$\xi(x)=K'x$$

holds for $x\ge x_0$. Here K' is a finite positive constant. Consequently,

$$E(\xi(\lambda_n))\le\xi(x_0)+K'E(\lambda_n)$$

and so

$$E(\Phi(X_n))\le\lambda_1\varphi(\lambda_1)+\xi(x_0)+K'E(\lambda_n).$$

Let us note that $\lambda_1\varphi(\lambda_1)$ is a constant and put

$$K_\Phi=\max(\lambda_1\varphi(\lambda_1)+\xi(x_0),K').$$

Then with this K_Φ we finally obtain

$$E(\Phi(X_n))\le K_\Phi(E(\lambda_n)+1).$$

This proves the first part of the assertion.

Conversely, suppose that the inequality

$$E(\Phi(X_n))\le K_\Phi[E(\lambda_n)+1]$$

holds, where $K_\Phi>1$ is a constant depending only on Φ. Then we necessarily have

$$\xi(x)=0(x)$$

as $x\to+\infty$. In fact, for the nonnegative martingale (X_n, F_n) use the inequality of Doob: if $x>0$ then we have

$$xP(X_n^*>x)\le E(X_n\chi(X_n^*>x)).$$

For arbitrary $a>0$ put

$$X_k'=\min(X_k,a).$$

Then the maximum random variable $\max\limits_{1\le k\le n} X_k'=X_n^{**}$ has the property that

$$X_n^{**}=\min(X_n^*,a)$$

and from the above inequality of Doob we can deduce that

$$xP(X_n^{**}>x)\le E(X_n\chi(X_n^{**}>x)).$$

Integrate this on $(0,+\infty)$ with respect to the measure generated by the function $\varphi(x)$. Then we get

$$E(\xi(X_n^{**}))\le bE(\frac{X_n}{b}\varphi(X_n^{**})),$$

where $0<b<1$ is an arbitrary constant. Applying the Young inequality to the right hand side and remarking that $\Psi(\varphi(x))=\xi(x)$, from this we deduce that

$$E(\xi(X^{**}))\le b[E(\Phi(\frac{X_n}{b}))+E(\xi(X_n^{**}))]$$

or after rearranging,

$$(1-b)E(\xi(X_n^{**}))\le bE(\Phi(\frac{X_n}{b})).$$

$\xi(x)$ being an increasing and continuous function we get by the monotone convergence theorem that

$$(1-b)E(\xi(X_n^*))\le bE(\Phi(\frac{X_n}{b})),$$

since

$$X_n^{**}\uparrow X_n^*$$

as $a\to+\infty$.

The random variable X/b is also predictable. Consequently, by the above supposition we get

$$(1-b)E(\xi(X_n^*))\le bK_\Phi[E(\frac{\lambda_n}{b})+1].$$

This proves the assertion if we take $X \equiv x$. In fact, in this case $X_n^* \equiv x$ and λ_n can be taken to be x. Finally, put, e.g., $b=\frac{1}{2}$.

Theorem 2 characterizes the class of the nonnegative random variables which are predictable in L_1. Namely, they belong to all the spaces L^Φ for which

$$\xi(x)=x\varphi(x)-\Phi(x)=O(x) \quad (x\to+\infty).$$

A little more general setting of Theorem 2 is the following

Theorem 3. If the nonnegative random variable is predictable in L^g, where g is a Young function with finite power, then with some constant $K_{\Phi,g}$ we have

$$E(\Phi(g(X_n)))\leq K_{\Phi,g}[E(g(\lambda_n))+1]$$

if and only if

$$x\varphi(g(x))\hat{g}(x)-\Phi(g(x))=O(g(x)) \text{ as } x\to+\infty,$$

where

$$\varphi(g(x))\hat{g}(x)$$

denotes the "density function" of the composed Young function $\Phi(g(x))$, and $\hat{g}(x)$ is the same for $g(x)$.

REFERENCES

[1] A. M. GARSIA: Martingale inequalities. Reading, Massachusetts, 1973.

[2] M. A. KRASNOSELSKII and Ya.B. RUTICKII: Convex functions and Orlicz spaces. Transl. from Russian by L. F. Boron. Noordhoff, Groningen, 1961.

[3] J. MOGYORÓDI and T. MÓRI: A necessary and sufficient condition for the validity of the maximal inequality for nonnegative submartingales. Accepted for publication in Acta Math. Szeged, 1982.

[4] J. MOGYORÓDI: On a problem of R. F. Gundy. Annales Univ. Sci. Budapest, Sectio Mathematica, Vol. XXV (1982), 273-278.

[5] R. F. GUNDY: On the class LlogL, martingales and singular integrals. Studia Mathematica, 33 (1969), 109-118.

[6] P. A. MEYER: Martingales and stochastic integrals. Lecture Notes in Mathematics, 284, Springer, Berlin, 1972.

[7] J. NEVEU: Discrete parameter martingales. North Holland, Amsterdam, 1975.

[8] D. L. BURKHOLDER: Distribution function inequalities for martingales. The Annals of Probability. 1 (1973), 19-42.

Proc. of the 3rd Pannonian Symp. on Math. Stat., Visegrád, Hungary 1982
J. Mogyoródi, I. Vincze, W. Wertz, eds

I-DIVERGENCE GEOMETRY OF DISTRIBUTIONS AND STOCHASTIC GAMES

TAMÁS F. MÓRI

Department of Probability Theory,
Eötvös Loránd University, Budapest, Hungary

STOCHASTIC GAMES

Consider a gambling house where one can participate in d different games at the same time. The games are not supposed to be independent of each other (you can think the various kinds of betting in roulette). If the gambler stakes a sum of money c_i in the ith game, then he wins $c_i X_i$ where X_i is a random variable, not less than -1. So the system of games can be fully described by a d-dimensional random vector $\underline{X}$.

Consider a sequence of independent repetitions of the games, i.e. a sequence of i.i.d. random vectors

$$\underline{X}_n=(X_{n,1},X_{n,2},\ldots,X_{n,d})' \qquad n=1,2,\ldots$$

(by vector a column vector is meant, the corresponding transpose is denoted by '). Sometimes $\underline{X}_n$ is called the nth round.

In every round the gambler may decide about the proportion of his fortune to be staked in the different games. To do this he can make use of earlier observations. Thus a non-anticipating strategy is a sequence of random vectors

$$\underline{a}_n=(a_{n,1},\ldots,a_{n,d})' \qquad n=1,2,\ldots$$

with the following properties:

(I) $\underline{a}_n$ is measurable to the σ-field $\mathcal{F}_{n-1}$ generated by the first n-1 rounds,

(II) $\underline{a}_n$ takes its values from the simplex

$$\Delta=\{\underline{a}\in\mathbb{R}^d:\ a_i\geqq 0\ \ 1\leqq i\leqq d,\ \ \sum_{i=1}^{d} a_i\leqq 1\}.$$

Demonstratively, $a_{n,i}$ shows, what proportion of his actual fortune the gambler stakes in the ith game of the nth round. Denote the gambler's fortune after n rounds by T_n and let the initial fortune T_o be equal to 1. Then we have

$$T_n=T_o\prod_{j=1}^{n}(1+\underline{a}_j'\underline{X}_j).$$

The gambler's aim is to find the strategy that provides the highest rate of growth for his fortune. It was observed even by D. Bernoulli that the best strategy is a constant non-random one, which maximizes a logarithmic type expectation instead of the expected gain. More precise meaning of this optimality was given by Breiman (1961), also in Móri-Székely (1982). According to their results, the optimal strategy $\underline{a}_n\equiv\underline{a}^*$ can be evaluated by maximizing the expectation

$$E[\log(1+\underline{a}'\underline{X})-\log(1+\sum_{i=1}^{d}(1+X_i))],\ \underline{a}\in\Delta.$$

(The only role of the negative term is to supersede a restriction on the finiteness of $E(\log(1+X_i))$.)

To this moment the favourability of the games was not used. But if none of the games is favourable (i.e. they are of non-positive expectation) then the optimal strategy turns out to be 0, which means that the best you can do is not to play at all. This case is out of interest, of course.

In a system of games having at least one favourable component the optimal strategy provides an exponential rate of growth for the fortune. The aim of the present note is to characterize this optimal rate in an information theoretical manner. Further results concerning favourable stochastic games can be found in Móri (1981), (1982).

INFORMATIONAL DISTANCE OF DISTRIBUTIONS

Let $(S, \mathfrak{S})$ be a measurable space and consider two probability measures P and Q given on it. The informational divergence of P and Q is defined as follows:

$$D(P||Q)=\begin{cases}\int\limits_S \log(dP/dQ)dP & \text{if } P\ll Q,\\ +\infty & \text{otherwise.}\end{cases}$$

This quantity measures how much the distribution Q differs from P in the sense of statistical distinguishability. Hence $D(P||Q)$ is also called the mean information for discrimination between Q and P. It was introduced by Kullback and Leibler (1951).

Denote the n-fold direct product of the above probability space with itself by (S^n, $\mathfrak{S}^n$, P^n and Q^n, resp.). The following assertion is known as

STEIN's LEMMA (See e.g. Csiszár-Körner (1981))

For any $0<c<1$

$$\lim \frac{1}{n} \log(\inf\{Q^n(A):A\in\mathfrak{S}^n, P^n(A)\geq c\})=-D(P||Q).$$

Though the informational divergence is not symmetric, it is an analogue of the squared Euclidean distance in some respect (see Csiszár (1975) for details).

A GEOMETRIC PROPERTY OF THE OPTIMAL RATE OF GROWTH

Consider the measurable space

$$S=\{\underline{x}\in \mathbb{R}^d: x_i \geq -1, \quad 1\leq i\leq d\}$$

with the Borel σ-field $\mathcal{S}$ and let P be the distribution of our system of games $\underline{X}$. Denote by $\mathcal{Q}$ the set of all nonfavourable probability distributions given on this measurable space, i.e.

$$\mathcal{Q}=\{Q : \int_S x_i Q(d\underline{x})\leq 0, \quad 1\leq i\leq d\}.$$

Now we are in position to state our

THEOREM

$$\max_{\{\underline{a}_n\}}(\limsup \frac{1}{n} \log T_n)=\min_{Q\in\mathcal{Q}} D(P||Q) \quad \text{a.s.}$$

(on the left-hand side the maximum is taken over all nonanticipating strategies).

This theorem asserts that the optimal growth exponent of the gambler's fortune is equal to the informational distance of the given system of games $\underline{X}$ from those ones having no favourable components.

Proof: Denote the fortune obtained with the optimal strategy by T_n^*. By the strong law of large numbers

$$\lim \frac{1}{n} \log T_n=E(\log(1+\underline{a}^{*\prime}\underline{X}))=E^*.$$

In virtue of Móri-Székely (1982), Theorem 1, for any strategy $\{\underline{a}_n\}$ $\lim(T_n/T_n^*)$ exists and it is finite with probability 1. Consequently

$$\limsup \frac{1}{n} \log T_n\leq E^*,$$

thus

$$\max_{\{\underline{a}_n\}}(\limsup \frac{1}{n} \log T_n)=E^*.$$

It remains to show that $\min_{Q\in\mathcal{Q}} D(P||Q)=E^*$.

First define Q by

$$Q(B)=\int_B \frac{1}{1+\underline{a}^{*\prime}\underline{x}} P(d\underline{x})+C\chi_B(-\underline{1}), \quad B\in\mathcal{S},$$

where $\underline{1}\in\mathbb{R}^d$ denotes the vector with coordinates 1, χ_B stands for the indicator function of B and

$$C=E\left(\frac{\underline{a}^{*\prime}\underline{X}}{1+\underline{a}^{*\prime}\underline{X}}\right).$$

Since for arbitrary $\underline{a}\in\Delta$

$$E\left(\frac{(\underline{a}-\underline{a}^*)'\underline{X}}{1+\underline{a}^{*\prime}\underline{X}}\right)\leq 0$$

holds (cf. Móri-Székely (1982)), substitution $\underline{a}=\underline{0}$ gives $C\geq 0$, therefore Q is a probability measure. Further, if all coordinates of $\underline{a}$ equal 0 except the ith one which is 1, we obtain

$$\int_S x_i Q(d\underline{x})=\int_S \frac{x_i-\underline{a}^{*\prime}\underline{x}}{1+\underline{a}^{*\prime}\underline{x}} P(d\underline{x})\leq 0,$$

consequently $Q\in\mathcal{Q}$.

Clearly $P\ll Q$ and for $\underline{x}\neq-\underline{1}$

$$\frac{dP}{dQ}(\underline{x})=1+\underline{a}^{*\prime}\underline{x}.$$

Supposing $P(\underline{X}=-\underline{1})>0$, from the definition of $\underline{a}^*$ we can deduce $\underline{1}'\underline{a}^*<1$, hence for $\underline{a}=\frac{1}{\underline{1}'\underline{a}^*}\underline{a}^*$ the cited inequality gives $C=0$. Thus the above formula for the Radon-Nikodym derivative $\frac{dP}{dQ}$ remains valid also at $\underline{x}=-\underline{1}$.

By all means

$$D(P||Q)=\int_S \log(1+\underline{a}^{*\prime}\underline{x})P(d\underline{x})=E^*,$$

thus $\min\leq E^*$.

For the proof of the converse inequality first suppose $E^*<+\infty$. Let $\varepsilon>0$ and consider the Borel set

$$A_n(\varepsilon)=\{(\underline{x}_1,\underline{x}_2,\ldots,\underline{x}_n)\in S^n:\ \prod_{j=1}^{n}(1+\underline{a}^{*\prime}\underline{x}_j)>2^{n(E^*-\varepsilon)}\}.$$

By the law of large numbers $\lim P^n(A_n(\varepsilon))=1$. Hence Stein's lemma implies

$$\liminf \frac{1}{n}\log Q^n(A_n(\varepsilon))\geq -D(P||Q)$$

for arbitrary $Q\in\mathcal{Q}$. On the other hand

$$\int_{S^n}\prod_{j=1}^{n}(1+\underline{a}^{*\prime}\underline{x}_j)Q^n(d\underline{x}_1,\ldots,d\underline{x}_n)\leq 1.$$

Applying the Markov inequality we obtain

$$Q^n(A_n(\varepsilon))\leq 2^{-n(E^*-\varepsilon)},$$

whence

$$E^*\leq D(P||Q)+\varepsilon,\quad Q\in\mathcal{Q},$$

which completes the proof in the finite case.

In case of infinite E^* we have to proceed in a similar way, using the Borel set

$$A_n(N)=\{(\underline{x}_1,\ldots,\underline{x}_n)\in S^n:\ \prod_{j=1}^{n}(1+\underline{a}^{*\prime}\underline{x}_j)>2^{nN}\}$$

instead of $A_n(\varepsilon)$, then letting N tend to infinity.

Remark

Our Theorem can be generalized for the case when the successive rounds form a stationary ergodic process. For non-i.i.d. rounds the optimal strategy $\underline{a}_n^*$ is defined as an $\mathcal{F}_{n-1}$-measurable random vector with values from the simplex Δ, such that it maximizes a.s. the conditional expectation

$$E[\log(1+\underline{a}'\underline{X}_n)-\log(1+\sum_{i=1}^{d}(1+X_{n,i}))|\mathcal{F}_{n-1}]$$

(cf. Móri (1981)). For a stationary ergodic sequence of rounds the optimal growth exponent of the fortune equals the minimal average distance of our sequence from those ones which are stationary, ergodic and absolutely non-favourable. This assertion can be formulated as follows:

A sequence of rounds $\mathbb{Y}=(\underline{Y}_1,\underline{Y}_2,\dots)$ is called absolutely non-favourable if

$$E(Y_{n,i}\mid \mathcal{F}_{n-1})\leq 0 \text{ a.s. for } n=1,2,\dots,\ 1\leq i\leq d.$$

Let $\mathcal{Q}$ denote the set of all absolutely non-favourable stationary ergodic sequences of rounds.

The average divergence $\overline{D}(\mathbb{X}\|\mathbb{Y})$ of two stationary sequences $\mathbb{X}=(\underline{X}_1,\underline{X}_2,\dots)$ and $\mathbb{Y}=(\underline{Y}_1,\underline{Y}_2,\dots)$ is defined as

$$\lim_{n\to\infty}\frac{1}{n}\overline{D}(\underline{X}_1,\dots,\underline{X}_n\|\underline{Y}_1,\dots,\underline{Y}_n).$$

The limit above always exists, it can be shown in a similar way as it is done when defining the entropy rate of a stationary source (Csiszár-Körner (1981), 1.§4.).

With these notations we have the following

THEOREM

$$\max_{\{\underline{a}_n\}}\left(\limsup\frac{1}{n}\log T_n\right)=\min_{\mathbb{Y}\in\mathcal{Q}}\overline{D}(\mathbb{X}\|\mathbb{Y}) \quad \text{a.s.}$$

REFERENCES

[1] BREIMAN, L. (1961). Optimal gambling systems for favourable games. In: Proc. 4th Berkeley Symp. on Math. Statist. and Probability, Univ. California Press, Berkeley. Vol.1, 65-78.

[2] CSISZÁR, I. (1975). I-divergence geometry of probability distributions and minimization problems. Ann. Probability 3, 146-158.

[3] CSISZÁR, I. and KÖRNER, J. (1981). Information Theory: Coding Theorems for Discrete Memoryless Systems. Akadémiai Kiadó - Academic Press, Budapest - New York.

[4] KULLBACK, S. and LEIBLER, R.A. (1951). On information and sufficiency. Ann. Math. Statist. 22, 79-86.

[5] MÓRI, T.F. (1981). On favourable stochastic games. Annales Univ. Sci. Budapest, Sectio Comp.

[6] MÓRI, T.F. (1982). Asymptotic properties of the empirical strategy in favourable stochastic games. In: Limit Theorems in Probability and Statistics, Coll. Math. Soc. J. Bolyai. North-Holland, Amsterdam.

[7] MÓRI, T.F. and SZÉKELY, G.J. (1982). How to win if you can? In: Limit Theorems in Probability and Statistics, Coll. Math. Soc. J. Bolyai. North-Holland, Amsterdam.

Proc. of the 3rd Pannonian Symp. on Math. Stat., Visegrád, Hungary 1982
J. Mogyoródi, I. Vincze, W. Wertz, eds

SOME RESULTS ON COMMUTATION MATRICES WITH STATISTICAL APPLICATIONS

H. NEUDECKER*, T. WANSBEEK**

*University of Amsterdam

**Central Bureau of Statistics, Voorburg,
The Netherlands

ABSTRACT

The commutation matrix P_{mn} changes the order of multiplication of a Kronecker matrix product. The vec operator stacks columns of a matrix one under another in a single column. It is possible to express the vec of a Kronecker matrix product in terms of a Kronecker product of vecs of matrices. The commutation matrix plays an important role here.
'Super' vec operators like vec $A \otimes$ vec A, vec$(A \otimes A)$ and vec$\{(A \otimes A)P_{nn}\}$ are very convenient. Several of their properties are being studied.
Both the traditional commutation matrix and vec operator and the newer concepts developed from these are applied to multivariate statistical and related problems.

1. THE COMMUTATION MATRIX

In the last few years, there has been an upsurge of interest for the Kronecker product permutation matrix in the statistical literature. This matrix changes the order of multiplication of a Kronecker matrix product. It is alternatively known as permuted identity matrix, commutation matrix, vec-permutation matrix or universal flip matrix.

Following Magnus and Neudecker (1979) we use the name commutation matrix.According to Henderson and Searle (1981) early use of this matrix was by Ledermann (1936) and Murnaghan

(1938), but it has been developed independently by numerous other researches, like Searle (1966), Tracy and Dwyer (1969), Malinvaud and Milleron (1970) and Browne (1974).

Its properties are extensively discussed by Hartwig and Morris (1975), Balestra (1976) and Magnus and Neudecker (1979). We use similar notation to Balestra (1976), writing $P_{m,n}$ as P_{mn} but retaining the comma when deemed necessary for clarity.

The reader is assumed to be familiar with the basic properties of Kronecker products. See e.g. Neudecker (1969) and Magnus and Neudecker (1979). We want to mention

$$\mathrm{tr}(A \otimes B) = \mathrm{tr}\, A . \mathrm{tr}\, B \tag{1.1}$$

for square matrices A and B of any order and the fundamental connection between vecs and Kronecker products

$$\mathrm{vec}\, ABC = (C' \otimes A)\mathrm{vec}\, B. \tag{1.2}$$

Further

$$a \otimes b = \mathrm{vec}\, ba' , \tag{1.3}$$

a special case of (1.2).

The basic properties of the commutation matrix are

$$P_{nm}\, \mathrm{vec}\, A = \mathrm{vec}\, A' \tag{1.4}$$

$$P_{mp}(A \otimes B)P_{qn} = B\otimes A,\ P_{mp}\, A\otimes B = (B\otimes A)P_{nq}, \tag{1.5}$$

A and B being (m,n) and (p,q) matrices, respectively. In fact (1.5) follows from the operational definition of P_{nm} (1.4), through (1.2).

A descriptive definition of the commutation matrix P_{nm} of order (mn,mn) is as follows. It consists of an array of m×n blocks each of n×m matrix. The $(i,j)^{th}$ block has a unit element in the $(j,i)^{th}$ position and zeros elsewhere, and will be written as $H_{ij}' = e_j^n(e_i^m)'$, where e_i^m is the ith unit column vector of order m.

Hence

$$P_{nm} = \sum_{i=1}^{m} \sum_{j=1}^{n} (H_{ij} \otimes H_{ij}'). \tag{1.6}$$

The following properties of P_{nm} are of interest:

$$P_{nm}' - P_{mn},\ P_{nm} P_{mn} = I_{mn},\ P_{n1} = P_{1n} = I_n. \tag{1.7}$$

It is obvious that commutation matrices of three indices can be used to reverse the orders of Kronecker products of three matrices. We mention:

$$P_{m,rs} = (I_r \otimes P_{ms})(P_{mr} \otimes I_s). \tag{1.8}$$

Clearly

$$P_{m,rs} P_{r,ms} = P_{mr,s}. \tag{1.9}$$

Interchanging indices yields additional equalities in (1.8) and (1.9). (We can interchange the indices r and s, m and s, m and r, respectively.)

Further

$$\begin{aligned}(\text{vec } B)'\text{vec } A &= \text{tr } AB' = \text{tr } P_{nm}(B' \otimes A) = \text{tr}(A \otimes B')P_{mn} \\ &= \text{tr}(B \otimes A')P_{mn} = \text{tr } P_{nm}(A' \otimes B),\end{aligned} \tag{1.10}$$

where A and B are (m,n)-matrices. For proofs see Balestra (1976, Section 2.3), Henderson and Searle (1979) or Magnus and Neudecker (1979).

2. STATISTICAL RESULTS SOME OF WHICH INVOLVE THE COMMUTATION MATRIX

We start with two results of an elementary nature.

Lemma 2.1

Let the (n,1)-vector x and the (m,1)-vector y be jointly distributed with $E(x) = \mu_1$, $E(y) = \mu_2$ and $E\{(y-Ey)(x-Ex)'\} = V_{21}$.

Then

$$E(x \otimes y) = \text{vec } V_{21} + \mu_1 \otimes \mu_2 \; .$$

Proof: $E(x \otimes y) = E(\text{vec } yx') = \text{vec } E(yx') = \text{vec}(V_{21} + \mu_2\mu_1')$

$$= \text{vec } V_{21} + \mu_1 \otimes \mu_2 \text{ , by virtue of (1.3).}$$

Note: Magnus and Neudecker (1979) derived the result for jointly normally distributed x and y. (Theorem 4.3i).

Lemma 2.2

Let the (n,1)-vector u have components u_i, with $E(u_i) = 0$, $E(u_i^2) = \sigma^2$ and $E(u_i^4) = \varphi^4$, u_i and u_j, $j \neq i$ being stochastically independent. Then

(i) $E(uu' \otimes uu') = \sigma^4\{I_{n^2} + P_{nn} + \text{vec } I_n(\text{vec } I_n)'\} + (\varphi^4 - 3\sigma^4) \sum_i (E_{ii} \otimes E_{ii})$.

(ii) $\mathfrak{D}(u \otimes u) - \sigma^4(I_{n^2} + P_{nn}) + (\varphi^4 - 3\sigma^4) \sum_i (E_{ii} \otimes E_{ii})$

(iii) the rank of $\mathfrak{D}(u \otimes u)$ is $\frac{1}{2}\, n(n + 1)$

(iv) the trace of $\mathfrak{D}(u \otimes u)$ is $n^2\sigma^4 + n(\varphi^4 - 2\sigma^4)$.

Here $E_{ij} = e_i^n(e_j^n)'$, where e_i^n is the ith unit column vector of order n.

Proof: (i) $E(u_i^2 uu') = \sigma^4 I_{n^2} + (\varphi^4 - \sigma^4)E_{ii}$

$E(u_i u_j uu') = \sigma^4(E_{ij} + E_{ji})$ for $i \neq j$

Hence the result follows immediately.

(ii)
$$\begin{aligned}\mathfrak{D}(u \otimes u) &= E\{(u \otimes u)(u \otimes u)'\} - E(u\otimes u)\{E(u\otimes u)\}' \\ &= E(uu' \otimes uu') - \text{vec } E(uu')\{\text{vec } E(uu')\}' \\ &= \sigma^4\{I_{n^2}+P_{nn}+\text{vec } I_n(\text{vec } I_n)'\} + \\ &+ (\varphi^4 - 3\sigma^4) \sum_i (E_{ii} \otimes E_{ii}) \\ &- \sigma^4 \text{ vec } I_n(\text{vec } I_n)' \; .\end{aligned}$$

(iii) Consider the matrix $I_{n^2} + P_{nn} + \lambda\sum_i(E_{ii} \otimes E_{ii})$.

It has the following set of eigenvalues and corresponding orthonormal eigenvectors:

$$2+\lambda\ ,\ e_i \otimes e_i \qquad (i=1,\ldots,n)$$

$$2\ ,\ \frac{1}{\sqrt{2}}(e_i \otimes e_j + e_j \otimes e_i) \qquad (i \neq j)$$

$$0\ ,\ \frac{1}{\sqrt{2}}(e_i \otimes e_j - e_j \otimes e_i) \qquad (i \neq j)$$

Its rank is thus $n + \frac{1}{2}n(n-1) = \frac{1}{2}n(n+1)$.

Hence $\mathfrak{D}(u \otimes u)$ has the same rank.

(iv) follows by direct verification.

Note: Magnus and Neudecker (1979) derived similar results to (i) and (ii) for normally distributed u. (Theorem 4.1i and Lemma 4.1.ii). These follow from the above given lemma for $\varphi^4 - 3\sigma^4$.

Finally we have

Lemma 2.3.

Let x and y be two stochastically independent vectors, of orders (p,1) and (q,1) respectively. Let further $E(xx') = V_1$, $E(yy') = V_2$, $E(x) = \mu_1$ and $E(y) = \mu_2$.

Then

(i) $E(xy' \otimes xy') = \text{vec}\, V_1(\text{vec}\, V_2)'$;

(ii) $E(xy' \otimes yx') = (V_1 \otimes V_2)P_{qp} = P_{qp}(V_2 \otimes V_1)$;

(iii) $E(x \otimes x \otimes y \otimes y) = \text{vec}\, V_1 \otimes \text{vec}\, V_2$;

(iv) $E(x \otimes y \otimes x \otimes y) = (I_p \otimes P_{pq} \otimes I_q)(\text{vec}\, V_1 \otimes \text{vec}\, V_2)$;

(v) $E(x \otimes y \otimes y \otimes x) = (I_p \otimes P_{p,qq})(\text{vec}\, V_1 \otimes \text{vec}\, V_2)$;

(vi) $\mathfrak{D}(x \otimes y) = \mathfrak{D}(x) \otimes \mathfrak{D}(y) + \mathfrak{D}(x) \otimes \mu_2\mu_2' + \mu_1\mu_1' \otimes \mathfrak{D}(y)$

$$= V_1 \otimes V_2 - \mu_1\mu_1' \otimes \mu_2\mu_2'\ .$$

Proof: (i) $E(xy'\otimes xy') = E\{(x\otimes x)(y\otimes y)'\} = E\{\text{vec}\, xx'(\text{vec}\, yy')'\}$

$$= \text{vec}\, E(xx')\{\text{vec}\, E(yy')\}' = \text{vec}\, V_1(\text{vec}\, V_2)';$$

(ii) $E(xy'\otimes yx') = E\{(x\otimes y)(y\otimes x)'\} = E\{(x\otimes y)(x\otimes y)'P_{qp}\}$

$$= E(xx'\otimes yy')P_{pq} = (V_1\otimes V_2)P_{qp} = P_{qp}(V_2\otimes V_1).$$

(iii) $E(x \otimes x \otimes y \otimes y) = E(\text{vec}\ xx' \otimes \text{vec}\ yy') =$

$= \text{vec}\ E(xx') \otimes \text{vec}\ E(yy') =$

$= \text{vec}\ V_1 \otimes \text{vec}\ V_2$;

(iv) $E(x \otimes y \otimes x \otimes y) = (I_p \otimes P_{pq} \otimes I_q)\ E(x \otimes x \otimes y \otimes y)$

$= (I_p \otimes P_{pq} \otimes I_q)(\text{vec}\ V_1 \otimes \text{vec}\ V_2)$;

(v) $E(x \otimes y \otimes y \otimes x) = (I_{pq} \otimes P_{pq})\ E(x \otimes y \otimes x \otimes y)$

$= (I_p \otimes I_q \otimes P_{pq})(I_p \otimes P_{pq} \otimes I_q)(\text{vec}\ V_1 \otimes \text{vec}\ V_2)$

$= (I_p \otimes P_{p,qq})(\text{vec}\ V_1 \otimes \text{vec}\ V_2)$;

(vi) $\mathfrak{D}(x \otimes y) = E\{(x \otimes y)(x \otimes y)'\} - E(x \otimes y)\{E(x \otimes y)\}'$

$= E(xx' \otimes yy') - (Ex \otimes Ey)(Ex \otimes Ey)'$

$= E(xx') \otimes E(yy') - (Ex \otimes Ey)(Ex \otimes Ey)'$

$= V_1 \otimes V_2 - \mu_1\mu_1' \otimes \mu_2\mu_2'$

$= (V_1 - \mu_1\mu_1') \otimes (V_2 - \mu_2\mu_2') + (V_1 - \mu_1\mu_1') \otimes \mu_2\mu_2'$

$+ \mu_1\mu_1' \otimes (V_2 - \mu_2\mu_2') = \mathfrak{D}(x) \otimes \mathfrak{D}(y) + \mathfrak{D}(x) \otimes \mu_2\mu_2'$

$+ \mu_1\mu_1' \otimes \mathfrak{D}(y)$.

We used (1.3), (1.5) and (1.8).

3. STACKING A KRONECKER MATRIX PRODUCT

The aim of this section is to express the vec of a Kronecker product of two matrices in terms of the Kronecker product of their vecs.

We shall consider vec $(A \otimes B)$, vec $(A' \otimes B)$ and vec $(A \otimes B')$. To accomplish this we need the following results on commutation matrices with four indices:

$$P_{mnp,q} = (P_{mq} \otimes I_{np})(I_m \otimes P_{nq} \otimes I_p)(I_{mn} \otimes P_{pq}); \qquad (3.1)$$

$$P_{mn,pq} = (I_p \otimes P_{mq} \otimes I_n)(P_{mp} \otimes P_{nq})(I_m \otimes P_{np} \otimes I_q) \qquad (3.2)$$

These results can be proved by applying the commutation matrices $P_{mnp,q}$ and $P_{mn,pq}$ to the Kronecker product $a \otimes b \otimes c \otimes d$, the column vectors being of orders m,n,p and q respectively. We give the detailed proof of (3.1):

Proof: $P_{mnp,q}(a \otimes b \otimes c \otimes d) = d \otimes a \otimes b \otimes c =$

$$= (P_{mq} \otimes I_{np})(a \otimes d \otimes b \otimes c)$$

$$= (P_{mq} \otimes I_{np})(I_m \otimes P_{nq} \otimes I_p)(a \otimes b \otimes d \otimes c)$$

$$= (P_{mq} \otimes I_{np})(I_m \otimes P_{nq} \otimes I_p)(I_{mn} \otimes P_{pq})(a \otimes b \otimes c \otimes d),$$

for which (3.1) follows.

Note: This proof is due to an unknown referee.

Let us partition $A = (a_1, \ldots, a_n)$, a_i being the ith column of A. Clearly $\text{vec}(A \otimes b) = \text{vec}(a_1 \otimes b \ldots a_n \otimes b) = \text{vec } A \otimes b$. However, $\text{vec } b \otimes A \neq b \otimes \text{vec } A$.

The stacking of Kronecker matrix products takes place according to the results contained in the following theorem.

Theorem 3.1.

Let A and B be arbitrary (m,n) and (p,q)-matrices respectively. Then

(i) $\text{vec}(A \otimes B) = (I_n \otimes P_{mq} \otimes I_p)(\text{vec } A \otimes \text{vec } B)$;

(ii) $\text{vec}(A' \otimes B) = (P_{n,mq} \otimes I_p)(\text{vec } A \otimes B) =$

$$= [P_{n,mq}(I_n \otimes P_{qp}) \otimes I_p]\text{vec}(A \otimes B);$$

(iii) $\text{vec}(A \otimes B') = (I_n \otimes P_{mq,p})(\text{vec } A \otimes \text{vec } B) =$

$$= [I_n \otimes P_{mq,p}(P_{qm} \otimes I_p)]\text{vec}(A \otimes B).$$

Proof: (i) Let again a_i be the ith column of A.

Then $\text{vec}(a_i \otimes B) = \text{vec}\{P_{pm}(B \otimes a_i)\} =$

$$= (I_q \otimes P_{pm})\text{vec}(B \otimes a_i)$$

$= (I_q \otimes P_{pm})(\text{vec } B \otimes a_i) = (I_q \otimes P_{pm})P_{m,pq}(a_i \otimes \text{vec } B)$

$= (P_{mq} \otimes I_p)(a_i \otimes \text{vec } B)$, by virtue of (1.5), (1.7), (1.2) and (1.8).

Further, $\text{vec}(A \otimes B) = \text{vec}(a_1 \otimes B \ldots a_n \otimes B)$

$= \text{vec}[\text{vec}(a_1 \otimes B) \ldots \text{vec}(a_n \otimes B)]$

$= \text{vec}[(P_{mq} \otimes I_p)(a_1 \otimes \text{vec } B) \ldots (P_{mq} \otimes I_p)(a_n \otimes \text{vec } B)]$

$= \text{vec}[(P_{mq} \otimes I_p)(a \otimes \text{vec } B \ldots a_n \otimes \text{vec } B)]$

$= (I_n \otimes P_{mq} \otimes I_p) \text{ vec}(a_1 \otimes \text{vec } B \ldots a_n \otimes \text{vec } B)$

$= (I_n \otimes P_{mq} \otimes I_p)(\text{vec } A \otimes \text{vec } B).$

(ii) $\text{vec}(A' \otimes B) = (I_m \otimes P_{nq} \otimes I_p)(\text{vec } A' \otimes \text{vec } B)$

$= (I_m \otimes P_{nq} \otimes I_p)(P_{nm} \otimes I_{pq})(\text{vec } A \otimes \text{vec } B)$

$= [(I_m \otimes P_{nq})(P_{nm} \otimes I_q) \otimes I_p](\text{vec } A \otimes \text{vec } B)$

$= (P_{n,mq} \otimes I_p)(\text{vec } A \otimes \text{vec } B)$

$= (P_{n,mq} \otimes I_p)(I_n \otimes P_{qm} \otimes I_p)\text{vec}(A \otimes B),$

by virtue of (1.10) and (i).

(iii) $\text{vec}(A \otimes B') = P_{mq,np} \text{ vec}(A' \otimes B) =$

$= P_{mq,np}(P_{n,mq} \otimes I_p)(\text{vec } A \otimes \text{vec } B)$

$= (I_n \otimes P_{mq,p})(P_{mq,n} \otimes I_p)(P_{n,mq} \otimes I_p)(\text{vec } A \otimes \text{vec } B)$

$= (I_n \otimes P_{mq,p})(\text{vec } A \otimes \text{vec } B) =$

$= [I_n \otimes P_{mq,p}(P_{qm} \otimes I_p)]\text{vec}(A \otimes B),$

by (1.4), (ii), (1.8) and (1.7).

Theorem 3.1 permits an interesting extension of (1.1) to <u>rectangular</u> A and B, but with $A \otimes B$ still square.

<u>Theorem 3.2.</u>

Let A and B be arbitrary (m,n) and (p,q)-matrices, respectively, such that mp = nq.

Then

$$\text{tr}(A \otimes B) = (\text{vec } I_n \otimes \text{vec } I_q)'(\text{vec } A \otimes \text{vec } B').$$

<u>Proof</u>: $\text{tr}(A \otimes B) = (\text{vec } I_{mp})' \text{ vec } (A \otimes B) =$

$= (\text{vec } I_{mp})'(I_n \otimes P_{mq} \otimes I_p)(\text{vec } A \otimes \text{vec } B)$

$=(\text{vec } I_{mp})'(I_n \otimes P_{mq} \otimes I_p)(I_{mn} \otimes P_{pq})(\text{vec } A \otimes \text{vec } B')$

$=(\text{vec } I_{mp})'(I_n \otimes P_{mp,q})(\text{vec } A \otimes \text{vec } B')$

$=(\text{vec } I_{nq})'(I_n \otimes P_{nq,q})(\text{vec } A \otimes \text{vec } B')$

$=(\text{vec } I_n \otimes \text{vec } I_q)'(\text{vec } A \otimes \text{vec } B')$, because mp = nq, and by virtue of (1.10), Theorem 2.1 (iii) and (i), (1.8) and (1.3).

<u>Note</u>: This theorem is due to Pietro Balestra, who suggested it to the second author.

It follows from Theorem 3.2 that

$$\mathrm{tr}(A \otimes B) = (\mathrm{vec}\ I_n \otimes \mathrm{vec}\ I_q)'(I_{mn} \otimes P_{qp})(\mathrm{vec}\ A \otimes \mathrm{vec}\ B), \tag{3.3}$$

in case A and B satisfy the conditions of the theorem.

For m=n and p=q (3.3) boils down to (1.1), through (1.10).

4. SUPER VEC OPERATORS

In this section we shall study three operators vec A ⊗ vec A, vec(A ⊗ A) and vec{(A ⊗ A)P_{nn}}, where A is an arbitrary (m,n)-matrix. These will be applied to the expectations of expressions like $x_i \otimes x_j \otimes x_k \otimes x_\ell$ with two pairs of identical indices, the vectors being independently $N_n(o,V)$ distributed.

We shall introduce the 'super' vec operators of order $(m^2n^2,1)$.

$$w_1(A) = \mathrm{vec}\ A \otimes \mathrm{vec}\ A = \mathrm{vec}\{\mathrm{vec}\ A(\mathrm{vec}\ A)'\}; \tag{4.1}$$

$$w_2(A) = \mathrm{vec}(A \otimes A); \tag{4.2}$$

$$w_3(A) = \mathrm{vec}\{(A \otimes A)P_{nn}\}. \tag{4.3}$$

We do not follow Bellman's (1960, p. 228) notation $A^{[k]} = A \otimes \ldots \otimes A$ (k times) for reasons of economy. It is then easy to see that

$$w_2(A) = (I_n \otimes P_{mn} \otimes I_m)\ w_1(A); \tag{4.4}$$

$$w_3(A) = (P_{nn} \otimes P_{m^2})\ w_2(A) = (I_{n^2} \otimes P_{mm})\ w_2(A). \tag{4.5}$$

Result (4.4) follows from Theorem 3.1(i).

A more intricate result is contained in the following lemma.

Lemma 4.1.

Let A be an (m,n)-matrix and B a (p,q)-matrix.

Then

(i) $\mathrm{vec}\{(A' \otimes B)P_{qm}\} = (P_{n,mq} \otimes I_p)\ \mathrm{vec}(A \otimes B)$

$= (I_q \otimes P_{nm} \otimes I_p)\mathrm{vec}\{(A \otimes B)P_{qn}\};$

(ii) $\mathrm{vec}\{(A' \otimes A)P_{nm}\} = (P_{n,mn} \otimes I_m)\ w_2(A)$

$= (I_n \otimes P_{nm} \otimes I_m)\ w_3(A);$

if A is a symmetric (m,m)-matrix, then

(iii) $(I_m \otimes P_{mm} \otimes I_m)\ w_3(A) = W_3(A).$

Proof: (i) $\{(P_{mq} \otimes I_n)P_{n,mq}(I_n \otimes P_{qm}) \otimes I_p\}\ \mathrm{vec}(A \otimes B)$

$= (P_{n,mq} \otimes I_p)\ \mathrm{vec}(A \otimes B) =$

$=\{(I_q \otimes P_{nm})(P_{nq} \otimes I_m) \otimes I_p\}\ \mathrm{vec}(A \otimes B)$

$= (I_q \otimes P_{nm} \otimes I_p)(P_{nq} \otimes I_{mp})\ \mathrm{vec}(A \otimes B)$

$= (I_q \otimes P_{nm} \otimes I_p)\ \mathrm{vec}\{(A \otimes B)P_{qn}\}$, by virtue of (1.2), Theorem 3.1(ii), (1.5) and (1.8);

(ii) Put B=A, p=m and q=n in (i);

(iii) Put A'=A and m=n in (ii).

5. STATISTICAL APPLICATIONS

We now present more applications, in addition to the ones given in Section 2.

Application 1.

Let vec X and vec Y be jointly distributed with

$$E(X) = M_1,\ E(Y) = M_2 \text{ and } \mathfrak{D}\{\mathrm{vec}(X:Y)\} = \begin{pmatrix} V_{11} & V'_{21} \\ V_{21} & V_{22} \end{pmatrix},$$

X and Y being (m,n) and (p,q)-matrices, respectively, where : joins the matrices X and Y

(i) $E\ \mathrm{vec}(X \otimes Y) = (I_n \otimes P_{mq} \otimes I_p)(\mathrm{vec}\ V_{21} + \mathrm{vec}\ M_1 \otimes \mathrm{vec}\ M_2)$;

(ii) $E\ \mathrm{vec}(X \otimes X) = (I_n \otimes P_{mn} \otimes I_m)(\mathrm{vec}\ V_{11} + \mathrm{vec}\ M_1 \otimes \mathrm{vec}\ M_1)$;

(iii) $E\ \mathrm{vec}(Y \otimes Y) = (I_q \otimes P_{pq} \otimes I_p)(\mathrm{vec}\ V_{22} + \mathrm{vec}\ M_2 \otimes \mathrm{vec}\ M_2)$.

Proof: Follows from Lemma 2.1 and Theorem 3.1(i).

Application 2.

Let vec X and vec Y be stochastically independent vectors, X and Y being (m,n) and (p,q)-matrices respectively. Let further $E(X) = M_1$, $E(Y) = M_2$, $\mathfrak{D}(\mathrm{vec}\ X) = V_{11}$ and $\mathfrak{D}(\mathrm{vec}\ Y) = V_{22}$.

Then

$$\mathfrak{D}\ \mathrm{vec}(X \otimes Y) = (I_n \otimes P_{mq} \otimes I_p\ \ V_{11} \otimes V_{22} + V_{11} \otimes \mathrm{vec}\ M_2(\mathrm{vec}\ M_2)' + \\ + \mathrm{vec}\ M_1(\mathrm{vec}\ M_1)' \otimes V_{22}).(I_n \otimes P_{qm} \otimes I_p).$$

Proof: Follows from Lemma 2.3(vi) and Theorem 3.1(i).

Application 3.

Let X be normally distributed, with $E(X) = M$ and $\mathfrak{D}(\mathrm{vec}\ X) = V$, X being an (m,n)-matrix. Then

$$\mathfrak{D}\ \mathrm{vec}(X \otimes X) = (I_{m^2n^2} + P_{nn} \otimes P_{mm})(I_n \otimes P_{mn} \otimes I_m) \\ (V \otimes V + V \otimes \mathrm{vec}\ M(\mathrm{vec}\ M\ ') + \\ + \mathrm{vec}\ M(\mathrm{vec}\ M)' \otimes V)(I_n \otimes P_{nm} \otimes I_m).$$

Proof: Follows from Theorem 3.1(i), Theorem 4.3(iv) in Magnus and Neudecker (1979) and (3.2).

Application 4.

Let x and y be stochastically independent $N_k(o,V)$-distributed vectors, with $|V| \neq 0$.

Then

(i) $E(x \otimes x \otimes x \otimes x) = w_1(V) + w_2(V) + w_3(V)$,

(ii) $E(x \otimes x \otimes y \otimes y) = w_1(V)$;

(iii) $E(x \otimes y \otimes x \otimes y) = w_2(V)$;

(iv) $E(x \otimes y \otimes y \otimes x) = w_3(V)$.

Proof: (i) $E(x \otimes x \otimes x \otimes x) = E(\text{vec } xx' \otimes \text{vec } xx') =$

$= (I_k \otimes P_{kk} \otimes I_k)\, E\, \text{vec}(xx' \otimes xx')$

$= (I_k \otimes P_{kk} \otimes I_k)\, \text{vec}\{(I_{k^2} + P_{kk})(V \otimes V) + \text{vec } V(\text{vec } V)'\}$

$= (I_k \otimes P_{kk} \otimes I_k)\{w_1(V) + w_2(V) + w_3(V)\}$

$= w_1(V) + w_2(V) + w_3(V)$, by virtue of Theorem 3.1(i), Theorem 4.3(iv) in Magnus and Neudecker (1979), (4.1), (4.2), (4.3), (4.4) and Lemma 4.1(iii);

(ii) follows from Lemma 2.3(iii) and (4.1);

(iii) follows from Lemma 2.3(iv) and (4.4);

(iv) follows from Lemma 2.3(v), (1.9), (4.4) and (4.5).

Application 5: (A result of Bentler and Lee (1978))

Let Z be a (p,q)-matrix. Let X and Y be (m,n) and (r,s)-matrices, respectively, each of whose elements is a differentiable function of all pq elements of Z.

Then

$$\frac{\partial(X \otimes Y)}{\partial Z} = \left\{\frac{\partial X}{\partial Z} \otimes (\text{vec } Y')' + (\text{vec } X')' \otimes \frac{\partial Y}{\partial Z}\right\}(I_m \otimes P_{rn} \otimes I_s)$$

$$= \left\{\frac{\partial X}{\partial Z}\,(I_{mn} \otimes \text{vec } Y')' + \frac{\partial Y}{\partial Z}(\text{vec } X' \otimes I_{rs})'\right\}(I_m \otimes P_{rn} \otimes I_s),$$

where $\frac{\partial X}{\partial Z} = \left\{\frac{\partial \text{ vec } X}{\partial \text{ vec } Z}\right\}' P_{mn}$ and $\frac{\partial \text{ vec } X}{\partial \text{ vec } Z}$ follows from the definition of $\frac{\partial x}{\partial z}$ whose $(i,j)^{th}$ element is $\frac{\partial x_i}{\partial z_j}$.

Proof: Following a procedure as given by the first author (1969) we get $d\ \text{vec}(X \otimes Y) = \text{vec}(dX \otimes Y) + \text{vec}(X \otimes dY)$

$$= (I_n \otimes P_{ms} \otimes I_r)(\text{vec}\ dX \otimes \text{vec}\ Y + \text{vec}\ X \otimes \text{vec}\ dY)$$

$$= (I_n \otimes P_{ms} \otimes I_r)\left\{\frac{\partial\ \text{vec}\ X}{\partial\ \text{vec}\ Z}\ d\ \text{vec}\ Z \otimes \text{vec}\ Y + \text{vec}\ X \otimes \frac{\partial\ \text{vec}\ Y}{\partial\ \text{vec}\ Z}\ d\ \text{vec}\ Z\right\}$$

$$= (I_n \otimes P_{ms} \otimes I_r)\left\{\frac{\partial\ \text{vec}\ X}{\partial\ \text{vec}\ Z} \otimes \text{vec}\ Y + \text{vec}\ X \otimes \frac{\partial\ \text{vec}\ Y}{\partial\ \text{vec}\ Z}\right\}\ d\ \text{vec}\ Z,$$

from which follows

$$\frac{\partial\ \text{vec}(X \otimes Y)}{\partial\ \text{vec}\ Z} = (I_n \otimes P_{ms} \otimes I_r)\ \left\{\frac{\partial\ \text{vec}\ X}{\partial\ \text{vec}\ Z} \otimes \text{vec}\ Y + \text{vec}\ X \otimes \frac{\partial\ \text{vec}\ Y}{\partial\ \text{vec}\ Z}\right\}$$

and

$$\frac{\partial(X \otimes Y)}{\partial Z} = \left[\left(\frac{\partial\ \text{vec}\ X}{\partial\ \text{vec}\ Z}\right)' \otimes (\text{vec}\ Y)' + (\text{vec}\ X)' \otimes \left(\frac{\partial\ \text{vec}\ Y}{\partial\ \text{vec}\ Z}\right)'\right](I_n \otimes P_{sm} \otimes I_r)P_{mr,n}$$

$$= \left\{\frac{\partial X}{\partial Z}\ P_{nm} \otimes (\text{vec}\ Y')'P_{st} + (\text{vec}\ X')'P_{nm} \otimes \frac{\partial Y}{\partial Z}\ P_{sr}\right\}(I_n \otimes P_{sm} \otimes I_r)\ P_{mr,ns}$$

$$= \left\{\frac{\partial X}{\partial Z} \otimes (\text{vec}\ Y')' + (\text{vec}\ X')' \otimes \frac{\partial Y}{\partial Z}\right\}\ (P_{nm} \otimes P_{sr})(I_n \otimes P_{sm} \otimes I_r)P_{mr,ns}$$

$$= \left\{\frac{\partial X}{\partial Z} \otimes (\text{vec}\ Y')' + (\text{vec}\ X')' \otimes \frac{\partial Y}{\partial Z}\right\}\ (I_m \otimes P_{rn} \otimes I_s)\ .$$

The last line then follows from $\frac{\partial X}{\partial Z} \otimes (\text{vec}\ Y')' = \frac{\partial X}{\partial Z}(I_{pq} \otimes \text{vec}\ Y')'$ and

$$(\text{vec}\ X')' \otimes \frac{\partial Y}{\partial Z} = \frac{\partial Y}{\partial Z}\ (I_{pq} \otimes \text{vec}\ X')'\ P_{mn,pq} = \frac{\partial Y}{\partial Z}(\text{vec}\ X' \otimes I_{rs})'$$

We applied Theorem 3.1(i), (3.2) and (1.4). Bentler and Lee (1978) applied their Theorem 4 to a specific statistical problem which we shall not go into. In fact our Application 5 replaces Lemmas 2 through 6 and Theorem 4 of Bentler and Lee. A similar result can be obtained more easily by using two alternative definitions of matrix derivatives, viz. $\frac{\partial Y}{\partial Z} = \left[\frac{\partial y_{kl}}{\partial Z}\right]$ or $\frac{\partial Y}{\partial Z} = \left[\frac{\partial Y}{\partial z_{ij}}\right]$. We present without proof:

$$\frac{\partial(X \otimes Y)}{\partial Z} = (I_m \otimes P_{pr})\left(\frac{\partial X}{\partial Z} \otimes Y\right)(I_n \otimes P_{sq}) + X \otimes \frac{\partial Y}{\partial Z}$$

$$\frac{\partial(X \otimes Y)}{\partial Z} = \frac{\partial X}{\partial Z} \otimes Y + (I_p \otimes P_{rm})\left(\frac{\partial Y}{\partial Z} \otimes X\right)(I_q \otimes P_{ns}).$$

Proofs can be found in Neudecker (1981).

Application 6.

Let the (n,k)-matrix Y have columns $y_1 \ldots y_k$, being stochastically independent $N_n(0,V)$-distributed vectors.

Then $S = YY' \sim W_n(k,V)$, $E(S) = kV$ and $\mathcal{D}(\text{vec } S) = n(I_{n^2} + P_{nn})(V \otimes V)$ from Corollary 4.2 of Magnus and Neudecker (1979).

The following result can then be proved:

$$E(Y \otimes Y \otimes Y \otimes Y) = (\text{vec } V \otimes \text{vec } V)(\text{vec } I_k \otimes \text{vec } I_k)'$$
$$+ \text{vec}(V \otimes V)\{\text{vec}(I_k \otimes I_k)\}' + \text{vec}\{P_{nn}(V \otimes V)\}(\text{vec } P_{kk})'.$$

Proof: We shall first develop $E\{\text{vec}(Y \otimes Y)\}$ and $\mathcal{D}\{\text{vec}(Y \otimes Y)\}$.

$$E\{\text{vec}(Y \otimes Y)\} = (I_k \otimes P_{nk} \otimes I_n)E(\text{vec } Y \otimes \text{vec } Y)$$
$$= (I_k \otimes P_{nk} \otimes I_n) \text{ vec } E\{\text{vec } Y(\text{vec } Y)'\}$$
$$=(I_k \otimes P_{nk} \otimes I_n)\text{vec}\mathcal{D}(\text{vec } Y)=(I_k \otimes P_{nk} \otimes I_n) \text{ vec}(I_k \otimes V)$$
$$= \text{vec } I_k \otimes \text{vec } V.$$ Further

$$\mathcal{D}\{\text{vec}(Y \otimes Y)\} =(I_k \otimes P_{nk} \otimes I_n)\mathcal{D}(\text{vec } Y \otimes \text{vec } Y)(I_k \otimes P_{kn} \otimes I_n)$$
$$=(I_k \otimes P_{nk} \otimes I_n)(I_{n^2k^2}+P_{nk,nk})(I_k \otimes V \otimes I_k \otimes V)(I_k \otimes P_{kn} \otimes I_n)$$
$$= (I_{n^2k^2} + P_{kk} \otimes P_{nn})(I_{k^2} \otimes V \otimes V).$$

Therefore $\text{vec } E(Y \otimes Y \otimes Y \otimes Y) = (I_{k^2} \otimes P_{n^2,k^2} \otimes I_{n^2})E\{\text{vec}(Y \otimes Y) \otimes \text{vec}(Y \otimes Y)\}$

$$= (I_{k^2} \otimes P_{n^2,k^2} \otimes I_{n^2}) \text{ vec } \mathcal{D}\{\text{vec}(Y \otimes Y)\}$$
$$+ (I_{k^2} \otimes P_{n^2,k^2} \otimes I_{n^2}) \text{ vec}[E\{\text{vec}(Y \otimes Y)\}E\{\text{vec}(Y \otimes Y)\}']$$
$$= (I_{k^2} \otimes P_{n^2,k^2} \otimes I_{n^2}) \text{ vec}(I_k \otimes I_k \otimes V \otimes V)$$
$$+ (I_{k^2} \otimes P_{n^2,k^2} \otimes I_{n^2}) \text{ vec}\{(P_{kk} \otimes P_{nn})(I_{k^2} \otimes V \otimes V)\}$$

$$+ (I_{k^2} \otimes P_{n^2,k^2} \otimes I_{n^2})(\text{vec } I_k \otimes \text{vec } V \otimes \text{vec } I_k \otimes \text{vec } V)$$

$$= \text{vec}(I_k \otimes I_k) \otimes \text{vec}(V \otimes V) + \text{vec } P_{kk} \otimes \text{vec } P_{nn}(V \otimes V)$$

$$+ \text{vec } I_k \otimes \text{vec } I_k \otimes \text{vec } V \otimes \text{vec } V.$$

Hence $E(Y \otimes Y \otimes Y \otimes Y) = \text{vec}(V \otimes V)\{\text{vec}(I_k \otimes I_k)\}' +$

$$+ \text{vec}\{P_{nn}(V \otimes V)\}(\text{vec } P_{kk})' +$$

$$+ (\text{vec } V \otimes \text{vec } V)(\text{vec } I_k \otimes \text{vec } I_k)'.$$

We used Theorem 3.1(i), (3.2), (1.5) and Theorem 4.3(iv) and (2.2) from Magnus and Neudecker (1979).

ACKNOWLEDGEMENT

We are grateful to Pietro Balestra for suggesting Theorem 3.2 and supplying a (different) proof.
We want to express our gratitude to an unknown referee whose valuable comments inspired us to improve our paper in content and presentation.

REFERENCES

BALESTRA, P. (1976). La dérivation matricielle. Collection de l'IME, no. 12. Sirey, Paris.

BELLMAN, R. (1960). Introduction to Matrix Analysis. McGraw-Hill, New York.

BENTLER, P.M. and LEE, SIK-YUM (1978). Matrix Derivatives with Chain Rule and Rules for Simple, Hadamard, and Kronecker Products. Journal of Mathematical Psychology 17, 255-262.

BROWNE, M.W. (1974). Generalized Least-Squares Estimators in the Analysis of Covariance Structures, South African Statistical Journal 8, 1-24 .

HARTWIG, R.E. and S. BRENT MORRIS (1975). The Universal Flip Matrix and the Generalized Faro-Shuffle. Pacific Journal of Mathematics 50, 445-455.

HENDERSON, H.V. and S.R. SEARLE (1981). The Vec-Permutation Matrix, the Vec Operator and Kronecker Products: A Review. Linear and Multilinear Algebra, 9, 271-288.

LEDERMANN, W. (1936). On singular pencils of Zehfuss, compound, and Schläflian matrices. Proceedings of the Royal Society of Edinburgh, LVI, 50-89.

MAGNUS, J.R. and H. NEUDECKER (1979). The Commutation Matrix: Some Properties and Applications. The Annals of Statistics 7, 381-394.

MALINVAUD, E. and J.C. MILLERON (1970). Généralisation des modèles d'analyse de la variance dans le cas de séries temporelles de coupe instantanées. Conditions d'identifiabilité et méthodes d'estimation. INSEE Report 348/930. Paris.

MURNAGHAN, F.D. (1938). The Theory of Group Representations. Johns Hopkins Press, Baltimore.

NEUDECKER, H. (1969). Some Theorems on Matrix Differentiation with Special Reference to Kronecker Matrix Products. Journal of the American Statistical Association 64, 953-963.

NEUDECKER, H. (1981). Two germane matrix derivatives: properties and relationships. Paper no. AE 21/81, Faculty of Actuarial Sciences and Econometrics, University of Amsterdam.

SEARLE, S.R. (1966). Matrix Algebra for the Biological Sciences. Wiley, New York.

TRACY, D.S. and P.S. DWYER (1969). Multivariate Maxima and Minima with Matrix Derivatives. Journal of the American Statistical Association 64, 1576-1594.

Proc. of the 3rd Pannonian Symp. on Math. Stat., Visegrád, Hungary 1982
J. Mogyoródi, I. Vincze, W. Wertz, eds

OPTIMAL MULTIPLIERS FOR LINEAR CONGRUENTIAL PSEUDO-RANDOM NUMBERS: THE DECIMAL CASE

H. NIEDERREITER

Johann Wolfgang Goethe-Universität,
Frankfurt, FRG
Österreichische Akademie der Wissenschaften,
Vienna, Austria

1. INTRODUCTION

Linear congruential pseudo-random numbers (abbreviated LCPRN) are the most commonly used uniform random number generators. They are produced by the recursion

$$y_{n+1} \equiv \lambda y_n + r \bmod m \text{ for } n=0,1,\ldots, \tag{1}$$

where λ, r, m, and y_0 are integers, $\gcd(\lambda,m)=1$, and $0 \le y_n < m$, and they are normalized by setting $x_n = y_n/m \in [0,1]$. For the case where the modulus m is a prime or a prime power, effective criteria for the choice of the multiplier λ in order to guarantee desirable distribution and statistical independence properties of the pseudo-random numbers have already been established (see [2, Ch. 3], [4], [5]). In the present paper we discuss another case of practical importance, namely that of m being a (large) power of 10. As usual, we will have to distinguish the homogeneous case $r \equiv 0 \bmod m$ and the inhomogeneous case $r \not\equiv 0 \bmod m$.

The distribution and statistical independence properties of the LCPRN are measured according to a univariate resp. multivariate Kolmogorov test. For $s \ge 1$ and $N \ge 1$ let $\underline{t}_0, t_1, \ldots, t_{N-1}$ be points of $[0,1]^s$ and set

$$D_N(\underline{t}_0,\underline{t}_1,\ldots,\underline{t}_{N-1})=\sup|F_N(J)-V(J)|,$$

where the supremum is extended over all subintervals J of $[0,1]^s$, $F_N(J)$ is N^{-1} times the number of terms among $\underline{t}_0,\underline{t}_2,\ldots,\underline{t}_{N-1}$ falling into J, and V(J) denotes the volume of J. This test is applied to the points

$$\frac{1}{m}\underline{y}_n=\frac{1}{m}(y_n,y_{n+1},\ldots,y_{n+s-1}),$$

and in this case we write

$$D_N^{(s)}=D_N(\frac{1}{m}\underline{y}_0,\frac{1}{m}\underline{y}_1,\ldots,\frac{1}{m}\underline{y}_{N-1}).$$

For s=1 the quantity $D_N^{(s)}$ measures the deviation of the empirical distribution of the LCPRN from the uniform distribution on [0,1], and for $s\geq 2$ it measures the deviation from statistical independence among s successive LCPRN.

For moduli which are powers of 10, effective estimates will be given for $D_N^{(s)}$ in terms of s-dimensional "figures of merit" $\rho^{(s)}(\lambda,m)$, where λ is the multiplier and m is either the original modulus or a somewhat reduced modulus. These estimates lead to the criterion that a sequence of LCPRN has the property that any s successive terms are statistically almost independent if and only if the figure of merit $\rho^{(s)}(\lambda,m)$ is large. We use the definition of $\rho^{(s)}(\lambda,m)$ introduced in [7], which is slightly different from that in [5]. For any integer h set $r(h)=\max(1,2|h|)$, and for any lattice point $\underline{h}=(h\ ,\ldots,h_s)$, i.e. any point with integer coordinates, set

$$r(\underline{h})=r(h_1)\ldots r(h_s).$$

Then we define

$$\rho^{(s)}(\lambda,m)=\min r(\underline{h}), \qquad (2)$$

where the minimum is extended over all nonzero lattice points $\underline{h}$ with

$$\underline{h}\cdot\underline{\lambda}=h_1+h_2\lambda+\ldots+h_s\lambda^{s-1}\equiv 0 \bmod m. \qquad (3)$$

Here we use $\underline{\lambda}$ to denote the lattice point $(1,\lambda,\ldots,\lambda^{s-1})$.

For $m\geq 2$ let $C(m)$ denote the set of nonzero lattice points $\underline{h}$ with $-m/2<h_i\leq m/2$ for $1\leq i\leq s$. Then for $s\geq 2$ it suffices to extend the minimum in (2) over all $\underline{h}\in C(m)$ with $\underline{h}\cdot\underline{\lambda}\equiv 0 \bmod m$. This is due to the fact that for $s\geq 2$ we have $\rho^{(s)}(\lambda,m)\leq 2m$, since the condition (3) can be satisfied by choosing $h_2=1$, $h_3=\ldots=h_s=0$, and $h_1\equiv-\lambda \bmod m$, $-m/2<h_1\leq m/2$, whereas $r(\underline{h})\geq 2m$ for all nonzero $\underline{h}\notin C(m)$ satisfying (3).

For $s\geq 1$, $m\geq 2$, and a positive integer d we define

$$R^{(s)}(\lambda,m,d)=\sum_{\substack{\underline{h}\in C(m)\\ \underline{h}\cdot\underline{\lambda}\equiv 0 \bmod d}}\frac{1}{r(\underline{h})}.$$

We recall some notation and a result from [4]. For $0<|h|\leq m/2$ we write $r(h,m)=m\sin\pi|h/m|$, and we put $r(0,m)=1$. Then for $\underline{h}=(h_1,\ldots,h_s)\in C(m)$ we set

$$r(\underline{h},m)=r(h_1,m)\ldots r(h_s,m).$$

We write $e(x)=e^{2\pi ix}$ for real x. The following lemma is a special case of Lemma 2.2 in [4].

Lemma 1. With the notation above we have for any $s\geq 1$ and $N\geq 1$,

$$D_N^{(s)}\leq\frac{s}{m}+\sum_{\underline{h}\in C(m)}\frac{1}{r(\underline{h},m)}\left|\frac{1}{N}\sum_{n=0}^{N-1}e(\underline{h}\cdot\underline{y}_n/m)\right|.$$

Lemma 2. For all $\underline{h}\in C(m)$ we have $r(\underline{h},m)\geq r(\underline{h})$.

Proof. From $\sin\pi x\geq 2x$ for $0\leq x\leq\frac{1}{2}$ it follows that $r(h,m)\geq r(h)$ for $|h|\leq m/2$, and this implies already the result. □

For a sequence of integers which is periodic when considered mod m we write $p(m)$ for the (least) period of the sequence of least residues mod m. For $\gcd(\lambda,m)=1$ let $\mathrm{ord}(\lambda,m)$ denote the least positive integer n with $\lambda^n \equiv 1 \bmod m$. We note that under the condition $\gcd(\lambda,m)=1$ the sequence $y_0, y_1, \ldots$ generated by (1), and thus the sequence $x_0, x_1, \ldots$ of LCPRN, is purely periodic. For the applications of LCPRN it is important to make the period as large as possible. The performance of the LCPRN under statistical tests is only of interest when considered over the full period or over parts of the period.

2. THE HOMOGENEOUS CASE

In this section we consider the case $r \equiv 0 \bmod m$. We may assume w. l. o. g. that $\gcd(y_0,m)=1$, for otherwise the same sequence of LCPRN could be produced with a smaller modulus. For the sequence $y_0, y_1, \ldots$ we have then $y_n \equiv \lambda^n y_0 \bmod m$ and $p(m)=\mathrm{ord}(\lambda,m)$. If now $m=10^a$, $a \geq 4$, then the largest value of $p(m)$ can be obtained by choosing λ in such a way that both $\mathrm{ord}(\lambda,2^a)$ and $\mathrm{ord}(\lambda,5^a)$ are maximal, i.e.

$$\lambda \equiv \pm 3 \bmod 8,\ \lambda \equiv \pm 2,\ \pm 3,\ \pm 8,\ \pm 12 \bmod 25. \tag{4}$$

In this case

$$p(m)=\mathrm{lcm}(\mathrm{ord}(\lambda,2^a),\mathrm{ord}(\lambda,5^a))=\mathrm{lcm}(2^{a-2},4\cdot 5^{a-1})=2^{a-2}.5^{a-1}=\frac{m}{20}.$$

The following result is basic.

<u>Lemma 3.</u> If $m=2^b 5^c$ with integers $b,c \geq 0$ is not a divisor of 80, λ satisfies (4), and $p=\mathrm{ord}(\lambda,m)$, then

$$\sum_{n=0}^{p-1} e(k\lambda^n/m)=0 \text{ for all integers } k \text{ with } \mathrm{gdc}(k,m)=1.$$

<u>Proof.</u> The condition on m implies that either $b \geq 5$ or $c \geq 2$. With $\emptyset$ denoting Euler's totient function, we have in the first case

$$p=\mathrm{lcm}(\mathrm{ord}(\lambda,2^b),\ \mathrm{ord}(\lambda,5^c))=\mathrm{lcm}(2^{b-2},\emptyset(5^c))=2^{b-2}.5^d$$

for some integer $d \geq 0$ and

$$Q:=\mathrm{ord}(\lambda,\tfrac{m}{2})=\mathrm{lcm}(2^{b-3},\emptyset(5^c))=2^{b-3}.5^d=\frac{p}{2}.$$

For $\gcd(k,m)=1$ the sequence $(k\lambda^n)$, $n=0,1,\ldots$, satisfies $p(m)=p$ and $p(\frac{m}{2})=Q$. If $a_1,\ldots,a_Q$ are the least residues mod $\frac{m}{2}$ of the sequence, then the least residues mod m of the sequence are necessarily among the values $a_1,\ldots,a_Q,a_1+\frac{m}{2},\ldots,a_Q+\frac{m}{2}$. But these are $2Q=p$ distinct least residues mod m, and so they are exactly the least residues mod m of the sequence.

Therefore,

$$\sum_{n=0}^{p-1} e(k\lambda^n/m)=\sum_{j=0}^{Q-1}(e(\frac{a_j}{m})+e(\frac{a_j}{m}+\frac{1}{2}))=0.$$

In the case $c\geq 2$ we have

$$p=\operatorname{lcm}(2^f,4\cdot 5^{c-1})=2^{\max(f,2)}\cdot 5^{c-1}$$

for some integer $f\geq 0$ and

$$q:=\operatorname{ord}(\lambda,\frac{m}{5})=\operatorname{lcm}(2^f,4\cdot 5^{c-2})=2^{\max(f,2)}\cdot 5^{c-2}=\frac{p}{5}.$$

The sequence $(k\lambda^n)$ from above satisfies $p(\frac{m}{5})=q$. If $b_1,\ldots,b_q$ are the least residues mod $\frac{m}{5}$ of the sequence, then the least residues mod m of the sequence are necessarily among the arithmetic progressions $\{b_j+\frac{m}{5}t\}_{t=0}^{4}$, $1\leq j\leq q$. These yield $5q=p$ distinct least residues mod m, and so they produce exactly all the least residues mod m of the sequence. Therefore,

$$\sum_{n=0}^{p-1} e(k\lambda^n/m)=\sum_{j=0}^{q-1}\sum_{t=0}^{4} e(\frac{b_j}{m}+\frac{t}{5})=\sum_{j=0}^{q-1} e(\frac{b_j}{m})\sum_{t=0}^{4} e(\frac{t}{5})=0. \qquad \square$$

Corollary 1. If $m=10^a$, $a\geq 4$, λ satisfies (4), and $p=\operatorname{ord}(\lambda,m)$, then

$$\sum_{n=0}^{p-1} e(k\lambda^n/m)=0 \text{ for } k\not\equiv 0 \bmod \frac{m}{80}.$$

Proof. Let $d=\operatorname{gdc}(k,m)$, so that $k=k_1d$, $m=m_1d$ with $\gcd(k_1,m_1)=1$. Then

$$\sum_{n=0}^{p-1} e(k\lambda^n/m)=\sum_{n=0}^{p-1} e(k_1\lambda^n/m_1)=\frac{p}{p_1}\sum_{n=0}^{p_1-1} e(k_1\lambda^n/m_1),$$

where $p_1=\mathrm{ord}(\lambda,m_1)$. Now $k\not\equiv 0 \bmod \frac{m}{80}$ implies $d\not\equiv 0 \bmod \frac{m}{80}$, and so $m_1=\frac{m}{d}$ is not a divisor of 80. Thus the result follows from Lemma 3. □

Theorem 1. If $m=10^a$, $a\geq 4$, λ satisfies (4), and $p=\mathrm{ord}(\lambda,m)=m/20$, then for any $s\geq 1$ we have

$$D_p^{(s)}\leq\frac{s}{m}+R^{(s)}(\lambda,m,\frac{m}{80}). \tag{5}$$

Proof. From Lemmas 1 and 2 we get

$$D_p^{(s)}\leq\frac{s}{m}+\sum_{\underline{h}\in C(m)}\frac{1}{r(\underline{h})}\left|\frac{1}{p}\sum_{n=0}^{p-1} e(\underline{h}\cdot\underline{y}_n/m)\right|.$$

From $\underline{y}_n=(y_n,y_{n+1},\ldots,y_{n+s-1})$ and $y_n\equiv\lambda^n y_0 \bmod m$ it follows that

$$\sum_{n=0}^{p-1} e(\underline{h}\cdot\underline{y}_n/m)=\sum_{n=0}^{p-1} e(y_0(\underline{h}\cdot\underline{\lambda})\lambda^n/m).$$

Using $\gcd(y_0,m)=1$ and Corollary 1, we see that this sum vanishes whenever $\underline{h}\cdot\underline{\lambda}\not\equiv 0 \bmod (m/80)$. For $\underline{h}\cdot\underline{\lambda}\equiv 0 \bmod (m/80)$ we use the trivial upper bound p for the absolute value of the sum. Consequently,

$$D_p^{(s)}\leq\frac{s}{m}+\sum_{\substack{h\in C(m)\\ \underline{h}\cdot\underline{\lambda}\equiv 0 \bmod \frac{m}{80}}}\frac{1}{r(\underline{h})}=\frac{s}{m}+R^{(s)}(\lambda,m,\frac{m}{80}).$$ □

Theorem 2. Under the conditions of Theorem 1 we have

$$D_p^{(s)}\leq\frac{C_s(\log 2m)^s}{\rho^{(s)}(\lambda,m/80)} \text{ for any } s\geq 1,$$

with an effective constant C_s only depending on s.

Proof. The method in the proof of Theorem 3.8 in [4] and of Theorem 4 in [8] shows that

$$R^{(s)}(\lambda,m,\frac{m}{80})<\frac{(2\log 2m)^s+2^s(2\log 2m)^{s-1}}{(\log 2)^{s-1}\rho^{(s)}(\lambda,m/80)}. \tag{6}$$

For the first term on the right-hand side of (5) we use the $\rho^{(s)}(\lambda,m/80)\leq m/40$. □

Theorems 1 and 2 pertain to the behavior of the sequence of LCPRN over the full period. The method also yields results for parts of the period.

<u>Theorem 3.</u> Under the conditions of Theorem 1 we have for any $s\geq 1$ and $N\geq 1$,

$$D_N^{(s)} < \frac{s}{m} + R^{(s)}(\lambda, m, \frac{m}{80}) + \frac{m^{1/2}}{N}\left(\frac{2}{\pi}\log p + \frac{7}{5}\right)\left(\frac{2}{\pi}\log m + \frac{7}{5}\right)^s.$$

<u>Proof.</u> Using Lemma 1 and the fact that $\underline{y}_n=(y_n,y_{n+1},\ldots,y_{n+s-1})$ with $y_n\equiv\lambda^n y_0 \bmod m$, we obtain

$$D_N^{(s)} \leq \frac{s}{m} + \sum_{\underline{h}\in C(m)} \frac{1}{r(\underline{h},m)}\left|\frac{1}{N}\sum_{n=0}^{N-1} e(y_0(\underline{h}\cdot\underline{\lambda})\lambda^n/m)\right|. \tag{7}$$

For an integer k put $d=\gcd(k,m)$, so that $k=k_1d$, $m=m_1d$ with $\gcd(k_1,m_1)=1$. If $p_1=\mathrm{ord}(\lambda,m_1)$, then for given N we write $N=qp_1+N_1$ with $0\leq N_1<p_1$. Thus

$$\sum_{n=0}^{N-1} e(k\lambda^n/m) = \sum_{n=0}^{N-1} e(k_1\lambda^n/m_1) = q\sum_{n=0}^{p_1-1} e(k_1\lambda^n/m_1) + \sum_{n=0}^{N_1-1} e(k_1\lambda^n/m_1).$$

If $k\not\equiv 0 \bmod \frac{m}{80}$, then m_1 is not a divisor of 80, and so the first sum in the last expression vanishes because of Lemma 3. An application of Lemma 7 in [3] yields

$$\left|\sum_{n=0}^{N-1} e(k\lambda^n/m)\right| = \left|\sum_{n=0}^{N_1-1} e(k_1\lambda^n/m_1)\right| < m_1^{1/2}\left(\frac{2}{\pi}\log p_1 + \frac{7}{5}\right) \leq m^{1/2}\left(\frac{2}{\pi}\log p + \frac{7}{5}\right).$$

For $k=y_0(\underline{h}\cdot\underline{\lambda})$ with $\underline{h}\in C(m)$ this estimate is applicable whenever $\underline{h}\cdot\underline{\lambda}\not\equiv 0 \bmod \frac{m}{80}$. In the remaining case we use the trivial bound N for the absolute value of the exponential sum. Then from (7),

$$D_N^{(s)} < \frac{s}{m} + \sum_{\substack{\underline{h}\in C(m)\\ \underline{h}\cdot\underline{\lambda}\equiv 0 \bmod \frac{m}{80}}} \frac{1}{r(\underline{h},m)} + \frac{m^{1/2}}{N}\left(\frac{2}{\pi}\log p + \frac{7}{5}\right) \sum_{\substack{\underline{h}\in C(m)\\ \underline{h}\cdot\underline{\lambda}\not\equiv 0 \bmod \frac{m}{80}}} \frac{1}{r(\underline{h},m)}$$

$$< \frac{s}{m} + R^{(s)}(\lambda, m, \frac{m}{80}) + \frac{m^{1/2}}{N}\left(\frac{2}{\pi}\log p + \frac{7}{5}\right) \sum_{\underline{h}\in C(m)} \frac{1}{r(\underline{h},m)},$$

where we used Lemma 2 in the last step. The inequality

$$\sum_{\underline{h}\in C(m)} \frac{1}{r(\underline{h},m)} < \left(\frac{2}{\pi}\log m+\frac{7}{5}\right)^s \tag{8}$$

obtained from Lemma 2.3 in [4] yields the desired result. □

The quantity $D_N^{(s)}$ can be estimated in terms of the figure of merit $\rho^{(s)}(\lambda,m/80)$ by applying (6) in Theorem 3. Lower bounds for $D_N^{(s)}$ may also be established in terms of figures of merit. In analogy with a lemma in [6] one shows first the following result.

Lemma 4. For any points $\underline{t}_0,\underline{t}_1,\ldots,\underline{t}_{N-1}$ in $[0,1]^s$, any nonzero lattice point $\underline{h}$, and any real Θ we have

$$\left|\frac{1}{N}\sum_{n=0}^{N-1}\cos 2\pi(\underline{h}\cdot\underline{t}_n-\Theta)\right| < \frac{2}{\pi}\left(\frac{\pi}{2}+\frac{1}{4}\right)^s r(\underline{h})D_N(\underline{t}_0,\underline{t}_1,\ldots,\underline{t}_{N-1}).$$

Theorem 4. For any even $m\geq 2$ we have in the homogeneous case

$$D_N^{(s)} > \frac{C_s'}{\rho^{(s)}(\lambda,m/2)} \quad \text{for all } s\geq 1 \text{ and } N\geq 1, \text{ where } C_s' = \frac{\pi}{2}\left(\frac{\pi}{2}+\frac{1}{4}\right)^{-s}.$$

Proof. By the definition of $\rho^{(s)}(\lambda,m/2)$ there exists a nonzero lattice point $\underline{h}$ with $\underline{h}\cdot\underline{\lambda}\equiv 0 \bmod \frac{m}{2}$ and $r(\underline{h})=\rho^{(s)}(\lambda,m/2)$. For $\underline{t}_n=\frac{1}{m}\underline{y}_n$ we have

$$\underline{h}\cdot\underline{t}_n \equiv \frac{1}{m}(\underline{h}\cdot\underline{\lambda})\lambda^n y_0 \bmod 1.$$

If $\underline{h}\cdot\underline{\lambda}\equiv 0 \bmod m$, then $\underline{h}\cdot\underline{t}_n\equiv 0 \bmod 1$ for all n. Otherwise $\underline{h}\cdot\underline{\lambda}\equiv\frac{m}{2} \bmod m$, and since λ^n and y_0 are odd, we get $\underline{h}\cdot\underline{t}_n\equiv\frac{1}{2} \bmod 1$ for all n. In any case, there exists Θ such that $\underline{h}\cdot\underline{t}_n\equiv\Theta \bmod 1$ for all n. From Lemma 4 we obtain

$$1 < \frac{2}{\pi}\left(\frac{\pi}{2}+\frac{1}{4}\right)^s r(\underline{h})D_N^{(s)},$$

whence the desired inequality. □

By combining Theorems 2 and 4, we see that under the conditions of Theorem 1 we have

$$\frac{C'_s}{\rho^{(s)}(\lambda,m/2)} < D_p^{(s)} \le \frac{C_s(\log 2m)^s}{\rho^{(s)}(\lambda,m/80)} \quad \text{for any } s\ge 1.$$

Thus, the order of magnitude of $D_p^{(s)}$ is essentially governed by these figures of merit.

3. THE INHOMOGENEOUS CASE

We take $m=10^a$, $a\ge 2$, and $r\not\equiv 0 \bmod m$. According to a well-known criterion (see [2, Ch. 3], [5, Sec. 7]), the maximal period in this case is m and it is attained precisely for $\lambda\equiv 1 \bmod 20$ and $\gcd(r,m)=1$. We need information on the period $p(d)$ of sequences of integers $z_0, z_1, \ldots$ considered modulo divisors d of m and satisfying a recursion $z_{n+1}=\lambda z_n+r'$. For a prime q and integer g we write $q^k \| g$ if q^k is the largest power of q dividing g, where $k=\infty$ for $g=0$.

<u>Lemma 5.</u> Let $z_0, z_1, \ldots$ be a sequence of integers with $z_{n+1}=\lambda z_n+r'$ for all $n\ge 0$. Let q be a prime and $q^w \| ((\lambda-1)z_0+r')$. If $\lambda\equiv 1 \bmod q$ and $\lambda\equiv 1 \bmod 4$ if $q=2$, then $p(q^a)=q^{a-\min(a,w)}$ for all $a\ge 0$.

<u>Proof.</u> If $w\ge a$, then $(\lambda-1)z_0+r'\equiv 0 \bmod q^a$, hence $z_1\equiv z_0 \bmod q^a$, and so $p(q^a)=1=q^{a-\min(a,w)}$. If $\lambda=1$, then $p(q^a)=q^a/\gcd(r',q^a)=$ $=q^{a-\min(a,w)}$. Thus we may assume $w<a$ and $\lambda\ne 1$. In this case, Lemma 8 in [3] implies that $p(q^a)=\operatorname{ord}(\lambda,q^{a-w+k})$, where k is defined by $q^k \| (\lambda-1)$. The condition on λ shows that $k\ge 1$, and $k\ge 2$ if $q=2$. We have $\lambda=1+tq^k$ with $\gcd(t,q)=1$. We show by induction on j that

$$\lambda^{q^j}=1+t_j q^{k+j} \quad \text{for all } j\ge 0, \text{ where } \gcd(t_j,q)=1. \tag{9}$$

The statement (9) is true for $j=0$. Assuming its truth for some $j\ge 0$, we have

$$\lambda^{q^{j+1}}=(1+t_j q^{k+j})^q=1+t_j q^{k+j+1}+\sum_{i=2}^{q}\binom{q}{i}t_j^i q^{(k+j)i} .$$

If $q>2$, then $k\geq 1$, and so $(k+j)i\geq 3k+j\geq k+j+2$ for $i\geq 3$, whereas for $i=2$ we have $(k+j)i\geq 2k+j\geq k+j+1$ and $\binom{q}{i}\equiv 0 \bmod q$, so that

$$\sum_{i=2}^{q} \binom{q}{i} t_j^i q^{(k+j)i} \equiv 0 \bmod q^{k+j+2}.$$

Consequently, the statement (9) holds with j replaced by j+1. If q=2, then $k\geq 2$, hence $(k+j)i\geq 2k+j\geq k+j+2$ for i=2, so that (9) holds again for j+1.

For fixed $j\geq 1$ it follows from (9) that

$$\lambda^{q^j}\equiv 1 \bmod q^{k+j},\ \lambda^{q^{j-1}}=1+t_{j-1}q^{k+j-1}\not\equiv 1 \bmod q^{k+j}.$$

Incorporating the trivial case j=0, we get

$$\mathrm{ord}(\lambda,q^{k+j})=q^j \quad \text{for } j\geq 0, \tag{10}$$

hence $p(q^a)=\mathrm{ord}(\lambda,q^{a-w+k})=q^{a-w}=q^{a-\min(a,w)}$, and so the formula for $p(q^a)$ is shown in all cases. □

Lemma 6. Let the sequence $z_0,z_1,\ldots$ be as in Lemma 5 with $\lambda\equiv 1 \bmod 20$, and let $m=2^b5^c$ with integers $b,c\geq 0$. Then

$$p(m)=\frac{m}{\gcd((\lambda-1)z_0+r',m)}.$$

Proof. Let $2^v\,||\,((\lambda-1)z_0+r')$, $5^w\,||\,((\lambda-1)z_0+r')$. Then by Lemma 5,

$$p(m)=\mathrm{lcm}(p(2^b),\ p(5^c))=\mathrm{lcm}(2^{b-\min(b,v)},\ 5^{c-\min(c,w)})=$$

$$=\frac{2^b5^c}{2^{\min(b,v)}5^{\min(c,w)}}=\frac{m}{\gcd((\lambda-1)z_0+r',m)}. \qquad □$$

Lemma 7. If $m=10^a$, $a\geq 1$, and the sequence $z_0,z_1,\ldots$ is as in Lemma 5 with $\lambda\equiv 1 \bmod 20$, then for $p=p(m)$ we have

$$\sum_{n=0}^{p-1} e(z_n/m)=0 \text{ whenever } (\lambda-1)z_0+r'\not\equiv 0 \bmod m.$$

Proof. If $(\lambda-1)z_0+r'\not\equiv 0 \bmod m$, we have $\gcd((\lambda-1)z_0+r',m)=2^b5^c$ with $0\leq b,\ c\leq a$ and either $b<a$ or $c<a$. Suppose $b<a$; then by Lemma 6,

$$p\left(\frac{m}{2}\right) = \frac{m/2}{\gcd((\lambda-1)z_0+r', m/2)} = \frac{m/2}{\gcd((\lambda-1)z_0+r', m)} = \frac{p}{2}.$$

The identity $\sum_{n=0}^{p-1} e(z_n/m)=0$ can now be shown by the method in the proof of Lemma 3. If $c<a$, we consider $m/5$ instead of $m/2$. □

Theorem 5. If $m=10^a$, $a\geq 2$, $\lambda\equiv 1 \bmod 20$, and $\gcd(r,m)=1$, then for any $s\geq 1$ we have

$$D_m^{(s)} \leq \frac{s}{m} + R^{(s)}(\lambda, m, m).$$

Proof. From Lemmas 1 and 2 we get

$$D_m^{(s)} \leq \frac{s}{m} + \sum_{\underline{h}\in C(m)} \frac{1}{r(\underline{h})} \left| \frac{1}{m} \sum_{n=0}^{m-1} e(\underline{h}\cdot\underline{y}_n/m) \right|. \tag{11}$$

For fixed $\underline{h}\in C(m)$ put $u_n=\underline{h}\cdot\underline{y}_n$ for all $n\geq 0$. Then

$$u_{n+1}=h_1 y_{n+1}+h_2 y_{n+2}+\ldots+h_s y_{n+s} \equiv h_1(\lambda y_n+r)+h_2(\lambda y_{n+1}+r)+$$

$$+\ldots+h_s(\lambda y_{n+s-1}+r)\equiv \lambda u_n+r' \bmod m$$

with $r'=(h_1+h_2+\ldots+h_s)r$. Let $z_0, z_1, \ldots$ be the sequence with $z_0=u_0$ and $z_{n+1}=\lambda z_n+r'$ for all $n\geq 0$. Then $z_n\equiv u_n \bmod m$ for all $n\geq 0$. Furthermore,

$$(\lambda-1)z_0+r' \equiv u_1-u_0 \equiv h_1(y_1-y_0)+h_2(y_2-y_1)+\ldots+h_s(y_s-y_{s-1}) \bmod m.$$

An easy induction shows $y_{n+1}-y_n\equiv \lambda^n(y_1-y_0) \bmod m$ for all $n\geq 0$, hence

$$(\lambda-1)z_0+r' \equiv (\underline{h}\cdot\underline{\lambda})(y_1-y_0) \equiv (\underline{h}\cdot\underline{\lambda})((\lambda-1)y_0+r) \bmod m.$$

Since $\gcd((\lambda-1)y_0+r, m)=1$, we have $(\lambda-1)z_0+r'\not\equiv 0 \bmod m$ precisely if $\underline{h}\cdot\underline{\lambda}\not\equiv 0 \bmod m$. Moreover, if $p=p(m)$ refers to the sequence $z_0, z_1, \ldots$, then p divides m by Lemma 6, and for $\underline{h}\cdot\underline{\lambda}\not\equiv 0 \bmod m$ we get

$$\sum_{n=0}^{m-1} e(\underline{h}\cdot\underline{y}_n/m) = \sum_{n=0}^{m-1} e(z_n/m) = \frac{m}{p}\sum_{n=0}^{p-1} e(z_n/m) = 0$$

by Lemma 7. It follows then from (11) that

$$D_m^{(s)} \le \frac{s}{m} + \sum_{\substack{\underline{h}\in C(m) \\ \underline{h}\cdot\underline{\lambda}\equiv 0 \bmod m}} \frac{1}{r(\underline{h})} = \frac{s}{m} + R^{(s)}(\lambda,m,m).$$

Theorem 6. Under the conditions of Theorem 5 we have

$$D_m^{(s)} \le \frac{C_s(\log 2m)^s}{\rho^{(s)}(\lambda,m)} \quad \text{for any } s\ge 1,$$

with an effective constant C_s only depending on s.

Proof. This follows from Theorem 5 by using an analog of (6), namely

$$R^{(s)}(\lambda,m,m) < \frac{(2\log 2m)^s + 2^s(2\log 2m)^{s-1}}{(\log 2)^{s-1}\rho^{(s)}(\lambda,m)} \tag{12}$$

and the fact that $\rho^{(s)}(\lambda,m)\le 2m$. □

Theorems 5 and 6 pertain to the behavior of the sequence of LCPRN over the full period. To get optimal results for parts of the period, we not only require that $\lambda\equiv 1 \bmod 20$, but also that $\mathrm{ord}(\lambda,m)$ be as large as possible under this condition. This leads to the additional conditions $\lambda\equiv 5 \bmod 8$ and $\lambda\not\equiv 1 \bmod 25$.

Theorem 7. If $m=10^a$, $a\ge 2$, $\lambda\equiv 21 \bmod 40$, $\lambda\not\equiv 1 \bmod 25$, and $\gcd(r,m)=1$, then for any $s\ge 1$ and $N\ge 1$ we have

$$D_N^{(s)} < \frac{s}{m} + R^{(s)}(\lambda,m,m) + \frac{1}{N}(20m)^{1/2}\left(\frac{2}{\pi}\log m + \frac{7}{5}\right)^{s+1}.$$

Proof. For fixed $\underline{h}\in C(m)$, let the sequence $z_0, z_1, \ldots$ and $p=p(m)$ be defined as in the proof of Theorem 5. If $\underline{h}\cdot\underline{\lambda}\not\equiv 0 \bmod m$, then we have seen in that proof that $(\lambda-1)z_0+r'\not\equiv 0 \bmod m$. It follows then from Lemma 7 that $\sum_{n=0}^{p-1} e(z_n/m)=0$. For given N we write $N=qp+N_1$ with $0\le N_1<p$. Then

$$\sum_{n=0}^{N-1} e(\underline{h}\cdot\underline{y}_n/m) = \sum_{n=0}^{N-1} e(z_n/m) = q\sum_{n=0}^{p-1} e(z_n/m) + \sum_{n=0}^{N_1-1} e(z_n/m) = \sum_{n=0}^{N_1-1} e(z_n/m).$$

Since $2^2 || (\lambda-1)$ and $5 || (\lambda-1)$, the formula (10) yields $\mathrm{ord}(\lambda, 2^a) = 2^{a-2}$ and $\mathrm{ord}(\lambda, 5^a) = 5^{a-1}$, hence

$$\mathrm{ord}(\lambda, m) = \mathrm{lcm}(\mathrm{ord}(\lambda, 2^a), \mathrm{ord}(\lambda, 5^a)) = 2^{a-2} \cdot 5^{a-1} = \frac{m}{20}$$

By Theorem 8.1 in [5] we get then

$$\left| \sum_{n=0}^{N-1} e(\underline{h} \cdot \underline{y}_n / m) \right| = \left| \sum_{n=0}^{N_1-1} e(z_n / m) \right| < (20p)^{1/2} \left(\frac{2}{\pi} \log p + \frac{2}{5}\right) + \frac{N_1}{p} \left(\frac{mp-p^2}{m/20}\right)^{1/2}$$

$$\leq (20m)^{1/2} \left(\frac{2}{\pi} \log m + \frac{7}{5}\right)$$

whenever $\underline{h} \cdot \underline{\lambda} \not\equiv 0 \bmod m$. Thus Lemmas 1 and 2 yield

$$D_N^{(s)} < \frac{s}{m} + \sum_{\substack{\underline{h} \in C(m) \\ \underline{h} \cdot \underline{\lambda} \equiv 0 \bmod m}} \frac{1}{r(\underline{h}, m)} + \frac{1}{N} (20m)^{1/2} \left(\frac{2}{\pi} \log m + \frac{7}{5}\right) \sum_{\substack{\underline{h} \in C(m) \\ \underline{h} \cdot \underline{\lambda} \not\equiv 0 \bmod m}} \frac{1}{r(\underline{h}, m)}$$

$$\leq \frac{s}{m} + R^{(s)}(\lambda, m, m) + \frac{1}{N} (20m)^{1/2} \left(\frac{2}{\pi} \log m + \frac{7}{5}\right) \sum_{\underline{h} \in C(m)} \frac{1}{r(\underline{h}, m)} \cdot$$

An application of (8) concludes the argument.

The quantity $D_N^{(s)}$ can be estimated in terms of the figure of merit $\rho^{(s)}(\lambda, m)$ by applying (12) in Theorem 7. A lower bound for $D_N^{(s)}$ in terms of $\rho^{(s)}(\lambda, m)$ can also be given.

<u>Theorem 8.</u> If $m = 10^a$, $a \geq 1$, $\lambda \equiv 1 \bmod 20$, and r is arbitrary, then

$$D_N^{(s)} > \frac{C_s'}{\rho^{(s)}(\lambda, m)} \quad \text{for all } s \geq 1 \text{ and } N \geq 1, \text{ where } C_s' = \frac{\pi}{2} \left(\frac{\pi}{2} + \frac{1}{4}\right)^{-s}.$$

<u>Proof.</u> By the definition of $\rho^{(s)}(\lambda, m)$ there exists a nonzero lattice point $\underline{h}$ with $\underline{h} \cdot \underline{\lambda} \equiv 0 \bmod m$ and $r(\underline{h}) = \rho^{(s)}(\lambda, m)$. With this $\underline{h}$ we consider the sequence $z_0, z_1, \ldots$ as in the proof of Theorem 5. The argument there shows $(\lambda-1) z_0 + r' \equiv 0 \bmod m$, so that $p(m) = 1$ by Lemma 6. Consequently, there exists an integer k with $\underline{h} \cdot \underline{y}_n \equiv k \bmod m$ for all n. Applying Lemma 4 with $\underline{t}_n = \frac{1}{m} \underline{y}_n$ and $\theta = \frac{k}{m}$, we obtain

$$1<\frac{2}{}(\frac{\pi}{2}+\frac{1}{4})^{s} r(\underline{h}) D_N^{(s)}$$

whence the desired inequality. □

By combining Theorems 6 and 8, we see that under the conditions of Theorem 5 we have

$$\frac{C'_s}{\rho^{(s)}(\lambda,m)} < D_m^{(s)} \leq \frac{C_s(\log 2m)^s}{\rho^{(s)}(\lambda,m)} \quad \text{for any } s \geq 1.$$

Thus, the order of magnitude of $D_m^{(s)}$ is essentially governed by $\rho^{(s)}(\lambda,m)$. We note that for $s=2$ and given m, values of λ for which $\rho^{(s)}(\lambda,m)$ is large can be found by the search method in [1].

REFERENCES

1 I. BOROSH and H. NIEDERREITER, Optimal multipliers for pseudo-random number generation by the linear congruential method, Nordisk Tidskr. for Informationsbehandling (BIT), to appear.

2 D. E. KNUTH, The Art of Computer Programming, Vol. 2: Seminumerical Algorithms, 2nd ed., Addison-Wesley, Reading, Mass., 1981.

3 H. NIEDERREITER, On the distribution of pseudo-random numbers generated by the linear congruential method. III. Math. Comp. 30, 571-597(1976).

4 H. NEIDERREITER, Pseudo-random numbers and optimal coefficients, Advances in Math. 26, 99-181(1977).

5 H. NIEDERREITER, Quasi-Monte Carlo methods and pseudo-random numbers, Bull. Amer. Math. Soc. 84, 957-1041(1978).

6 H. NIEDERREITER, Statistical tests for linear congruential pseudo-random numbers, COMPSTAT 1978: Proceedings in Computational Statistics (Leiden, 1978). pp. 398-404, Physica-Verlag, Vienna, 1978.

7 H. NIEDERREITER, Nombres pseudo-aléatoires et équirépartition, Astérisque, no. 61, pp. 155-164, Soc. Math. France, Paris, 1979.

8 H. NIEDERREITER, Statistical tests for Tausworthe pseudo-random numbers, Probability and Statistical Inference (W. Grossmann, G. C. Pflug, and W. Wertz, eds.), pp. 265-274, D.Reidel, Dordrecht, 1982.

Proc. of the 3rd Pannonian Symp. on Math. Stat., Visegrád, Hungary 1982
J. Mogyoródi, I. Vincze, W. Wertz, eds

STOCHASTIC OPTIMIZATION METHODS FOR SOLVING MATHEMATICAL PROGRAMMING PROBLEMS*

J. PINTÉR

Computing Center, Eötvös Loránd University,
Budapest, Hungary

ABSTRACT

Convergence properties of stochastic optimization procedures are investigated when convex and/or differentiable structure of the mathematical programming problems to be solved is not supposed. Some general stochastic optimization schemes and stochastically combined methods are introduced, their global and local convergence characteristics are presented in a unified framework.

Key words: Non-convex and/or Non-differentiable Optimization Problems, Stochastic Optimization Methods, Random Search, Hybrid Algorithms.

1. INTRODUCTION

The planning and management of engineering or economic systems frequently lead to optimization problems without global convexity and/or differentiability properties. In such cases it is desirable to apply optimization strategies which are able to find the solution of possibly non-convex or non-smooth problems. As it is known, a broad class of stochastic optimization techniques, including random searches, is convergent under mild assumptions; for example we refer to the respective papers in Dixon and Szegő (1975, 1978), or Devroye (1978), Pintér (1979), Marti (1980), Solis and Wets (1981), Baba (1981). At the same

*This paper was presented at the 7th Symposium on Operations Research (Hochschule St. Gallen, Aug. 19-21., 1982).

time, most random search methods become rather inefficient in the neighbourhood of (local or global) optima. If there the objective function can be locally approximated quadratically, then random search ought to be "switched over" to some locally more efficient procedure, e.g. to some method of conjugate directions (Hestenes (1980)).

In Pintér (1981a, 1982a) a general class of stochastically combined (hybrid) optimization procedures was investigated and convergence properties of such methods were proved. Here we present the mentioned results in a more general, unified framework. Firstly, some global and local convergence properties of stochastic optimization methods are displayed, then the way of extending these results to hybrid procedures for solving optimization problems in presence of noise is highlighted. For the sake of brevity only the main results are exposed here, more details can be found in Pintér (1981a, b, 1982a, b).

2. CONVERGENCE PROPERTIES OF STOCHASTIC OPTIMIZATION METHODS

Denote by C the set of feasible solutions to an optimization problem, where C is a subset of the real Euclidean n-space E^n, further define the set of optimal solutions $C^* \subset C$. To find points of C^*, a stochastic sequence $\{x_k\}$ is generated according to the rule

$$x_{k+1} = D_k(z_k, \bar{\omega}_k) \in C$$

$$z_k = (x_0, x_1, \ldots, x_k)$$

$$\bar{\omega}_k = (\omega_0, \omega_1, \ldots, \omega_k).$$

This means that x_{k+1} is selected according to some conditional probability measure μ_k defined on the measurable space $(C, \mathcal{C})$

(e.g. $\mathcal{C}$ is the σ-algebra of Borel-measurable subsets of the Borel-measurable set C), i.e. the decision rule D_k depends on the search information z_k and the k-th projection of the sequence of random factors $\omega=(\omega_0,\omega_1,\omega_2,\ldots)$, influencing the decisions D_k k=0,1,2,... . Let $\omega\in\Omega$, we suppose that to the above stochastic decision scheme corresponds a probability measure P defined on the measurable space $(\Omega,\mathcal{A})$. We want to find sufficient conditions which assure that for the generated sequence $\{x_k\}$ there holds

$$P(\lim_{k\to\infty} \rho(x_k,\, C^*)=0)=1 \qquad \rho(x,C^*)=\inf_{x^*\in C^*}||x-x^*||.$$

Let $\{A_k\}$ $A_k\in\mathcal{A}$ a sequence of random events, define $A_\infty=\bigcap_{n=0}^{\infty}\bigcup_{k=n}^{\infty} A_k$.

First we give a simple but useful generalization of the Borel-Cantelli lemma.

Lemma 2.1: Assume that for every pair of indices k,n k≥n there holds

$$P(A_k|\bigcap_{j=n}^{k-1}\bar{A}_j)\geq p_k \quad 0\leq p_k\leq 1, \text{ where } \sum_{k=0}^{\infty} p_k=\infty; \text{ then } P(A_\infty)=1.$$

On the other hand, if for k,n=0,1,2,... k≥n we have

$$P(A_k|\bigcap_{j=n}^{k-1}\bar{A}_j)\leq p_k \quad 0\leq p_k\leq 1 \quad \text{and} \quad \sum_{k=0}^{\infty} p_k<\infty, \quad \text{then} \quad P(A_\infty)=0.$$

The first assertion of Lemma 2.1 states that - under the given assumptions - among the events $\{A_k\}$ infinitely many take place with probability 1 (w.p.1), while its second assertion implies that only a finite number of A_k's occurs w.p.1. Therefore, with properly defined A_k's, Lemma 2.1 can serve as a base of necessary and sufficient convergence conditions for different types of adaptive stochastic algorithms. We mention here only some special results of Pintér (1981b, 1982b).

We consider the problem of minimizing a measurable function $f:C\to E^1$ on the measurable set $C\subseteq E^n$. The following conceptual algorithm can be applied to find points of the set C^* of optimal solutions (we assume that $f(x)>-\infty$ $x\in C$, and $C^*\neq\emptyset$).

Step 0. Let $x_0\in C$, set $k:=0$.

Step 1. Select a new point $y=D_k(z_k, \bar{\omega}_k)\in C$.

Step 2. Set $x_{k+1}=\begin{cases} y & \text{if } f(y)<f(x_k), \\ x_k & \text{otherwise;}\end{cases}$

then (setting $k:=k+1$) return to step 1.

For arbitrary $\hat{\delta}>0$ and $\hat{\varepsilon}>0$, we define the sets

$$G(C^*,\hat{\delta})=\{y\in C:\rho(y,C^*)<\hat{\delta}\},\quad \hat{C}=\hat{C}(\hat{\delta})=C\setminus G(C^*,\hat{\delta})=\{y\in C:\rho(y,C^*)\geq\hat{\delta}\},$$

and the random events

$$F_k=F_k(\hat{\delta}):\{x_k\in\hat{C}\},\quad A_k=A_k(\hat{\varepsilon}):\{f(x_{k+1})\leq f(x_k)-\hat{\varepsilon}\}\quad k=0,1,2,\ldots$$

<u>Theorem 2.1:</u> Suppose that for every $\hat{\delta}>0$ there exist $\hat{\varepsilon}=\hat{\varepsilon}(\hat{\delta})>0$ and $\{\hat{p}_k\}$ $\hat{p}_k=\hat{p}_k(\hat{\delta})$ $0\leq\hat{p}_k\leq 1$ such that for every pair of indices n,k $0\leq n\leq k$ we have

$$P\{A_k\cup\bar{F}_k \mid \bigcap_{j=n}^{k-1}(\bar{A}_j\cap F_j)\}\geq\hat{p}_k,\quad \text{where}\quad \sum_{k=0}^{\infty}\hat{p}_k=\infty.$$

Then $P(\liminf_{k\to\infty}\rho(x_k, C^*)=0)=1$.

<u>Corollary 2.1:</u> If - besides the assumptions of Theorem 2.1 - suppose that f is uniformly continuous on C, and for every $\delta>0$ there exists $\varepsilon=\varepsilon(\delta)>0$ such that

$$x\in C\setminus G(C^*,\delta)=\{x\in C:\rho(x,C^*)\geq\delta\}\quad\text{implies}\quad f(x)\geq f^*+\varepsilon,$$

then

$$P(\lim_{k\to\infty}\rho(x_k,C^*)=0)=P(\lim_{k\to\infty}f(x_k)=f^*)=1\qquad (f^*=\min_{x\in C}f(x)).$$

Consider next the (more special) global optimization problem

$$\min_{x \in C} f(x),$$

where $C \subset E^n$ is a (non-empty) compact set and $f(x)$ is a continuous function. From these assumptions there follows that the set of global solutions

$$C^* = \{x^* \in C : f(x^*) = \min_{x \in C} f(x)\}$$

is non-empty. Suppose that the Lebesgue-measure $\mu^{(n)}$ of the set

$$G(C^*, \delta) = \{x \in C : \rho(x, C^*) < \delta\}$$

is positive for every $\delta > 0$. For solving the problem a conceptual stochastic algorithm can be applied, which consists of the following steps:

Step 0. Let $x_0 \in C$, set $k:=0$.

Step 1. Generate a search direction $s_k = s_k(z_k, \bar{\omega}_k)$ according to some probability measure μ_k given in E^n. We assume that the sequence $\{\mu_k\}$ is defined in accordance with the following requirement: if $C_1 \subset C$, $C_1 \in \mathcal{C}$, $\mu^{(n)}(C_1) > 0$, then for every $\hat{\delta} > 0$ there exist $\hat{\varepsilon} = \hat{\varepsilon}(\hat{\delta}) > 0$, $\hat{p} = \hat{p}(C_1, \hat{\delta}) > 0$ such that

$$P\{B_k \cup \bar{F}_k \mid \bigcap_{j=n}^{k-1} (\bar{A}_j \cap F_j)\} \geq \hat{p} \qquad k,n = 0,1,2,\ldots \quad k \geq n,$$

where the random events A_k, B_k, F_k are defined as follows:

$$A_k = A_k(\hat{\varepsilon}) : \{f(x_{k+1}) \leq f(x_k) - \hat{\varepsilon}\}, \; B_k = B_k(C_1) : \{\exists \gamma \geq 0 : x_k + \gamma\, s_k \in C_1\},$$

$$F_k = F_k(\hat{\delta}) : \{\rho(x_k, C^*) \geq \hat{\delta}\} .$$

Step 2. Having selected s_k, the step-size $\gamma_k = \gamma_k(z_k, \bar{\omega}_k)$ is computed by univariate global search on C:

$$f(x_{k+1}) = f(x_k + \gamma_k s_k) = \min_{\substack{\gamma \geq 0 \\ x_k + \gamma s_k \in C}} f(x_k + \gamma s_k),$$

then (setting $k:=k+1$) return to step 1.

Theorem 2.2: For the generated sequence $\{x_k\}$ we have

$$P(\lim_{k\to\infty} \rho(x_k,C^*)=0)=P(\lim_{k\to\infty} f(x_k)=f^*)=1.$$

The proof of this theorem is based on a rather straightforward argumentation which shows that the assumptions of Theorem 2.1 and Corollary 2.1 are satisfied.

Consider now the problem of finding local solutions. Assume that $f:C\to E^1$ ($f(x)>-\infty$ $x\in C$) is a measurable function, $C\subseteq E^n$ is the closure of an open set, and the set of local solutions $C^*\subset C$ is non-empty. The following conceptual algorithm can be applied to find points of C^*.

Step 0. Let $x_0\in C$, set $k:=0$.

Step 1. Generate a search direction $s_k=s_k(z_k,\bar{\omega}_k)$ $||s_k||=1$, according to some probability measure μ_k defined on the surface of the unit sphere in E^n.

Step 2. Having selected s_k, the step-size $\gamma_k=\gamma_k(z_k,\bar{\omega}_k)$ results from univariate local search on C:

$$f(x_{k+1}) = f(x_k + \gamma_k s_k) = \min_{\substack{\gamma\in[0,h_k] \\ x_k+\gamma s_k\in C}} f(x_k + \gamma s_k) \quad 0<h_k\le\bar{h},$$

then return to step 1 ($k:=k+1$).

(The figuring measures $\{\mu_k\}$ and maximal step-size $\bar{h}$ have to satisfy the requirements of the theorem below.)

For arbitrary $\hat{\delta}>0$ and $\{\hat{\varepsilon}_k\}>0$ define the sets

$$G(C^*,\hat{\delta})=\{x\in C:\rho(x,C^*)<\hat{\delta}\},\quad \hat{C}=\hat{C}(\hat{\delta})=C\backslash G(C^*,\hat{\delta}),$$

and the random events

$$F_k = F_k(\hat{\delta}):\{x_k\in\hat{C}\},\quad A_k = A_k(\hat{\varepsilon}_k):\{f(x_{k+1})\le f(x_k) - \hat{\varepsilon}_k\}\ .$$

Theorem 2.3.: Suppose that for every $\hat{\delta}>0$ there exist $\{\hat{\varepsilon}_k\}$ $\hat{\varepsilon}_k = \hat{\varepsilon}_k(\hat{\delta})>0$ and $\{\hat{p}_k\}$ $\hat{p}_k = \hat{p}_k(\hat{\delta})$ $0\le\hat{p}_k\le 1$ such that for every pair of indices n,k $0\le n\le k$ we have

$$P\{A_k \cup \overline{F}_k \mid \bigcap_{j=n}^{k-1} (\overline{A}_j \cap F_j)\} \ge \hat{p}_k \ , \quad \text{where} \quad \sum_{k=0}^{\infty} \hat{p}_k = \infty,$$

and for every subsequence of indices $\{k_i\} \subseteq \{k\}$ there holds

$$\sum_{i=0}^{\infty} \hat{\varepsilon}_{k_i} = \infty.$$

Then $P(\liminf_{k\to\infty} \rho(x_k, C^*) = 0) = 1$.

Assume now that $C^* = \{x_j^*\}_{j\in J}$ $J=\{1,2,\ldots,r\}$, and there exists a system of disjoint sets $\{C_j\}_{j\in J}$ such that

$$x_j \in \text{int } C_j, \ \min_{x\in C_j} f(x) = f(x_j^*), \quad \bigcup_{j\in J} C_j = C.$$

We denote by int C_j the interior of C_j, while $\overline{C}_j$ - its closure.

Corollary 2.3: If besides the assumptions of Theorem 2.3 suppose that f is uniformly continuous on C, for every $\delta>0$ there exists $\varepsilon=\varepsilon(\delta)>0$ such that

$$x\in D_j = \{x\in\overline{C}_j : \rho(x, x_j^*) \ge \delta\}$$

implies $f(x)\ge f(x_j^*)+\varepsilon$, and $\liminf_{k\to\infty} \rho(x_k, x_j^*) = 0$ implies that $x_k \in \overline{C}_j$ $k\ge n=n(\omega)$, then $P(\lim_{k\to\infty} \rho(x_k, C^*) = 0) = 1$.

It is not difficult to see that the assumptions of Theorems 2.2 and 2.3 about the described conceptual algorithms can be satisfied in a number of ways. E.g. the directions s_k can be generated according to different random search strategies, while the step lengths γ_k can be computed respectively by global or local univariate minimization techniques.

3. HYBRID OPTIMIZATION METHODS IN THE PRESENCE OF NOISE

The above convergence results can be extended in two ways: instead of pure stochastic optimization strategies for solving optimization problems with exactly computable functions, we consider now hybrid methods which are applicable also to "noisy" problems. As an example of "noisy" problems we mention stochastic programming models (see e.g. Kall and Prékopa (1980)), in which some or all of the figuring functions cannot be evaluated analytically, and their values are estimated by Monte Carlo techniques. We outline here only the basic ideas of the extension, for more details see Pintér (1981a, 1982 a,b).

As in Section 2, let $C \subseteq E^n$ be the set of feasible solutions; a sequence of points $\{x_k\}$ is sought, which converges w.p.1 to a subset $C^* \neq \emptyset$ of C. To generate $\{x_k\}$, the following stochastically combined scheme is applied. Let $x_0 \in C$ arbitrary, k:=0; for k=0,1,2,... the next point is defined by the relations

$$x_{k+1} = H_k(z_k, \bar{\omega}_k),$$

$$z_k = (x_0, x_1, \ldots, x_k),$$

$$\bar{\omega}_k = (\omega_0, \omega_1, \ldots, \omega_k),$$

$$H_k : \alpha_k G_k(z_k, \bar{\omega}_k) + (1-\alpha_k)\, L_k(z_k, \bar{\omega}_k) \qquad 0 \le \alpha_k \le 1.$$

This means that at step k - independently from the results of the previous iterations - either subprocedure G_k is accomplished with probability α_k, or subprocedure L_k is accomplished with probability $(1-\alpha_k)$; both G_k and L_k usually depend on the search information z_k and also on random factors represented by $\bar{\omega}_k$. Denote $\omega=(\omega_0, \omega_1, \omega_2, \ldots)$ $\omega \in \Omega$. As earlier, we suppose that there exists a probability measure P, defined on the measurable space $(\Omega, \mathcal{A})$, which corresponds to the described stochastic decision scheme.

In connection with the convergence properties of hybrid optimization methods, first we present a simple extension of the first part of Lemma 2.1. Let $\{A_k\}$ $A_k \in \mathcal{A}$ $k=0,1,2,\ldots$ a sequence of random events, where A_k depends only on $\bar{\omega}_k$.

<u>Lemma 3.1:</u> If the application of subalgorithm G_k assures that for every pair of non-negative integers k,n $k \geq n$ we have

$$P(A_k \mid \bigcap_{j=n}^{k-1} \bar{A}_j;\ G_k) \geq p_k \quad 0 \leq p_k \leq 1 \quad k \geq n \quad k,n=0,1,2,\ldots$$

and

$$\sum_{k=0}^{\infty} \alpha_k p_k = \infty, \quad \text{then} \quad P(A_\infty) = P(\bigcap_{n=0}^{\infty} \bigcup_{k=n}^{\infty} A_k) = 1.$$

On the base of the results of Section 2 and Lemma 3.1 now we can briefly describe the way of extensions. Assume that instead of the "imbedded" deterministic problem

$$\min_{x \in C} f(x),$$

(where $C \subseteq E^n$, and $f: C_0 \to E^1$ is defined on a suitable open set C_0, $C \subseteq C_0$), only its "noisy" estimates can be obtained. In other words, for the sequentially generated approximate solutions x_k $k=0,1,2,\ldots$, we have estimates of the objective function and the feasible set in the form $f(x_k) + \xi(x_k)$, $C + \eta(x_k)$.

Here f and C have respectively analogous properties to those described in connection with Theorems 2.2 and 2.3.

The random variables $\xi(x_k) = \xi(x_k, \bar{\omega}_k)$ and random point-to-set mappings $\eta(x_k) = \eta(x_k, \bar{\omega}_k)$ satisfy uniform convergence conditions of the following form: there exist (uniformly in ω) constants $D>0$, $T>0$ such that we have

$$P(\mid \sum_{k=0}^{\infty} \xi(x_k) \mid \leq D) = 1, \quad P(\sum_{k=0}^{\infty} \|\eta(x_k)\| \leq T) = 1, \quad \|\eta(x_k)\| = \underset{z \in \eta(x_k)}{\text{vrai sup}} \|z\|;$$

moreover, for every $\mu>0, \nu>0$ there exist respective indices $k(\mu)$, $k(\nu)$ such that for $n\geq k(\mu)$ and $n\geq k(\nu)$ hold respectively

$$P(|\sum_{k=n}^{\infty} \xi(x_k)|\leq\mu)=1, \qquad P(\sum_{k=n}^{\infty} ||\eta(x_k)||\leq\nu)=1.$$

Finally, we suppose that the basic components of the hybrid procedure (denoted by $\{G_k\}$ in Lemma 3.1) have analogous properties to those described in the assumptions of Theorems 2.2 and 2.3. Then the presented convergence results can be generalized in a rather straightforward manner, for details see Pintér (1981a, 1982a,b).

4. CONCLUDING REMARKS

As it can be seen from the assumptions of the presented results, they can be satisfied by a very broad class of algorithms. As we noted earlier, e.g. random search methods satisfy the mentioned requirements and can serve as basic subprocedures in hybrid optimization schemes. Numerical experiments with simple realizations of locally and globally convergent hybrid algorithms (see Pintér (1980, 1981b)) indicate that they can provide a useful tool for solving optimization problems with complicated structure.

REFERENCES

1 BABA, N., Convergence of a random optimization method for constrained optimization problems. Journal of Optimization Theory and Applications 33 (1981) 451-461.

2 DEVROYE, L.P., Progressive global random search of continuous functions. Mathematical Programming 15 (1978) 330-342.

3 DIXON, L.C.W. and SZEGŐ, G.P., eds., Towards Global Optimisation, Vol. 1.-2., North-Holland, Amsterdam, 1975, 1978.

4 HESTENES, M.R., Conjugate Direction Methods in Optimization. Springer-Verlag, New York, 1980.

5 KALL, P. and PRÉKOPA, A., eds., Recent Results in Stochastic Programming. Lecture Notes in Economics and Operations Research 179. Springer-Verlag, Berlin-Heidelberg, 1980.

6 MARTI, K., Random search in optimization problems as a stochastic decision process (adaptive random search). Methods of Operations Research 36 (1980) 223-234.

7 PINTÉR, J., On the convergence and computational efficiency of random search optimization. Methods of Operations Research 33 (1979) 347-362.

8 PINTÉR, J. On a method of random search for unconstrained optimization, Avtomatika i Telemehanika 41 (1980) No.12., 76-85., (in Russian, translated as: Automation and Remote Control).

9 PINTÉR, J., Stochastically combined optimization procedures, their convergence and numerical performance, Methods of Operations Research 43 (1981a) 143-150.

10 PINTÉR, J., Stochastic procedures for solving optimization problems, Alkalmazott Matematikai Lapok 7(1981b) 217-252 (in Hungarian).

11 PINTÉR, J., Hybrid procedures for the solution of non-smooth constrained stochastic problems. Vestnik Moskovskogo Universiteta, Ser. 15., (1982a) No.1. 39-49., (in Russian, translated as: Moscow University Computational Mathematics and Cybernetics).

12 PINTÉR, J., Convergence properties of stochastic optimization procedures, (submitted for publication (1982b)).

13 SOLIS, F.J. and WETS, R.J-B., Minimization by random search techniques. Mathematics of Operations Research 6 (1981) 19-30.

*Proc. of the 3rd Pannonian Symp.
on Math. Stat., Visegrád, Hungary 1982
J. Mogyoródi, I. Vincze, W. Wertz, eds*

LOCAL RESISTANCE IN POLYNOMIAL REGRESSION AND RESTRICTED LEAST SQUARES

W. POLASEK

University of Vienna*, Austria

ABSTRACT

Using the dual derivative concepts of Balestra (1976) and Polasek (1980) we derive the local sensitivities for special linear models. These sensitivity results can be used for the sensitivity analysis in cubic spline models, using the estimation method of Buse and Lim (1977). First of all the local resistance, i.e. the sensitivity of the OLS-estimate with respect to data-variation is derived for the polynomial regression model. In the second part of the paper the sensitivities of mixed regression and restricted least squares with respect to the restriction matrix R are derived.

Keywords: Matrix by matrix derivatives, Polynomial Regression, Mixed Regression, Constrained Least Squares, Local Sensitivity Analysis.

0. INTRODUCTION

When I tried to analyse the problem of the sensitivities of connection points in cubic spline models, I had to derive first some results on related problems. These problems are the local resistance analysis of polynomial regression models (note that the term "local resistance" refers to local sensitivity derivatives of estimators with respect to changes in the data matrix), mixed regression and constrained least squares estimates. To see the connections, I have briefly outlined the cubic spline methods using constrained least squares estimates in the appendix.

---

*Institute of Statistics and Computer Science, A-1090 Vienna Rooseveltplatz 6, AUSTRIA

Section 1 introduces briefly into the theory of matrix by matrix derivatives. Details can be found in Balestra (1976), Polasek (1980) or Neudecker (1969).

In Section 2 we derive the "eigenderivatives", i.e. $\partial X/\partial x$-derivatives of polynomial regression matrices. The B-type derivative is easier to obtain for that case. The local resistance result for polynomial regression is given in Section 3. In Section 4 the sensitivities of the mixed regression estimator (Theil and Goldberger 1961) are derived with respect to the restriction matrix R. Similar results are obtained in Section 5 for the constrained least squares estimator. Note that all these results are sensitivity analyses for special linear models. More general aspects and results for sensitivity analysis in the general linear model can be found in Polasek (1980).

1. MATRIX DERIVATIVES

Local sensitivity analysis investigates the behaviour of estimates (or parameters of the posterior distribution) if certain "input" parameters for the calculation of these estimates are changed. Using matrix notation it is therefore desirable to extend the multivariate differential calculus for functions with matrices as arguments. If the functions can be arranged in matrices, i.e. $a_{ij}=a_{ij}(B)$, $i=1,\dots,n$, $j=1,\dots,m$, $B=(p \times q)$ then we can define matrix derivatives, where the arrangement of the derivatives implies certain ways of interpretation.

The idea of matrix to matrix derivatives dates back to Dwyer and MacPhail. They recognized the importance of the arrangement of matrix by matrix derivatives by developing an elementwise approach of the form $[\partial a_{ij}/\partial B]=\partial <A>/\partial B$ and $\partial A/\partial b_{kl}=A/\partial <B>$.

Balestra (1976) developed an overall calculus for the non-elementwise matrix by matrix derivatives, but only for the $\partial a_{ij}/\partial B$-type derivative, based on the work of MacRae (1974). Polasek (1980) showed that the second type of derivative $[\partial A/\partial b_{kl}]=\partial A//\partial B$, where the double-bar notation is preferred to the $\partial A/\langle\partial B\rangle$-notation e.g. for typing convenience, implies a dual type of derivative algebra, analogous to Balestra (1976). Fortunately, both types of derivatives can be easily transformed to each other by the use of permutation matrices $P_{m,p}=P'_{p,m}$:

$$\partial B/\partial A=P_{m,p}\partial R//\partial A\ P_{q,n}, \tag{1.1}$$

$$\partial B//\partial A=P_{q,m}\partial B/\partial A\ P_{n,q}. \tag{1.2}$$

For definition and properties of permutation matrices see Balestra (1976) or Magnus and Neudecker (1979). The transformation of derivatives is usually easily obtained, since permutation matrices exchange the variables in a Kronecker-product by the relationship

$$A\otimes B=P_{p,m}(B\otimes A)P_{n,q}. \tag{1.3}$$

For the cases of sensitivity analysis discussed in the following sections, the $\partial B//\partial A$-type derivative is preferred. The $\partial B//\partial A$-derivatives involve fewer permutation matrices and allow big matrices of derivative results to be broken up more easily for better interpretable results for statistical purposes. The translation into a dual form $\partial B/\partial A$ is possible by relation (2.1).

2. EIGENDERIVATIVES OF POLYNOMIAL REGRESSION MATRICES

In the following Section we derive the A- and the B-type eigenderivatives of the matrix of polynomial variables. They form the basis of the local resistance results for polynomial regression estimators.

2.1 B-type matrix derivatives

Let $x'=(x_1,\ldots,x_T)$ be a (Tx1) vector of (time series) observations and from this a (TxK) polynomial regression matrix can be formed by

$$X=\begin{pmatrix} x_1 & x_1^2 & \ldots & x_1^K \\ x_2 & x_2^2 & \ldots & x_2^K \\ \vdots & \vdots & & \vdots \\ x_T & x_T^2 & \ldots & x_T^K \end{pmatrix} \quad (2.1)$$

where K denotes the order of the polynomial approximation. Then we can obtain the following derivative result:

Lemma 2.1

The B-type eigenderivative of the polynomial regression matrix (2.1) is given by the (T^2xK) derivative matrix

$$\partial X/\partial x=(\text{vec}I_T, 2\text{vec}\tilde{X}, 3\text{vec}\tilde{X}^2, \ldots, K\text{vec}\tilde{X}^{K-1}) \quad (2.2)$$

where I_T is the identity matrix of order T and vec (.) denotes the usual vectorisation process of a matrix by columns. $\tilde{X}$ denotes a (TxT) diagonal matrix formed by the individual observations of the vector $x=(x,\ldots,x_T)$:

$$\tilde{X}=\text{diag}(x_1,\ldots,x_T)=\begin{pmatrix} x_1 & & \sigma \\ & \ddots & \\ \sigma & & x_T \end{pmatrix}. \quad (2.3)$$

Proof: The B-type derivative of Balestra (1976) is given by the definition

$$\partial X/\partial x=[\partial x_{ij}/\partial x],\ i=1,\ldots,T,\ j=1,\ldots,K. \quad (2.4)$$

If we partition the polynomial matrix (2.1) columnwise by $X=(X_{\cdot 1},X_{\cdot 2},\ldots,X_{\cdot K})$ we obtain immediately

$$\partial X_{\cdot 1}/\partial x = \begin{pmatrix} 1 \\ 0 \\ \cdots \\ 0 \\ 1 \\ \cdots \\ 0 \\ 1 \end{pmatrix} = \begin{pmatrix} e_1 \\ \vdots \\ e_T \end{pmatrix} = \mathrm{vec} I_T$$

$$\partial X_{\cdot 2}/\partial x = \begin{pmatrix} x_1 \\ 0 \\ \vdots \\ 0 \\ x_T \end{pmatrix} = \mathrm{vec}\ \tilde{X} \qquad (2.5)$$

$$\partial X_{\cdot K}/\partial x = \begin{pmatrix} K\ x_1^{K-1} \\ 0 \\ \vdots \\ 0 \\ K\ x_T^{K-1} \end{pmatrix} = K\ \mathrm{vec}(\tilde{X}^K).$$

For further calculations and also for interpretation purposes it is useful to look at the structure of the elementwise derivatives of (2.2). They are given by

$$\partial X_{tj}/\partial x = (j-1) e_t\ x_t^{j-1}, \quad t=1,\dots,T,\ j=1,\dots,K, \qquad (2.6)$$

where e_t denotes the t-th unity vector of order T. □

Note also that the derivative of the transposed polynomial matrix is given by the (KTxT) matrix

$$\partial X'/\partial x = \begin{pmatrix} \partial X_{\cdot 1}'/\partial x \\ \vdots \\ X_{\cdot T}'/\partial x \end{pmatrix} = \begin{pmatrix} I_T \\ \vdots \\ K\ \tilde{X}^{K-1} \end{pmatrix}. \qquad (2.7)$$

3. LOCAL RESISTANCE OF POLYNOMIAL REGRESSION

Let $y'=(y_1,\dots,y_T)$ and $x'=(x_1,\dots,x_T)$ be two statistical variables, then the polynomial regression of order K of the variable x on the variable y is given by the linear model

$$y = X\beta + u \qquad (3.1)$$

where X is the (TxK) polynomial regression matrix given in (2.1). The OLS-estimator is given by the usual formula

$$\hat{b}=(X'X)^{-1}X'y. \tag{3.2}$$

Theorem 3.1

The local resistance of the OLS-estimator (3.2) of the polynomial regression model (3.1) is given by (KxT) B-type derivative matrix

$$\partial\hat{b}/\partial x=((X'X)^{-1}\otimes I_T)\left[\begin{pmatrix}\hat{e}\\ 2\tilde{X}\hat{e}\\ \vdots\\ K\tilde{X}^{K-1}\hat{e}\end{pmatrix}-\sum_{k=1}^{K}k\hat{b}_k \operatorname{vec}X^{(k-1)}\right] \tag{3.3}$$

where $\hat{e}=y-X\hat{b}$ is the estimated residual and $\tilde{X}$ is given in (2.3). $X^{(n)}$ is a (TxK) special "superpolynomial" matrix created by columnwise exponential powers of matrix X.

$$X^{(n)}=\begin{pmatrix}x_1^{n+1} & \dots & x_1^{n+K}\\ x_2^{n+1} & \dots & x_2^{n+K}\\ \vdots & & \vdots\\ x_T^{n+1} & \dots & x_T^{n+K}\end{pmatrix},\ n\in N. \tag{3.4}$$

Proof: Let us split up the OLS-estimate (3.2) into $\hat{b}=H^{-1}g$ where the two components are given by $H=X'X$ and $g=X'y$. In the first step we derive the local resistance for the variance matrix H^{-1}.

$$\begin{aligned}\partial H^{-1}/\partial x&=-(H^{-1}\otimes I_T)(\partial X'X/\partial x)(H^{-1}\otimes I_1)\\ &=-(H^{-1}\otimes I_T)[(\partial X'/\partial x)X+(X'\otimes I_T)(\partial X/\partial x)]H^{-1}.\end{aligned} \tag{3.5}$$

Finally we obtain the local resistance for $\hat{b}$ by

$$\partial\hat{b}/\partial x=H^{-1}/\partial x=\partial H^{-1}/\partial x(g\otimes I_1)+(H^{-1}\otimes I_T)\partial g/\partial x$$

$$=(H^{-1}\otimes I_T)[-((\partial X'/\partial x)X+(X'\otimes I_T)(\partial X/\partial x))H^{-1}g+(\partial X'/\partial x)y]$$

$$=(H^{-1}\otimes I_T)[(\partial X'/\partial x)(y-Xb)-(X'\otimes I_T)(\partial X/\partial x)b]$$

$$=(H^{-1}\otimes I_T)\left[\begin{pmatrix}\hat{e}\\ 2\tilde{X}\hat{e}\\ \vdots\\ K\tilde{X}^{K-1}\hat{e}\end{pmatrix}-\sum_{k=1}^{K} k\, b_k \mathrm{vec}X^{(k-1)}\right]. \tag{3.6}$$

We used the transposed derivative (2.7) to simplify the first term in (3.5), i.e.

$$(\partial X'/\partial x)\hat{e}=\begin{pmatrix}I_T\\ \vdots\\ K\ \tilde{X}^{K-1}\end{pmatrix}\hat{e}=\begin{pmatrix}\hat{e}\\ 2\tilde{X}e\\ \vdots\\ K\tilde{X}^{K-1}\hat{e}\end{pmatrix} \tag{3.7}$$

and the second term in (3.5) is simplified by

$$(X'\otimes I_T)(\partial X/\partial x)\hat{b}=\begin{pmatrix}x_1\ I_T & \dots & x_T\ I_T\\ \vdots & & \vdots\\ x_1{}^K I_T & \dots & x_T{}^K I_T\end{pmatrix}(\mathrm{vec}I_T, 2\mathrm{vec}\tilde{X},\dots,\dots,K\mathrm{vec}\tilde{X}^{K-1})\hat{b}$$

$$=(\mathrm{vec}X,\ 2\mathrm{vec}X^{(1)},\dots,K\mathrm{vec}X^{(K-1)})\hat{b} \tag{3.8}$$

$$=\sum_{k=1}^{K} k\hat{b}_k \mathrm{vec}X^{(k-1)}$$

where $X^{(k-1)}$ is defined as in (3.4). □

Note that the B-type elementwise resistances of (3.3) are given by the (Tx1) derivative vectors

$$\partial\hat{b}_j/\partial x=\sum_{k=1}^{K} h^{jk}(k\tilde{X}^{k-1}e-\sum_{i=1}^{K} i\hat{b}_i\cdot\tilde{X}_k{}^{(i-1)}),\ j=1,\dots,K \tag{3.9}$$

where the h^{jk} are the elements of the covariance matrix $(X'X)^{-1}$.

4. LOCAL SENSITIVITY OF RESTRICTIONS IN MIXED REGRESSION MODELS

Since cubic spline models can be written as restricted least squares estimates, we need some general results on the sensitivity of restrictions in restricted least squares, before we can discuss the cubic spline problem. Given the mixed regression set up as in Theil and Goldberger (1961) or Theil (1971)

$$\begin{pmatrix} y \\ r \end{pmatrix} = \begin{pmatrix} X \\ R \end{pmatrix} \beta + \begin{pmatrix} \varepsilon \\ v \end{pmatrix}, \quad \begin{pmatrix} \varepsilon \\ v \end{pmatrix} \sim N\left[\sigma, \begin{pmatrix} \Sigma & \sigma \\ \sigma & B \end{pmatrix} \right] \tag{4.1}$$

then the restricted least squares estimate (posterior mean in a Bayesian framework) is given by

$$\hat{\beta}(R,r)=(X'\Sigma^{-1}X+R'B^{-1}R)^{-1}(X'\Sigma^{-1}Xb+R'B^{-1}r). \tag{4.2}$$

Note that the estimator $\hat{\beta}$ is strict symmetric in the matrices X,Σ and R,B.

Theorem 4.1

The local sensitivity of the restricted least squares estimate (4.2) with respect to the (pxq) restriction matrix R is given by the A-type derivative

$$\partial\hat{B}//\partial R=(B^{-1}\otimes H^{-1})G \tag{4.3}$$

where the H and G matrices are given by

$$G=(r-R\hat{\beta})\otimes I_q-\mathrm{vec}R'\,\mathrm{vec}'\hat{\beta}, \tag{4.4}$$

$$H=X'\Sigma^{-1}X+R'B^{-1}R. \tag{4.5}$$

Proof: Let us split up the estimate $\hat{\beta}$ in (4.2) into $\hat{\beta}=H^{-1}g$ where H is given in (4.5) and $g=X'\Sigma^{-1}y+R'B^{-1}r$. As a preliminary result we derive the sensitivity of the covariance matrix H^{-1} with respect to the restrictions R

$$\partial H^{-1}//\partial R=-P_{p,q}(H^{-1}\otimes H^{-1}B^{-1}R)-\mathrm{vec}H^{-1}R'B^{-1}\mathrm{vec}'H^{-1}. \tag{4.6}$$

The proof of this result is quite similar to results in Polasek (1980). Now the final derivative is given by

$$\partial\hat{\beta}//\partial R=\partial H^{-1}g//\partial R=\partial H^{-1}//\partial R(I_q\otimes g)+(I_p\otimes H^{-1})\partial g//\partial R$$

$$=-(I_p\otimes H^{-1})[P_{p,q}(I_q\otimes B^{-1}R')+\text{vec}R'B^{-1}\text{vec}'I_q)]\cdot$$

$$\cdot(I_q\otimes H^{-1})(I_q\otimes g)+(I_p\otimes H^{-1})P_{p,q}(I_q\otimes B^{-1}r)$$

$$=-(I_p\otimes H^{-1})P_{p,q}(I_q\otimes B^{-1}R\hat{\beta})+(\text{vec}R'B^{-1}\text{vec}'I_q)(I_q\otimes\hat{\beta})- \tag{4.7}$$

$$-(B^{-1}r\otimes I_q)$$

$$=(I_p\otimes H^{-1})(B^{-1}r\otimes I_q)-(B^{-1}R^{-1}\hat{\beta}\otimes I_q)-\text{vec}R'B^{-1}\text{vec}'\hat{\beta}$$

$$=(I_p\otimes H^{-1})(B^{-1}\otimes I_q)[(r-R\hat{\beta})\otimes I_q-\text{vec}R'\text{vec}'\hat{\beta}]$$

$$=(B^{-1}\otimes H^{-1})G.$$ □

5. LOCAL RESISTANCE OF RESTRICTED LEAST SQUARES

A special case of stochastic constrained least squares is constrained least squares. It is also worthwhile to derive the sensitivity of the constraints in that case. The restricted least squares estimate of the linear model

$$Y=X\beta+u \tag{5.1}$$

subject to the nonstochastic restriction $r=R\beta$ is given by

$$b_R=b+(X'X)^{-1}R(X'X)^{-1}R'^{-1}(r-Rb). \tag{5.2}$$

This can be shown by Lagrangian multipliers (see e.g. Leamer (1978)). To facilitate the subsequent computation, we denote the middle term of (5.2) by the (pxp) matrix Z, i.e.

$$Z=(R(X'X)^{-1}R')^{-1}, \tag{5.3}$$

and derive first the sensitivity of Z with respect to R.

Lemma 5.1 The sensitivity of Z in (5.3) with respect to R is given by

$$\partial Z//\partial R=-\text{vec}Z\ \text{vec}'ZR(X'X)^{-1}-P_{p,p}(ZR(X'X)^{-1}\otimes Z). \tag{5.4}$$

Proof: A straightforward application of the derivative calculus yields

$$Z//\partial R = -(I_p \otimes Z)R(X'X)^{-1}(R'//\partial R)(I_K \otimes Z)$$

$$= -(I_p \otimes Z)\left[\frac{\partial R}{\partial R}(I_K \otimes (X'X)^{-1}R' + (I_p \otimes R(X'X)^{-1}\right]\frac{\partial R'}{\partial R}(I_K \otimes Z)$$

$$= -(I_p \otimes Z)[\text{vec}I_p \text{vec}' I_K (I_K \otimes (X'X)^{-1}R') + (I_p \otimes R(X'X)^{-1})$$

$$P_{p,K}(I_K \otimes Z) \tag{5.5}$$

$$= \left[-\text{vec}Z\text{vec}' R(X'X)^{-1} - (I_p \otimes ZR(X'X)^{-1}P_{p,K}\right](I_K \otimes Z)$$

$$= -\text{vec}Z\text{vec}' ZR(X'X)^{-1} - P_{p,p}(ZR(X'X)^{-1} \otimes Z).$$

Theorem 5.2

The local sensitivity of the restricted least squares estimate (5.2) with respect to R is given by the A-type derivative

$$\partial b_R // \partial R = \text{vec}(X'X)^{-1}Z(\text{vec}'b' \; RZR(X'X)^{-1} - \text{vec}'b) +$$
$$+ Z(r-Rb) \otimes (X'X)^{-1}Z(X'X)^{-1}. \tag{5.6}$$

Proof: Using the derivative result of Lemma 5.1 we obtain

$$\partial b_R // \partial R = -(X'X)^{-1}Z(r-Rb)//\partial R$$

$$= -(I_p \otimes (X'X)^{-1}[(\partial Z//\partial R)(I_K \otimes (r-Rb) + (I_p \otimes Z)(\partial Rb//\partial R)]$$

$$= -(I_p \otimes (X'X)^{-1}[(-\text{vec}Z\text{vec}' R(X'X)^{-1}Z +$$

$$+ P_{p,p}(Z(X'X)^{-1} \otimes Z))(I_K \otimes (r-Rb)) + (I_p \otimes Z)(\text{vec}I_p \text{vec}' I_n)(I_K \otimes b)]$$

$$= -(I_p \otimes (X'X)^{-1})[-\text{vec}Z\text{vec}' b' R' ZR(X'X)^{-1} -$$

$$- P_{p,p}(Z(X'X)^{-1} \otimes Z(r-Rb) + \text{vec}Z\text{vec}' b]$$

$$= \text{vec}(X'X)^{-1}Z\text{vec}' b' R' ZR(X'X)^{-1} + (I_p \otimes (X'X)^{-1})$$

$$(Z(r-Rb) \otimes Z(X'X)^{-1}) - \text{vec}(X'X)^{-1}Z\text{vec}' b$$

$$= \text{vec}(X'X)^{-1}Z(\text{vec}' b' R' ZR(X'X)^{-1} - \text{vec}' b) +$$

$$+ Z(r-Rb) \otimes (X'X)^{-1}Z(X'X)^{-1}. \quad \square$$

APPENDIX

Cubic Splines as Restricted Least Squares Model

The cubic spline model in K regimes, i.e.

$$y=f_j(x)=a_j+b_jx+c_jx^2+d_jx^3, \quad e_{j-1}\le x\le e_j, \quad j=1,\dots,K,$$

with the joining (continuity) restrictions at the knots $e_1,\dots,e_{K-1}$ up to the second order

$$\begin{aligned} &f_j(e_j)=f_{j+1}(e_j), \\ &f'_j(e_j)=f'_{j+1}(e_j), \qquad\qquad j=1,\dots,K-1, \\ &f''_j(e_j)=f''_{j+1}(e_j), \end{aligned}$$

and the endpoint restrictions

$$\begin{aligned} 2c_1+6d\,e_0 &= \pi_0(2c_1+6d_1e_1), \\ 2c_K+6d_Ke_K &= \pi_K(2c_K+6d_Ke_{K-1}) \end{aligned}$$

can be formulated as restricted least squares problem of the form

$$\begin{aligned} &Y=X\beta+u \\ &r=R\beta \end{aligned} \qquad (x)$$

or

$$\begin{pmatrix} y \\ \vdots \\ y_K \end{pmatrix} = \begin{pmatrix} X_1 & & \sigma \\ & \ddots & \\ \sigma & & X_K \end{pmatrix} \begin{pmatrix} \beta_1 \\ \vdots \\ \beta_K \end{pmatrix} + u$$

$$\begin{pmatrix} r_0 \\ r_1 \\ r_2 \\ r_3 \end{pmatrix} = \begin{pmatrix} R_0 \\ R_1 \\ R_2 \\ R_3 \end{pmatrix} \beta$$

where the restriction matrices R_0, R_1 and R_2 are structured as

$$R_i = (K-1)\times 4K \begin{bmatrix} -e_{1i} & e_{1i} & 0 & 0 & 0 & 0 \\ 0 \ldots 0 & & -e_{2i} & e_{2i} & 0 \ldots & 0 \\ & & & & \ddots & \\ 0 & 0 & 0 & 0 & -e_{K-1,i} & e_{K-1,i} \end{bmatrix}, \quad i=0,1,2,$$

and the $e._K$-vectors represent the derivative restrictions up to the order K

$$R_0 \rightarrow e_{j0} = (1, e_j, e_j^2, e_j^3),$$

$$R_1 \rightarrow e_{j1} = (0, 1, 2e_j, 3e_j^2), \quad j=(1,\ldots,K-1),$$

$$R_2 \rightarrow e_{j2} = (0, 0, 2, 6e_j),$$

and $R_3 = (2\times 4K)$ contains the endpoint restrictions

$$R_3 = \begin{bmatrix} 0 & 0 & 2(\pi_0 - 1) & 6(\pi_0 e_1 - e_0) & \ldots & 0 & 0 & 0 & 0 \\ 0 & 0 & 0 & 0 & \ldots & 0 & 0 & 2(\pi_K - 1) & 6(\pi_K e_K - e_{K-1}) \end{bmatrix}.$$

The restricted least squares estimate of the model (x) is given by (see e.g. Theil (1971))

$$b_R = b + (X'X)^{-1} R (X'X)^{-1} R'^{-1} (r - Rb)$$

where

$$b = (X'X)^{-1} X'y$$

is the OLS-estimator.

REFERENCES

BALESTRA, P., (1976) La Derivation Matricielle, Sirey, Paris.

BUSE, A. and L. LIM, (1977) Cubic Splines as a Special Case of Restricted Least Squares, JASA 72, No 357, 64-68.

HENDERSON, H.V. and S.R. SEARLE (1979) Vec and Vech Operators for Matrices with some Uses in Jacobians and Multivariate Statistics. The Canadian Journal of Statistics 7, 65-81.

LEAMER, E.E., (1978) Specification Searches, John Wiley, New York.

MacRAE, E.CH., (1974) Matrix Derivatives with an Application to an Adaptive Linear Decision Problem, Annals of Statistics, Vol.2,337-348.

MAGNUS, J.R. and H. NEUDECKER, (1979) The Commutation Matrix: Some Properties and Applications, Annals of Statistics, 381-394.

NEUDECKER, H., (1969) Some Theorems on Matrix Differentiation with Special Reference to Kronecker Matrix Products, JASA Vol.64, 953-963.

POLASEK, W., (1980) Local Sensitivity Analysis in the General Linear Model, University of Southern California, MRG no. 8006.

POLASEK, W., (1982) Local Resistance in Distributed Lag Models, Statistische Hefte 23(1), 44-51.

ROGERS, G.S.,(1980) Matrix Derivatives, Marcel Decker, New York and Basel.

THEIL, H. and A.S. GOLDBERGER,(1961) On Pure and Mixed Statistical Estimation in Economics, IER 2, 65-78.

THEIL, H., (1971) Principles of Econometrics, John Wiley, New York.

Proc. of the 3rd Pannonian Symp. on Math. Stat., Visegrád, Hungary 1982
J. Mogyoródi, I. Vincze, W. Wertz, eds

ON STOCHASTIC PROCESSES WITH LEARNING PROPERTIES IN A TIME-VARYING ENVIRONMENT AND SOME APPLICATIONS

L. RUTKOWSKI

Technical University of Częstochowa, Poland

Abstract: A general procedure for nonparametric learning in a time-varying environment is proposed and its asymptotic properties are investigated. Applications to nonparametric estimation of time-varying probability densities and nonparametric classification of nonstationary patterns are presented.

1. INTRODUCTION

Let (X_n, Y_n), $n=1,2,\ldots$, be a sequence of independent pairs of random variables; X_n takes values in $A \subset R$, Y_n takes values in $B \subset R$. Let K_n, $n=1,2,\ldots$, be a sequence of bivariate Borel-measurable functions (so-called kernel functions) defined on $A \times A$. For examples of different types of kernels K_n we refer to [3]. Define

$$E\, Y_n K_n(x, X_n) = r_n(x). \tag{1}$$

Let R_n be a sequence of Borel-measurable functions, $n=1,2,\ldots$, defined on A such that

$$\int_A (r_n(x) - R_n(x))^2 dx \xrightarrow{n} 0. \tag{2}$$

It is also assumed that

$$\int_A (R_{n+i}(x) - R_n(x))^2 dx \xrightarrow{n} 0 \text{ for any fixed } i \geq 0. \tag{3}$$

Let a_n be a sequence of numbers such that

$$a_n \xrightarrow{n} 0, \; a_n \geq 0, \; \sum_{n=1}^{\infty} a_n = \infty. \tag{4}$$

Define a random process

$$\hat{R}_{n+1}(x)=\hat{R}_n(x)+a_{n+1}(Y_{n+1}K_{n+1}(x,X_{n+1})-\hat{R}_n(x)). \tag{5}$$

It should be noted that procedure (5) is suggested by earlier papers treating on sequential (nonparametric) estimates of probability density and regression functions [2],[4],[7],[8],[11]. In section 2 we investigate asymptotic properties of procedure (5) whereas section 3 contains applications to nonparametric estimation of time-varying probability densities and nonparametric classification of nonstationary patterns.

2. CONVERGENCE THEOREMS

THEOREM 1. Assume that conditions (2), (3) and (4) are satisfied. If

$$a_n \int_A E[Y_nK_n(x,X_n)]^2dx \xrightarrow{n} 0, \tag{6}$$

$$a_n^{-2}\int_A (r_{n+1}(x)-r_n(x))^2dx \xrightarrow{n} 0, \tag{7}$$

then

$$E\int_A (\hat{R}_n(x)-R_{n+i}(x))^2dx \xrightarrow{n} 0 \tag{8}$$

for fixed $i\geq 0$.

THEOREM 2. Assume that conditions (2), (3) and (4) are satisfied. If

$$\sum_{n=1}^{\infty} a_n^2 \int_A E[Y_nK_n(x,X_n)]^2dx < \infty, \tag{9}$$

$$\sum_{n=1}^{\infty} a_n^{-1}\int_A (r_{n+1}(x)-r_n(x))^2dx < \infty, \tag{10}$$

then

$$\int_A (\hat{R}_n(x)-R_{n+i}(x))^2dx \xrightarrow{n} 0 \tag{11}$$

with probability one for fixed $i\geq 0$.

Proof of Theorems 1 and 2. Observe that

$$\int_A (\hat{R}_n(x)-R_{n+i}(x))^2dx \leq 3\int_A (\hat{R}_n(x)-r_n(x))^2dx$$

$$+ 3\int_A (r_n(x)-R_n(x))^2dx+3\int_A (R_{n+i}(x)-R_n(x))^2dx.$$

Next

$$E[\int_A (\hat{R}_{n+1}(x)-r_{n+1}(x))^2dx|X_1,Y_1,...,X_n,Y_n]$$

$$=(1-a_{n+1})^2\int_A (\hat{R}_n(x)-r_n(x))^2dx$$

$$+a_{n+1}^2 E\int_A [Y_{n+1}K_{n+1}(x,X_{n+1})-r_{n+1}(x)]^2dx$$

$$+(1-a_{n+1})^2\int_A (r_{n+1}(x)-r_n(x))^2dx$$

$$+2(1-a_{n+1})^2\int_A (r_{n+1}(x)-r_n(x))(R_n(x)-r_n(x))dx.$$

From (1) it follows that

$$E\int_A [Y_{n+1}K_{n+1}(x,X_{n+1})-r_{n+1}(x)]^2dx \leq \int_A E[Y_{n+1}K_{n+1}(x,X_{n+1})]^2dx.$$

Using the inequality $2ab \leq a^2k+b^2k^{-1}$ (holding true for every $k>0$) and setting $k=(a_{n+1}c_1)^{-1}$, $0<c_1<1$, one gets

$$2\int_A (r_{n+1}(x)-r_n(x))(\hat{R}_n(x)-r_n(x))dx$$

$$\leq c_1a_{n+1}\int_A (\hat{R}_n(x)-r_n(x))^2dx+c_1^{-1}a_{n+1}^{-1}\int_A (r_{n+1}(x)-r_n(x))^2dx.$$

Therefore,

$$E[\int_A (\hat{R}_{n+1}(x)-r_{n+1}(x))^2dx|X_1,Y_1,...,X_n,Y_n]$$

$$\leq (1-a_{n+1}(1-c_1))\int_A (\hat{R}_n(x)-r_n(x))^2dx \qquad (12)$$

$$+a_{n+1}^2\int_A E[Y_{n+1}K_{n+1}(x,X_{n+1})]^2dx+c_2a_{n+1}^{-1}\int_A (r_{n+1}(x)-r_n(x))^2dx$$

for sufficiently large n.

We shall now use the following lemma:

Lemma (Braverman and Rozonoer [1]). Let U_n be a sequence of random variables on a probability space (Ω,F,P). Let F_n be a sequence of Borel fields, $F_n \subset F_{n+1} \subset F$, where U_n are measurable with respect to F_n. Finally, let b_n, s_n and q_n be sequences of numbers.

Suppose that

(i) $U_n \geqq 0$ a.s., $n=1,2,\ldots,$

(ii) $EU_1 < \infty$,

(iii) $b_n \geqq 0$, $b_n \xrightarrow{n} 0$, $\sum_{n=1}^{\infty} b_n = \infty$.

(a) If

$$E[U_{n+1}|F_n] \leqq (1-b_n)U_n + b_n s_n,$$

where

$$s_n \xrightarrow{n} 0,$$

then

$$EU_n \xrightarrow{n} 0.$$

(b) If

$$E[U_{n+1}|F_n] \leqq (1-b_n)U_n + q_n,$$

where

$$\sum_{n=1}^{\infty} q_n < \infty,$$

then

$$U_n \xrightarrow{n} 0 \text{ with probability one.}$$

By a direct application of the Lemma we establish convergence (8) and (11).

3. APPLICATIONS

A) Nonparametric estimation of time-varying probability densities

Let X_n, n=1,2,..., be a sequence of independent random variables with unknown probability densities f_n. Define

$$K_n(x,t)=h_n^{-1}K((x-t)/h_n), \tag{13}$$

where h_n is a sequence of positive numbers such that

$$h_n \xrightarrow{n} 0, \tag{14}$$

and K is a Borel-measurable function defined on R such that

$$K(x)\geq 0,\ \sup_{x\in R}|K(x)|<\infty,\ \int_R K(x)dx=1 \tag{15}$$

and

$$\int_R xK(x)dx=0,\ \int_R x^2K(x)dx<\infty. \tag{16}$$

Examples of commonly used weighting functions K (and their properties) are given in [6].

For nonparametric estimation of time-varying probability densities we apply the following procedure

$$\hat{f}_{n+1}(x)=\hat{f}_n(x)+a_{n+1}(h_{n+1}^{-1}K((x-X_{n+1})/h_{n+1})-\hat{f}_n(x)). \tag{17}$$

Corollary 1. Let X_n, n=1,2,..., be a sequence of independent random variables taking values on the real line with twice continuously differentiable probability densities f_n such that

$$\int_R f_n^2(x)dx<\infty,\ \int_R [f_n''(x)]^2dx<\infty,\ n=1,2,\ldots. \tag{18}$$

Assume that conditions (4), (14), (15), and (16) hold. If

$$a_n h_n^{-1} \xrightarrow{n} 0, \tag{19}$$

$$a_n^{-2}h_n^4\int_R [f_n''(x)]^2dx \xrightarrow{n} 0, \tag{20}$$

$$a_n^{-2}\int_R (f_{n+1}(x)-f_n(x))^2dx \xrightarrow{n} 0, \tag{21}$$

then for tracking scheme (17) we have

$$E\int_R [\hat{f}_n(x)-f_n(x)]^2dx \xrightarrow{n} 0. \tag{22}$$

Proof: First we note that condition (7) is implied by a pair

$$a_n^{-2}\int_R (r_n(x)-R_n(x))^2dx \xrightarrow{n} 0, \qquad (23)$$

$$a_n^{-2}\int_R (R_{n+1}(x)-R_n(x))^2dx \xrightarrow{n} 0. \qquad (24)$$

It is easily seen that conditions (6) and (23) become

$$a_n h_n^{-1}\int_R\int_R K^2(u)f_n(x-uh_n)dxdu \xrightarrow{n} 0,$$

$$a_n^{-2}\int_R(\int_R K(u)(f_n(x-uh_n)-f_n(x))du)^2dx \xrightarrow{n} 0,$$

respectively. Obviously

$$\int_R\int_R K^2(u)f_n(x-uh_n)dxdu \leq \sup_{x\in R} K(u).$$

Using arguments of Wertz [10] we obtain

$$\int_R(\int_R K(u)(f_n(x-uh_n)-f_n(x))du)^2dx$$

$$\leq \frac{h_n^4}{4}\int_R[f_n''(x)]^2dx(\int_R x^2K(x)dx)^2.$$

The proof of (22) is complete.

Remark. Nonparametric estimates of nonstationary probability densities were also studied in [5] and [9] with the restrictive assumption that f_n converges to the finite limit, i.e.

$$\int_R (f_n(x)-f(x))^2dx \xrightarrow{n} 0.$$

B) Nonparametric classification of nonstationary patterns

Let (T_n,X_n), n=1,2,..., be a sequence of independent pairs of random variables; $P(T_n=k)=p_k$, $k=1,2$, $p_1+p_2=1$, X_n takes values on the real line. Let f_{kn} be a conditional density of X_n given $T_n=k$. When f_{kn} and p_k are known a Bayes discriminant function

$$D_n(x)=p_1f_{1n}(x)-p_2f_{2n}(x)$$

classifies $X_n=y_1$ as coming from class 1 if $D_n(y_1)\geq 0$ and from class 2 otherwise. Define

$$t_{kn}=\begin{cases}1 & \text{if } T_n=k\\ 0 & \text{otherwise.}\end{cases}$$

For nonparametric learning of time-varying discriminant functions D_n we use the following procedure

$$\hat{D}_n(x)=\hat{D}_{n-1}(x)+a_n((t_{1n}-t_{2n})(h_n^{-1}K((x-X_n)/h_n)-\hat{D}_{n-1}(x)) \quad (25)$$

which classifies $X_{n+1}=y_2$ as coming from class 1 if $\hat{D}_n(y_2)\geq 0$ and from class 2 otherwise. Observe that $\hat{D}_n$ is an approximation of D_{n+1}.

Corollary 2. If the class conditional densities f_{1n} and f_{2n} and sequences a_n and h_n satisfy conditions of Corollary 1 then

$$E\int_R(\hat{D}_n(x)-D_{n+1}(x))^2dx \xrightarrow{n} 0.$$

4. FINAL REMARKS

In section 3 we established convergence of the mean integrated square errors for estimates (17) and (25). The almost sure convergence can be proved analogously using theorem 2 in section 2.

REFERENCES

[1] E.M. BRAVERMAN and L.I. ROZONOER, "Convergence of random processes in machine learning theory", Automation and Remote Control, vol. 30, pp. 44-64, 1969.

[2] L.P. DEVROYE and T.J. WAGNER, "On the L_1 convergence of kernel estimators of regression functions with applications in discrimination", Z. Wahrscheinlichkeitstheorie verw. Gebiete, vol. 51, pp. 15-25, 1980.

[3] A. FÖLDES and P. RÉVÉSZ, "A general method for density estimation", Studia Sci. Math. Hungar., vol. 9, pp. 81-92, 1974.

[4] W. GREBLICKI and A. KRZYŻAK, "Asymptotic properties of kernel estimates of a regression function", J. Statist. Planning Inference, vol. 4, pp. 81-90, 1980.

[5] L. GYŐRFI, Z. GYŐRFI and I. VAJDA, "A strong law of large numbers and some applications", Studia Sci. Math. Hungar., vol. 12, pp. 233-244, 1977.

[6] E. PARZEN, "On estimation of a probability density function and mode", Ann. Math. Statist., vol. 33, pp. 1065-1076, 1962.

[7] L. REJTŐ and P. RÉVÉSZ, "Density estimation and pattern classification", Problems of Control and Information Theory, vol. 2, pp. 67-80, 1973.

[8] L. RUTKOWSKI, "Sequential estimates of a regression function by orthogonal series with applications in discrimination", in Lecture Notes in Statistics, vol. 8, pp. 236-244, Springer-Verlag, New York - Heidelberg - Berlin, 1981.

[9] L. RUTKOWSKI, "On Bayes risk consistent pattern recognition procedures in a quasi-stationary environment", IEEE Trans. Pattern Anal. Machine Intell., vol. PAMI-4, pp. 84-87, 1982.

[10] W. WERTZ, "Fehlerabschätzung für eine Klasse von nichtparametrischen Schätzfolgen", Metrika, vol. 19, pp. 131-139, 1982.

[11] C.T. WOLVERTON and T.J. WAGNER, "Asymptotically optimal discriminant functions for pattern classification", IEEE Trans. Inform. Theory, vol. IT-25, pp. 258-265, 1969.

Proc. of the 3rd Pannonian Symp. on Math. Stat., Visegrád, Hungary 1982
J. Mogyoródi, I. Vincze, W. Wertz, eds

THE DUAL SPACE OF MARTINGALE VMO SPACE

F. SCHIPP

Eötvös Loránd University, Budapest, Hungary

1. INTRODUCTION

C. Fefferman [1] discovered the remarkable fact that the dual space of the classical Hardy space (i.e. the space of all bounded linear functionals) is the BMO space (the space of functions with bounded mean oscillation). A function φ defined on the thorus T belongs to the space BMO if

$$||\varphi||_* = \sup_{|I|} \frac{1}{I} \int_I |\varphi - \frac{1}{|I|} \int_I \varphi \, dx| dx < \infty ,$$

where the sup is taken over all intervals of T and $|I|$ denotes the length of I. Further, the dual space of VMO (of the space of functions with vanishing mean oscillation) is the classical Hardy space (see e.g. [2]). A function $\varphi \in$BMO belongs to VMO, if

$$\lim_{|I| \to 0} \frac{1}{|I|} \int_I |\varphi - \frac{1}{|I|} \int_I \varphi \, dx| dx = 0.$$

The martingale Hardy and BMO spaces are investigated in [3]. To define these concepts let us fix a probability space (X,A,μ) and a non-decreasing sequence of sub-σ-fields of A:

$$A_0 = \{X_1 \emptyset\} \subseteq A_1 \subseteq A_2 \subseteq \dots \subseteq A .$$

The conditional expectation of $f \in L = L^1(X,A,\mu)$ with respect to A_n will be denoted by $E_n f$ and the maximal function of f by

$$E^*f := \sup_{n\in \mathbb{N}} |E_n f| ,$$

where $\mathbb{N} = \{0,1,2,\ldots\}$. Further, let $||f||_p$ $(1\le p\le\infty)$ the $L^p(X,A,\mu)$-norm of f.

The martingale Hardy-space $\mathbb{H}$ coincides with the space of all functions $f\in L$, for which

$$||f||_{\mathbb{H}} := ||E^*f||_1 < \infty .$$

The dual space of $\mathbb{H}$ is the martingale $\mathbb{BMO}$ space, which is the space of all $\varphi\in L$ with the property

$$||\varphi||_{\mathbb{BMO}} := \sup_n ||E_n|f-E_{n-1}f|||_\infty < \infty$$

(see e.g. [3]).

The martingale $\mathbb{VMO}$ space is the space of the function f such that

$$\lim_{n\to\infty} ||E_n|f-E_{n-1}f|||_\infty = 0 .$$

The set $\mathbb{VMO}$ endowed with the $\mathbb{BMO}$-norm is a Banach-subspace of $\mathbb{BMO}$. In this paper it will be proved that if the A_n's are generated by a finite set of atoms, then the dual of $\mathbb{VMO}$ is $\mathbb{H}$.

2. PRELIMINARIES

For a sequence of Banach-spaces $(X_i, ||\ ||_i)$ $(i\in\mathbb{N}^*)$ let us introduce the following notations:

$$X = X_1 \times X_2 \times \ldots$$

$$||x||^1_X := \sum_{i=1}^{\infty} ||x_i||_i,\quad ||x||^\infty_X := \sup_{i\in\mathbb{N}^*} ||x_i||_i \qquad (1)$$

$(x = (x_1,x_2,\ldots)\in X;\ \mathbb{N}^* = \mathbb{N}\setminus\{0\})$

and denote

$$X^1 := \{x \in X : \|x\|_X^1 < \infty\}, \quad X^\infty := \{x \in X : \|x\|_X^\infty < \infty\}$$
$$X_o^\infty := \{x \in X^\infty : \lim_{i\to\infty} \|x_i\|_i = 0\} . \tag{2}$$

There will be used later the following

Lemma 1. Let Y_i, $|\ |_i$ the dual space of X_i, $\|\ \|_i$ ($i \in \mathbb{N}^*$).

i) The dual space of $(X^1, \|\ \|_X^1)$ is isomorphic to $(Y^\infty, |\ |_Y^\infty)$ and an isomorphism can be given by the mapping

$$Y^\infty \ni y = (y_i,\ i \in \mathbb{N}^*) \to L_y$$

where

$$L_y(x) := \sum_{i=1}^{\infty} y_i(x_i) \qquad (x = x_1, x_2, \ldots) \in X^1) . \tag{3}$$

ii) The dual space of $(X_o^\infty, \|\ \|_X^\infty)$ is isomorphic to $(Y^1, |\ |_Y^1)$ and $Y^1 \ni y = (y_i,\ i \in \mathbb{N}^*) \to L_y$ is an isomorphism.

The proof of Lemma 1 is similar to the proof of the following well-known statements:

$$(\ell^1)^* = \ell^\infty, \quad C_o^* = \ell^1 , \tag{4}$$

where C_o^* is the set of sequences $x = (x_n,\ n \in \mathbb{N}) \in \ell^\infty$, for which $\lim_{n\to\infty} x_n = 0$ and as before, S^* denotes the dual space of the space S (see e.g. [4]).

Let $B \subseteq A$ a σ-field generated by a finite set of atoms and for a function belonging to $L^2 = L^2(X, A, \mu)$ we set

$$\|f\|_{(B,2,1)} := \|(E(|f|^2 | B))^{1/2}\|_1 ,$$
$$\|f\|_{(B,2,\infty)} := \|(E(|f|^2 | B))^{1/2}\|_\infty . \tag{5}$$

Lemma 2. The dual space of $(L^2, \|\ \|_{(B,2,1)})$ is isomorphic to $(L^2, \|\ \|_{(B,2,\infty)})$, and conversely the dual of $(L^2, \|\ \|_{(B,2,\infty)})$ is $(L^2, \|\ \|_{(B,2,1)})$, and an isomorphism can be given by

$$f \to \langle \cdot, f \rangle$$

where $\langle f,g\rangle := \int_X f\, g\, d\mu$ is the usual scalar product in L^2.

We remark that the first statement holds for every σ-algebra $B \subseteq A$. The proof of Lemma 2 is similar to the proof of (4) (see e.g. [5]).

3. THE DUAL OF $\mathbb{V}$MO

For every $f \in \mathbb{H}$ let

$$L_f(\varphi) := \lim_{n\to\infty} \langle E_n\varphi, E_n f\rangle \qquad (\varphi \in \mathbb{V}MO)\ . \tag{6}$$

On the basis of the Fefferman inequality

$$|L_f(\varphi)| < C\|\varphi\|_{\mathbb{B}MO}\|f\|_{\mathbb{H}} \quad (f\in\mathbb{H},\ \varphi\in\mathbb{B}MO)$$

the functional L_f is bounded and linear, moreover

$$\|L_f\| < C\|f\|_{\mathbb{H}} \qquad (f\in\mathbb{H}) \tag{7}$$

where C is an absolute constant.

The following will be proved.

<u>Theorem</u>. If the A_n's are generated by finite sets of atoms, then the mapping

$$\mathbb{H} \ni f \to L_f \in \mathbb{V}MO^*$$

is a bounded linear bijection.

On the basis of this theorem every bounded linear functional of $\mathbb{V}$MO can be written in the form (6), i.e. the space $\mathbb{V}MO^*$ and $\mathbb{H}$ can be identified and the $\mathbb{V}MO^*$-norm is equivalent to the $\mathbb{H}$-norm (by this identification).

<u>Proof of Theorem</u>. Let

$$X_n := L^2,\ \|\ \|_n := \|\ \|_{(A_n,2,\infty)} \quad (n\in\mathbb{N}^*)\ .$$

Then using the notations of our Lemmas, we have

$$Y_n := L^2,\ |\ |_n := \|\ \|_{(A_n,2,1)}$$

and the dual of $(X_o^\infty, \|\ \|_X^\infty)$ is isomorphic to $(Y^1, |\ |_Y)$ and the isomorphism can be given by

$$Y^1 = L^2 x L^2 x \ \ldots \qquad f = (f_1, f_2, \ldots) \to L_f,$$

where

$$L_f(\Psi) = \sum_{n=1}^{\infty} \langle f_n, \Psi_n \rangle$$

$$(\Psi = (\Psi_1, \Psi_2, \ldots) \in X_o^\infty) .$$

On the basis of the definition of the IBMO-norm and $\|\ \|_X^\infty$ norm the mapping

$$\mathrm{VMO} \ni \varphi \to \Phi(\varphi) := (\varphi - E_{n-1}\varphi,\ n \in \mathrm{IN}^*) \in X_o^\infty$$

is an isomorphism of VMO and let $\widetilde{X} := \Phi(\mathbb{V}\mathrm{MO})$. Let F: $\mathbb{V}\mathrm{MO} \to \mathrm{IR}$ a bounded linear functional. It will be proved that there exists an $g \in \mathrm{IH}$ such that

$$F = L_g \text{ and } \|g\|_{\mathrm{IH}} < 3\|F\|. \tag{8}$$

The - obviously bounded and linear - function $F \circ \Phi^{-1} : \widetilde{X} \to \mathrm{IR}$ can be extended to a bounded linear functional $\widetilde{F}$, defined on the space X_o^∞, and consequently there exists a $f \in Y^1$ with

$$\|f\|_Y^1 := \| \sum_{n-1}^{\infty} (E_n |f_n|^2)^{1/2} \|_1 < \infty$$

such that

$$\widetilde{F}(\Psi) = \sum_{n=1}^{\infty} \langle f_n, \Psi_n \rangle \quad (\Psi \in X_o^\infty),\ \|\widetilde{F}\| = |f|_Y^1 .$$

Applying this for $\Psi = \Phi(\varphi)$ ($\varphi \in$ VMO) (i.e. on $\widetilde{X}$) we get

$$F(\varphi) = (F \circ \Phi^{-1})(\Psi) = \sum_{n=1}^{\infty} \langle f_n, \varphi - E_{n-1}\varphi \rangle$$

$$= \sum_{n=1}^{\infty} \langle f_n - E_{n-1} f_n, \varphi \rangle =$$

$$= \lim_{n \to} \langle \sum_{k=1}^{n} (f_k - E_{k-1} f_K), \varphi \rangle$$

and $\|F\| = \|f\|_Y^1$.

Since

$$\sum_{n=1}^{\infty} ||E_{n-1}f_n||_1 \leq \sum_{n=1}^{\infty} ||f_n||_1 < \sum_{n=1}^{\infty} \int_X (E_n|f_n|^2)^{1/2}\, d\mu = |f|^1_Y < \infty\,, \tag{9}$$

the series $\sum_{n=1}^{\infty} (f_n - E_{n-1}f_n)$ converges in L-norm to a function $g \in L$. It will be proved that

$$E^*g := \sup_n |E_n g| \in L$$

and

$$||g||_{|H} = ||E^*g||_1 \leq 3|f|^1_Y = 3||F||\,. \tag{10}$$

For every $m \in \mathbb{N}$ we have

$$E_m g = E_m f_1 + \ldots + E_m f_m - (E_0 f_1 + \ldots + E_{m-1} f_m)\,,$$

and thereby

$$E^*g < \sum_{n=1}^{\infty} \sup_{k \geq n} |E_k f_n| + \sum_{n=1}^{\infty} |E_{n-1} f_n|\,.$$

Using (9) and the martingale maximal theorem (see [5])

$$E_n(\sup_{k\geq n}|E_k f_n|) < (E_n(\sup_{k\geq n} |E_k f_n|)^2)^{1/2} < 2(E_n|f_n|^2)^{1/2}\,,$$

we get

$$||E^*g|| \leq \sum_{n=1}^{\infty} \int_X E_n(\sup_{k\geq n}|E_k f_n|)\, d\mu + |f|^1_Y$$

$$< 2 \sum_{n=1}^{\infty} \int_X (E_n|f_n|^2)^{1/2}\, d\mu + |f|^1_Y = 3|f|^1_Y$$

and (10) is proved.

On the set $L^o := \bigcup_{n=1}^{\infty} L^1(X, A_n, \mu)$ $(\subset L^\infty(X, A, \mu))$

which is dense everywhere in VMO we have

$$F(\varphi) = \langle g,\varphi\rangle = \lim_{n\to\infty} \langle g, E_n\varphi\rangle = \lim_{n\to\infty} \langle E_n g, E_n \varphi\rangle = L_g(\varphi)$$

$$(\varphi \in L^o)\,,$$

thus $F = L_g$ everywhere and (8) is proved.

REFERENCES

1 C. FEFFERMAN, Characterization of bounded mean oscillation, Bull. Amer. Math. Soc., 77 (1971), 587-588.

2 R. COIFMAN, G. WEISS, Extensions of Hardy spaces and their use in analysis, Bull. Amer. Math. Soc., 83 (1977), 569-645.

3 A. GARSIA, Martingale inequalities, Seminar notes on recent progress, W.A. Benjamin, Reading, Massachusetts 1973.

4 L.A. LJUSTERNIK, W.I. SOBOLEW, Elemente der Funktionanalysis. Akademie-Verlag, Berlin, 1955.

5 J. NEVEU, Discrete-parameter martingales, North-Holland Math. Library, Amsterdam, Oxford, New York, 1975.

Proc. of the 3rd Pannonian Symp. on Math. Stat., Visegrád, Hungary 1982
J. Mogyoródi, I. Vincze, W. Wertz, eds

MAXIMAL INEQUALITIES FOR DEPENDENT RANDOM VARIABLES

Á. SOMOGYI

Department of Probability Theory,
Eötvös Loránd University, Budapest, Hungary

Let $X_1, X_2, \ldots$ be random variables, they need not be independent or identically distributed. Set

$$S_{m,n} = \sum_{i=m+1}^{m+n} X_i \quad \text{and}$$

$$M_{m,n} = \max_{1 \le k \le n} |S_{m,k}| .$$

$S_{m,n}$ is the sum of n consecutive X_i' s and $M_{m,n}$ is the maximal function of $S_{m,n}$. Let further $S_n = S_{o,n}$ and $M_n = M_{o,n}$.

Lai and Stout in [1] prove the following statement:

Theorem: Let $p>0$. Suppose $\Psi: \{1,2,\ldots\} \to (0,+\infty)$ satisfies the following conditions:

$$\forall \varepsilon > 0, \ \exists \rho < 1$$

$$\limsup_{n\to\infty} \left\{ \max_{\rho n \le i \le n} \frac{\Psi(i)}{\Psi(n)} \right\} < 1+\varepsilon \tag{1}$$

and

$$\liminf_{n\to\infty} \frac{\Psi(Kn)}{\Psi(n)} > K \quad \text{for some integer } K \ge 2. \tag{2}$$

If further

$$E|S_{m,n}|^p \le \Psi(n) \quad \forall m \ge 0 \quad \text{and} \quad n \ge 1$$

then there exists a positive constant C such that

$$EM_{m,n}^p \le C\Psi(n) \quad \forall m \ge 0 \quad \text{and} \quad n \ge 1.$$

First we give an extension of this theorem, in which the power function $x \to x^p$ $(x \geq 0, p > 0$ fixed) is replaced by a wider class of functions.

Let φ be a non-negative function defined on $[0, +\infty)$ such that the integral

$$\phi(x) = \int_0^x \varphi(t)\,dt \tag{3}$$

exists for all $x \geq 0$ and suppose that ϕ satisfies the growth condition

$$\phi(2x) \leq d\phi(x) \tag{4}$$

for all $x \geq 0$, where $d \geq 1$ is a constant, not depending on x.

Especially, if ϕ is a Young-function, i.e. if $\varphi(t)$ is non-decreasing, zero at the origin, $+\infty$ at $+\infty$ and left-continuous function then our condition (4) implies that

$$\phi(nx) \leq n^p \phi(x) \tag{5}$$

holds for all $x \geq 0$ and $n = 1, 2, \ldots$, where

$$p = \sup_{t>0} \frac{t\varphi(t)}{\phi(t)} .$$

It is well known that from condition (4) it follows that $p < +\infty$. In fact, it is clear that

$$\frac{\phi(2x) - \phi(x)}{x} = \frac{1}{x} \int_x^{2x} \varphi(t)\,dt \geq \varphi(x)$$

and

$$\frac{\phi(2x) - \phi(x)}{x} \leq (d-1)\frac{\phi(x)}{x} ,$$

from which we have $d - 1 \geq p$.

Now for every $n \geq 1$ and $x > 0$ it follows

$$\log\left(\frac{\phi(nx)}{\phi(x)}\right) = \int_x^{nx} \frac{\varphi(t)}{\phi(t)}\,dt \leq p \log n,$$

from which (5) follows.

Let us denote by $f(n)=\sup_{t>0} \frac{\varphi(nt)}{\varphi(t)}$.

We have

$$f(n)\le \sup_{t>0} \frac{nt\varphi(nt)}{n\phi(t)} \le \sup_{t>0} n^{p-1}\frac{nt\varphi(nt)}{\phi(nt)} = n^{p-1}p. \qquad (6)$$

This shows that for all $\varepsilon>0$ the series

$$\sum_{n=1}^{\infty} f(n^2)n^2 2^{-\varepsilon n}<+\infty. \qquad (7)$$

(7) also holds if ϕ is a concave Young function, i.e. if φ is non-increasing. In fact, in this case we have $f(n)\le 1$.

Now we shall formulate our statement.

Theorem 1

Suppose that ϕ is a function defined on $[0,+\infty]$ satisfying (3), (4) and (7). Suppose that Ψ satisfies (1) and (2). If further the inequality

$$E\phi(|S_{m,n}|)\le\Psi(n)$$

holds for all $m\ge0$ and $n\ge1$, then there exists a constant C (depending only on ϕ and Ψ) such that

$$E\phi(M_{m,n})\le C\Psi(n)$$

holds for all $m\ge0$ and $n\ge1$.

Proof.

Suppose that $m=0$. The general case $m\ge0$ can be proved in the same way.

Let n be a positive integer. Let us define k by the formula $2^k\le n<2^{k+1}$. Then we can write uniquely

$$n-2^k=\sum_{i=0}^{k-1} \delta_i^{(n)}2^{k-1-i},$$

where $\delta_i^{(n)}=0$ or 1 $(i=0,1,2,\dots,k-1)$. So for every n we have

$$S_n = S_{2^k} + \sum_{i=0}^{k-1} S_{2^k+\alpha_i,\ 2^{k-1-i}\delta_i(n)},$$

where the integers α_i are defined as follows

$$\alpha_o = 0$$

$$\alpha_i = \sum_{\ell=0}^{i-1} \delta_\ell(n) \cdot 2^{k-1-\ell} \qquad (i=1,2,\ldots,k-1)$$

(and $S_{m,o}=0$ by definition).

Let us define the events B_k and $B_k(i,j)$ similarly as given in the proof of Lemma 1 of [1], i.e. let

$$B_k = \{M_{2^k,2^k} \geq 2x\}; \quad k=1,2,\ldots$$

and

$$B_k(i,j) = \{|S_{2^k+j2^{i-1},2^{i-1}}| \geq \frac{x}{(k-i+1)^2}\};$$

$$k=1,2,\ldots\ ; \quad i=1,2,\ldots,k\ ; \quad j=0,1,2,\ldots,2^{k-i}\ ,$$

where $x>0$ is an arbitrary real number. Then we have

$$B_k \subset \bigcup_{i=1}^{k} \bigcup_{j=0}^{2^{k-i}} B_k(i,j). \tag{8}$$

Let now $n=2^\nu$ for some integer $\nu \geq 1$. Then by (8) for $x>0$ we can write

$$P(M_n \geq 4x) \leq \sum_{k=0}^{\nu} \{P(|S_{2^k}| \geq 2x) + P(M_{2^k,2^k} \geq 2x)\} \leq$$

$$\leq P(|X_1| \geq x) + P(|X_2| \geq x) + \sum_{k=1}^{\nu} P(|S_{2^k}| \geq 2x) + \tag{9}$$

$$+ \sum_{k=1}^{\nu} \sum_{i=1}^{\nu} \sum_{j=0}^{2^{k-i}} P(|S_{2^k+j2^{i-1},2^{i-1}}| \geq \frac{x}{(k-i+1)^2}).$$

First we consider $P(|S_{2^k}| \geq 2x)$. By our conditions we have

$$\sum_{k=1}^{\nu} \int_o^\infty P(|S_{2^k}| \geq x)\varphi(x)dx \leq \sum_{k=1}^{\nu} g(2^k).$$

As shown in [1], the assumptions (1) and (2) imply that

a, $\Psi(n)\to\infty$ as $n\to\infty$

b, there exists a constant C′ such that

$$\frac{\max\limits_{i\le n}\Psi(i)}{\Psi(n)} \le C' \qquad (n=1,2,\dots)$$

c, that given $1<\gamma<\frac{\log A}{\log K}$, where $A=\liminf\limits_{n\to\infty} \frac{\Psi(Kn)}{\Psi(n)}$,

there exists N such that

$$\frac{\Psi(m\cdot n)}{\Psi(n)}>m^{\gamma} \quad \text{for all} \quad m\ge N \quad \text{and} \quad n\ge N.$$

So we have

$$\frac{\Psi(2^j)}{\Psi(2^{\nu})}<2^{-(\nu-j)\gamma} \text{ for all } 2^j\ge N \text{ and } 2^{\nu-j}\ge N,$$

and so

$$\sum_{k=1}^{\nu} \Psi(2^k)= \sum_{\substack{k=2^k\ge N \text{ and}\\ 2^{\nu-k}\ge N}} \Psi(2^k)+ \sum_{k=2^k<N} \Psi(2^k)+ \sum_{k=2^{\nu-k}<N} \Psi(2^k) \le$$

$$\le\Psi(2^{\nu})\{ \sum_{k=1}^{\nu} 2^{-(\nu-k)\gamma}+2C'\log_2 N\} \le$$

$$\le\Psi(2^{\nu})\{ \sum_{k=1}^{\infty} 2^{-j\gamma}+C''\}=C_1\Psi(2^{\nu}).$$

Finally we have

$$\int_0^{\infty}\{P(|X_1|\ge x)+P(|X_2|\ge x)+ \sum_{k=1}^{\nu} P(|S_{2^k}|\ge x)\}\varphi(x)dx\le C_2\Psi(2^{\nu}).$$

The last member of the right-hand side of inequality (9) can be estimated similarly

$$\sum_{k=1}^{\nu}\sum_{i=1}^{k}\sum_{j=0}^{2^{k-i}}\int_{o}^{\infty}P\left(|S_{2^k+j2^{i-1},2^{i-1}}|\geq\frac{x}{(k-i+1)^2}\right)\varphi(x)dx\leq$$

$$\leq\sum_{k=1}^{\nu}\sum_{i=1}^{k}\sum_{j=0}^{2^{k-i}} f((k-i+1)^2)(k-i+1)^2\int_{o}^{\infty}P(|S_{2^k+j2^{i-1},2^{i-1}}|\geq y)\varphi(y)dy\leq$$

$$\leq\sum_{k=1}^{\nu}\sum_{i=1}^{k}2^{k-i+1}f((k-i+1)^2)(k-i+1)^2\Psi(2^{i-1})\leq$$

$$\leq C_1\Psi(2^{\nu})\sum_{k=1}^{\nu}\sum_{i=1}^{k}2^{(k-i+1)-\gamma(\nu-i+1)}f((k-i+1)^2)(k-i+1)^2\leq$$

$$\leq C_1\Psi(2^{\nu})\sum_{k=1}^{\nu}2^{-\delta(\nu-k+1)}\sum_{i=1}^{k}2^{(1-\eta)(k-i+1)}f((k-i+1)^2)(k-i+1)^2,$$

where we take $\delta>0$ and $\eta>1$ such that $\gamma=\delta+\eta$. Finally by our conditions given for ϕ we have from (9)

$$\int_{o}^{\infty}P(M_n\geq x)\varphi(x)dx\leq C_3\Psi(n),\text{ if } n=2^{\nu}.$$

In the general case let $2^{\nu}\leq n<2^{\nu+1}$ for any fixed $n\geq 1$. Since

$$E\phi(M_n)\leq C_4\{E\phi(M_{2^\nu})+E\phi(M_{2^\nu,2^\nu})\}\leq C_5\Psi(2^{\nu})\leq C\Psi(n)$$

with a $C>0$ not depending on n, we have our statement.

In the following statement we shall replace our condition $E\phi(|S_{m,n}|)\leq\Psi(n)$ with the condition $E\phi(|S_{m,n}|)\leq\Psi(g_{m,n})$ where $g_{m,n}$ is a functional depending on the joint distribution of $(X_{m+1}, X_{m+2}, \ldots, X_{m+n})$ such that $g_{m,n}$ satisfies the following condition:

$$g_{m,n}+g_{m+n,k}\leq g_{m,n+k}\qquad(\forall m\geq 0, n\geq 1, k\geq 1).\tag{10}$$

We shall suppose that $\phi(x)$ is a Young-function satisfying the condition

$$\sup_{t>0}\frac{t\varphi(t)}{\phi(t)}=p<+\infty\tag{11}$$

and that $\Psi(x)$ is also a Young-function and

$$\inf_{t>0} \frac{t\psi(t)}{\Psi(t)} = q > 1. \tag{12}$$

Now we can formulate the following statement:

<u>Theorem 2</u>

Suppose that $g_{m,n}$, ϕ and Ψ satisfy conditions (10), (11) and (12). If further

$$E\phi(|S_{m,n}|) \le \Psi(g_{m,n})$$

holds for all $m \ge 0$ and $n \ge 1$, then there exists a constant C (depending only on ϕ and Ψ) such that

$$E\phi(M_{m,n}) \le C\Psi(g_{m,n})$$

holds for all $m \ge 0$ and $n \ge 1$.

This statement is an extension of the following theorem of Móricz (see [3]).

<u>Theorem</u> Let $p>1$, $\alpha>1$ and suppose that

$$E(|S_{m,n}|^p) \le g^{\alpha}_{m,n} \qquad (\text{all } m \ge 0, \ n \ge 1),$$

where $g_{m,n}$ satisfies (10). Then

$$E(M^p_{m,n}) \le K_{p,\alpha} g^{\alpha}_{m,n} \qquad (\text{all } m \ge 0, \ n \ge 1).$$

(Móricz proves this statement for $p>0$)

Similar statements are proved in [2] and [5].

The proof goes by induction on u using the usual bisection technique. The proof is omitted since it is a straightforward modification of the proof given in [5].

Finally we show some consequences of our inequalities.

<u>Theorem 3</u>

Suppose that the conditions of Theorem 1 hold. Suppose further that $c_1, c_2, \ldots$ is a sequence of non-negative numbers tending decreasingly to 0. If the series

$$\sum_{k=1}^{\infty} (c_k - c_{k+1})\Psi(k)$$

converges, then $c_n\phi(|S_n|)\to 0$ a.e. as $n\to\infty$.

Proof

Let $\lambda>0$. Using Markov inequality, Lemma 2 of [4] and Theorem 1 we have

$$P(\sup_{k\geq n} c_k\phi(|S_k|)\geq\lambda)\leq\frac{1}{\lambda}E(\sup_{k\geq n} c_k\phi(|S_k|))\leq$$

$$\leq\frac{1}{\lambda}\sum_{k=n}^{\infty} (c_k-c_{k+1})E\phi(M_k)\leq\frac{C}{\lambda}\sum_{k=n}^{\infty} (c_k-c_{k+1})\Psi(k)\to 0$$

as $n\to\infty$. So

$$\lim_{n\to\infty} \text{st} \sup_{k\geq n} c_k\phi(|S_k|)=0$$

and this implies the a.e. convergence. Thus our statement is proved.

In the same way we can show the following statement:

Theorem 4

Suppose that the conditions of Theorem 2 hold. Suppose again that $c_1, c_2, c_3, \ldots$ is a sequence of numbers such that $c_n\downarrow 0$. If the series

$$\sum_{k=1}^{\infty} (c_k-c_{k+1})\Psi(g_{o,k})$$

converges, then $c_n\phi(|S_n|)\to 0$ a.e. as $n\to\infty$.

REFERENCES

1 LAI,T.L. and STOUT W.: Limit theorems for Sums of Dependent Random Variables. Z. Wahrscheinlichkeitstheorie verw. Gebiete, 51 (1980) 1-14

2 LOGNECKER,M. and SERFLING R.J.: General moment and probability inequalities for the maximum partial sum. Acta Math. Acad. Scient. Hungar., 30 (1977) 129-133

3 MÓRICZ F.: Moment inequalities and the strong laws of large numbers. Z. Wahrscheinlichkeitstheorie verw. Gebiete, 35 (1976) 299-314

4 SOMOGYI Á.: Maximal inequalities for not necessarily orthogonal random variables and some applications. Analysis Mathematica, 3 (1977) 131-139

5 SOMOGYI Á.: An asymptotically optimal maximal inequality. Analysis Mathematica, 4 (1978) 53-59

Proc. of the 3rd Pannonian Symp. on Math. Stat., Visegrád, Hungary 1982
J. Mogyoródi, I. Vincze, W. Wertz, eds

UNIFORM THEORY OF COMPARISON OF LINEAR MODELS

C. STĘPNIAK

Agricultural University, Lublin, Poland

ABSTRACT

Separate ideas and results from comparison of linear models are completed and presented in a uniform theory by a coordinate free approach.

1. INTRODUCTION

First idea in this subject, due to Ehrenfeld (1955), refers to the comparison of linear models with the identity covariance by variances of the least squares estimators. An extension of this idea for general linear model, including the variance component model, was presented by Stępniak (1977); see also Stępniak and Torgersen (1981).

A modification of this concept, including also biased linear estimation with squared risk, was considered by Stępniak (1982b). It appears that, under a weak condition, the orderings of linear models induced by the two ways of comparison, are equivalent. Another modification of the ordering, regarding the case when some parameters are nuisance, was examined by Kiefer (1959) and by Wang and Wu (1981).

A necessary and sufficient condition for one of two considered models to be at least as good as the second was obtained by Ehrenfeld (1955) for the models with the identity covariances. This result was extended by Torgersen (1980) for models with trivial deterministic parts, and next by Stępniak (1981) and by Wang and Wu (1981) for arbitrary models with known covariances.

Linear models with unknown variance components were considered by Stępniak and Torgersen (1981). They have shown that the comparison of such models may be reduced to the same problem for linear models with known covariances.

Independently of the comparison of linear models with respect to linear estimation the theory of comparison of linear normal models is developed, based on more stronger criterion of "being at least as informative as" (Kiefer (1959), Hansen and Torgersen (1974), Swensen (1980), Torgersen (1980) and Stępniak (1982a)). In the case of known covariances the ordering of linear normal models coincides with the ordering of linear models w.r.t. linear estimation. However, this is not true in general.

Torgersen (1980) considers also other orderings of linear normal models and shows that all the orderings are equivalent in the case when the covariances are known.

2. GENERAL FRAMEWORK

Consider models where the observable random vector X belongs to a finite dimensional inner product space $\{H,(\cdot,\cdot)\}$. Assuming $E(X,X)<\infty$ there is a unique vector $E_X \in H$ and a unique operator Σ_X on H such that

$$E(h,X) = (h,E_X)$$

and

$$\text{Cov}\{(h,X),(h',X)\} = (h, \Sigma_x h')$$

for $h, h' \in H$.

The vector E_x and the operator Σ_x are called the expectation and the covariance of X, respectively.

It will be assumed that E_x depends linearly on an unknown parameter α, while Σ_x depends (perhaps non-linearly) on an unknown parameter γ. The set of possible values of α and γ will be assumed to constitute, respectively, a set Ξ in a finite dimensional inner product space $\{K, \langle\cdot, \cdot\rangle\}$ and a set Γ. Without loss of generality we shall assume that Ξ is a spanning set of K.

The parameter sets Ξ and Γ are fixed but arbitrary, while the sample space H may vary from model to model. Our notation does not distinguish between the inner products in different sample spaces.

The basic assumptions can be written in

$$E_x = A\alpha, \tag{1}$$

and

$$\Sigma_x = V_\gamma, \tag{2}$$

where A is a known linear operator from K to H, while V_γ, for each $\gamma\in\Gamma$, is a known non negative definite selfadjoint (n.n.d.) operator on H. We do not assume that V_γ is non singular. Moreover, there is a possibility of a relationship between α and γ.

We shall denote this structure by $L(A,V_\gamma;\ \gamma\in\Gamma)$. Say that X is subject to linear model $L(A,V_\gamma;\ \gamma\in\Gamma)$ if (1) and (2) hold. The usual variance component model defines such a structure.

If $V_\gamma = V$ does not depend of γ, i.e. in the case of known covariance, we may write $L(A,V)$, instead of $L(A,V_\gamma;\ \gamma\in\Gamma)$.

If T is an operator then T^*, $N(T)$ and $R(T)$ will denote the adjoint, the null space and the range of T, respectively. For a selfadjoint operator T we shall write $T \geq 0$ if T is n.n.d.

Suppose X_i, $i=1,2$, is subject to a linear model $L_i = L(A_i, V_\gamma^{(i)}; \gamma \in \Gamma)$. We shall then say that L_1 is at least as good as L_2 with respect to unbiased estimation with squared risk if for any unbiased estimator (h_2, X_2) of an unknown parameter there is an unbiased estimator (h_1, X_1) of this parameter such that $\mathrm{Var}_\gamma(h_1, X_1) \leq \mathrm{Var}_\gamma(h_2, X_2)$ for each $\gamma \in \Gamma$.
If this condition is satisfied then we shall write $L_1 \succ L_2$ (cf. Stępniak and Torgersen (1981)).

Another way of ordering of linear models is presented by Stępniak (1982b). Namely, model L_1 is said to be at least as good as model L_2 with respect to linear estimation with squared risk if for any function Ψ on $\Xi \times \Gamma$ and for any estimator (h_2, X_2) there is an estimator (h_1, X_1) such that

$$E_{\alpha,\gamma}[(h_1, X_1) - \Psi]^2 \leq E_{\alpha,\gamma}[(h_2, X_2) - \Psi]^2$$

for all $\alpha \in \Xi$ and $\gamma \in \Gamma$.

It was shown by Stępniak (1982b) that L_1 is at least as good as L_2 w.r.t. linear estimation iff $L_1 \succ L_2$, providing that the set Ξ, of possible values of α, constitutes a linear space and there is no relationship between α and γ.

A useful tool in the comparison of linear models is

Theorem 1. (Torgersen (1980)). $L(A_1, V_\gamma^{(1)}; \gamma \in \Gamma) \succ L(A_2, V_\gamma^{(2)}; \gamma \in \Gamma)$ if there is a linear map T from H_1 to H_2 such that $A_2 = TA_1$ and $V_\gamma^{(2)} - TV_\gamma^{(1)}T^* \geq 0$ for all $\gamma \in \Gamma$.

This result was earlier obtained by Stępniak (1977) under the assumption that $V_\gamma^{(1)}$ and $V_\gamma^{(2)}$ are non singular (cf. also Stępniak (1982b)).

3. LINEAR MODELS WITH KNOWN COVARIANCES

Suppose a random vector X is subject to a linear model L(A,V). A function $\Psi(\alpha)$ possesses a linear unbiased estimator in this model iff $\Psi(\alpha)\underset{\alpha}{\equiv}\langle k,\alpha\rangle$ for some $k\in R(A^*)$. Thus $\langle k,\alpha\rangle$ is unbiasedly estimable with zero variance iff $k\in A^*[N(V)]$. Say that model L(A,V) has trivial deterministic part if $A^*[N(V)]=\{0\}$, i.e. when $R(A)\subseteq R(V)$.

The main result in this Section is

Theorem 2. (Torgersen (1980)). For any two models $L(A_1, V_1)$ and $L(A_2, V_2)$ with trivial deterministic parts, $L(A_1,V_1)\succ L(A_2,V_2)$ iff $A_1^*V_1^-A_1 - A_2^*V_2^-A_2\geq 0$, where V_i^-, i=1,2, is a generalized inverse of V_i.

We shall give an elementary proof of the result. The proof is based on

Lemma 1. For any two n.n.d. operators T_1 and T_2 on a finite dimensional inner product space $\{K,\langle\cdot,\cdot\rangle\}$, the condition

(i) $\quad T_1 - T_2 \geq 0$

is equivalent to the two conditions

(ii) $\quad R(T_1)\supseteq R(T_2)$

and

(iii) $\quad \langle k,T_2^-k\rangle \geq \langle k,T_1^-k\rangle \quad$ for all $k\in R(T_2)$,

where T_i^-, i=1,2, is a generalized inverse of T_i.

Remark 1. This Lemma, suggested by Stępniak (1981), was stated by Wang and Wu (1981) in a matrix form.

Proof of the Lemma. The implication (i)$\Rightarrow$(ii) is evident. Thus we only need to show the equivalence (i)$\Leftrightarrow$(iii) under (ii).

Define a quasi-inner product $\langle\cdot,\cdot\rangle_1$ on K by

$\langle k,k'\rangle_1=\langle k,T_1k'\rangle.$

Then there exists an orthonormal basis in $R(T_1)$, say $k_1,\dots,k_p$, such that

$$\langle k,T_2k\rangle = \sum_{i=1}^{p} \rho_i\langle k,k_i\rangle_1^2,$$

where ρ_i, $i=1,\dots,p$, are non negative. On the other hand

$$\langle k,T_1k\rangle = \langle k,T_1 \sum_{i=1}^{p} \langle k,k_i\rangle_1 k_i\rangle = \sum_{i=1}^{p} \langle k,k_i\rangle_1^2 .$$

Thus the condition (i) is equivalent to

$$\rho_i \leq 1, \quad i=1,\dots,p. \tag{3}$$

First we shall show that

$$T_2k = T_1\sum\rho_i \langle k,k_i\rangle_1 k_i.$$

Really,

$$\langle k,T_1\sum\rho_i\langle k,k_i\rangle_1 k_i\rangle = \sum\rho_i\langle k,k_i\rangle_1^2 = \langle k,T_2k\rangle .$$

To show (3)$\Longleftrightarrow$(iii) we note that

$$\begin{aligned}\langle T_2k,T_1^-T_2k\rangle &= \langle T_1\sum\rho_i\langle k,k_i\rangle_1 k_i, T_1^-T_1\sum\rho_i\langle k,k_i\rangle_1 k_i\rangle\\ &= \langle\sum\rho_i\langle k,k_i\rangle_1 k_i, T_1\sum\rho_i\langle k,k_i\rangle_1 k_i\rangle\\ &= \sum\rho_i^2\langle k,k_i\rangle_1^2.\end{aligned}$$

Moreover,

$$\langle T_2k,T_2^-T_2k\rangle = \langle k,T_2k\rangle = \sum\rho_i\langle k,k_i\rangle_1^2.$$

This implies the desired result. □

<u>Proof of the Theorem</u>. A parameter $\langle k,\alpha\rangle$ is unbiasedly estimable in $L(A_i,V_i)$ iff $k\in R(A_i^*)$, $i=1,2$. Moreover, under the assumption $R(A_i)\subseteq R(V_i)$ the variance of any best linear unbiased estimator of $\langle k,\alpha\rangle$ is $\langle k,(A_i^*V_i^-A_i)^-k\rangle$ (cf. Rao (1973), p. 301). Define

$$\Pi_i(k) = \begin{cases}\langle k,(A_i^*V_i^-A_i)^-k\rangle & \text{if} \quad k\in R(A_i^*)\\ \infty & \text{otherwise.}\end{cases}$$

Thus $L(A_1,V_1)\succ L(A_2,V_2)$ iff $\Pi_1(k)\leq\Pi_2(k)$, $k\in K$. Note also that that $R(A_i)\subseteq R(V_i)$ implies $R(A_i^*V_i^-A_i) = R(A_i^*)$. Using Lemma 1 we obtain the desired result. □

Now we shall show how the comparison of arbitrary linear models may be reduced to the comparison of models having trivial deterministic parts.

Lemma 2. $L(A_1,V_1) \succ L(A_2,V_2)$ iff $L(A_1,V_1+A_1A_1^*) \succ L(A_2,V_2+A_2A_2^*)$.

Proof. Suppose X_i is subject to $L(A_i,V_i)$ and Y_i is subject to $L(A_i,V_i+A_iA_i^*)$, i=1,2. Then, for any $h_i \in H_i$,

$$E(h_i,Y_i) = E(h_i,X_i)$$

and

$$Var(h_i,Y_i) = Var(h_i,X_i) + \langle k,k\rangle,$$

where $k = A^*h_i$, i=1,2.

This implies the desired result. □

Remark 2. The operators $A_iA_i^*$, i=1,2, in the Lemma may be replaced by $cA_iA_i^*$, where c is a positive number.

As the models $L(A_i, V_i+A_iA_i^*)$, i=1,2, have trivial deterministic parts, Theorem 2 is applicable and a consequence of this is

Theorem 3. (Wang and Wu (1981)). $L(A_1,V_1) \succ L(A_2,V_2)$ iff $A_1^*(V_1+A_1A_1^*)^-A_1 - A_2^*(V_2+A_2A_2^*)^-A_2$ is n.n.d.

Another way to get rid of the assumption $R(A_i) \subseteq R(V_i)$, i=1,2, is

Lemma 3. $L(A_1,V_1) \succ L(A_2,V_2)$ iff $L(A_1P,V_1) \succ L(A_2P,V_2)$, where P is the projector on the orthogonal complement of $A_1^*[N(V_1)]$.

Proof. The conditions $L(A_1,V_1) \succ L(A_2,V_2)$ and $L(A_1P,V_1) \succ L(A_2P,V_2)$ may be written, respectively, in the form

(a) for each $h_2 \in H_2$ there is an $h_1 \in H_1$ such that

$$A_1h_1 = A_2h_2 \text{ and } (h_1,V_1h_1) \le (h_2,V_2h_2)$$

and

(b) for each $h_2 \in H_2$ there is an $h_1' \in H_1$ such that

$$PA_1h_1' = PA_2h_2 \text{ and } (h_1',V_1h_1') \le (h_2,Vh_2).$$

(a)$\Rightarrow$(b) evidently by putting $h_1' = h_1$.

(b)$\Rightarrow$(a). Suppose $PA_1h_1' = PA_2h_2$. Then $P(A_1^*h_1' - A_2^*h_2) = 0$. Defining $g = A_1^*h_1' - A_2^*h_2$ we notice that $g \in A_1^*[N(V_1)]$. Therefore there is a vector $f \in H_1$ such that $V_1f = 0$ and $A_1^*f = g$. Putting $h_1 = h_1' - f$ we have (a). This completes the proof. □

<u>Theorem 4</u>. $L(A_1,V_1) \succ L(A_2,V_2)$ iff $A_1^*[N(V_1)] \supseteq A_2^*[N(V_2)]$ and $P(A_1^*V_1^- A_1 - A_2^*V_2^-A_2)P$ is n.n.d.

<u>Proof</u>. Consider the models $L(B_i,V_i)$, where $B_i = A_iP$, $i=1,2$. We notice that $B_1^*[N(V_1)] = PA_1^*[N(V_1)] = \{0\}$ and $B_2^*[N(V_2)] = PA_2^*[N(V_2)]$. Thus $B_i^*[N(V_i)] = \{0\}$ as soon as

$$A_1^*[N(V_1)] \supseteq A_2^*[N(V_2)]. \tag{4}$$

Suppose $L(A_1,V_1) \succ L(A_2,V_2)$. Then (4) is satisfied by the definition of the ordering $\succ$. Using Lemma 3 and Theorem 2 we obtain

$$P(A_1^*V_1^-A_1 - A_2^*V_2^-A_2)P \geq 0. \tag{5}$$

Conversely, if (4) and (5) hold then it follows directly from Theorem 2 and Lemma 3, $L(A_1,V_1) \succ L(A_2,V_2)$. This completes the proof. □

4. LINEAR MODELS WITH NUISANCE PARAMETERS

Consider models where the expectation E_X of the observable vector X is of the form

$$E_X = A\alpha + B\beta,$$

where A and B are linear operators from $\{K,\langle\cdot,\cdot\rangle\}$ and $\{K_1,\langle\cdot,\cdot\rangle_1\}$, respectively, to H, while α and β are unknown parameters. We are interested only in estimating the parameters $\langle k,\alpha\rangle$, $k \in K$, i.e. β is assumed to be nuisance.

Such a linear model will be denoted by $L([A,B],V)$ or by $L([A,B],V_\gamma;\ \gamma \in \Gamma)$ according to whether the covariance is known or unknown.

Suppose X_i, i=1,2, is subject to a linear model L_i with the expectation $E_{X_i} = A_i\alpha + B_i\beta$. We shall write $L_1 \overset{A}{\succ} L_2$ if for any unbiased estimator (h_2,X_2) of a parameter $\langle k,\alpha\rangle$ there is an unbiased estimator (h_1,X_1) of this parameter such that $Var(h_1,X_1) \leq Var(h_2,X_2)$.

Let $L_i = L([A_i,B_i], V_i)$, i=1,2. Then (h_i,X_i), i=1,2, is an unbiased estimator of a parameter $\langle k,\alpha\rangle$ if and only if $(h_i,B_i\beta) \underset{\beta}{\equiv} 0$, i.e. when $h_i \in N(B_i^*)$.

Denote by Q_i, i=1,2, the orthogonal projector on $N(B_i^*)$. Then $L([A_1,B_1], V_1) \overset{A}{\succ} L([A_2,B_2], V_2)$ iff $L(Q_1A_1, Q_1V_1Q_1) \succ L(Q_2A_2, Q_2V_2Q_2)$. Moreover, by Theorem 1, model $L_i = L(Q_iA_i,Q_iV_iQ_i)$ is equivalent to the model $L_i' = L(Q_iA_i,V_i)$ in the sense that $L_i \succ L_i'$ and $L_i' \succ L_i$. Now, by transitivity of the ordering $\succ$ we obtain.

<u>Lemma 4</u>. $L([A_1,B_1],V_1) \overset{A}{\succ} L([A_2,B_2],V_2)$ iff $L(Q_1A_1,V_1) \succ L(Q_2A_2,V_2)$ where Q_i, i=1,2, is the orthogonal projector on $N(B_i^*)$.

<u>Remark 3.</u> The projector Q_i, i=1,2, may be presented in the form $Q_i = I - B_i(B_i^*B_i)^-B_i^*$.

By Lemma 1 and Theorem 2, noting that $Q_iB_iB_i^*Q_i=0$, i=1,2, this Lemma implies

<u>Theorem 5</u>. $L([A_1,B_1],V_1) \overset{A}{\succ} L([A_2,B_2],V_2)$ iff $A_1^*Q_1(V_1 + A_1A_1^*)^-Q_1A_1 - A_2^*Q_2(V_2 + A_2A_2^*)^-Q_2A_2$ is n.n.d.

<u>Remark 4</u>. If $R(Q_1A_1) \subseteq R(V_1)$ and $R(Q_2A_2) \subseteq R(V_2)$ then the operator $V_i + A_iA_i^*$, i=1,2, in this Theorem may be replaced by the operator V_i.

<u>Corollary 1</u>. (Kiefer (1959)). $L([A_1,B_1], I_{n_1}) \overset{A}{\succ} L([A_2,B_2], I_{n_2})$ iff $A_1^*A_1 - A_1^*B_1(B_1^*B_1)^-B_1^*A_1 - A_2^*A_2 + A_2^*B_2(B_2^*B_2)^-B_2^*A_2$ is n.n.d.

Another result concerning the comparison of linear models

with nuisance parameters was given by Wang and Wu (1981) in the form

Theorem 6. $L([A_1,B_1], V_1) \overset{A}{\succ} L([A_2,B_2], V_2)$ iff

$$A_1^*T_1^-A_1 - A_1^*T_1^-B_1(B_1^*T_1^-B_1)^-B_1^*T_1^-A_1$$
$$- A_2^*T_2^-A_2 + A_2^*T_2^-B_2(B_2^*T_2^-B_2)^-B_2^*T_2^-A_2 \text{ is n.n.d.,}$$

where $T_i = V_i + A_iA_i^* + B_iB_i^*$, $i=1,2$.

5. LINEAR MODELS WITH UNKNOWN VARIANCE COMPONENTS

Stępniak and Torgersen (1981) have shown that the problem of comparison for such models may be reduced to the same problem for known covariances by

Theorem 7. $L(A_1, V_\gamma^{(1)}; \gamma\in\Gamma) \succ L(A_2, V_\gamma^{(2)}; \gamma\in\Gamma)$ iff $L(A_1,V_\lambda^{(1)}) \succ L(A_2,V_\lambda^{(2)})$ for each probability distribution λ on Γ with finite support. Here $V_\lambda^{(1)} = \sum_\gamma \lambda(\gamma)V_\gamma^{(1)}$ and $V_\lambda^{(2)} = \sum_\gamma \lambda(\gamma)V_\gamma^{(2)}$.

Corollary 2. Assuming Γ is a subset of a linear space and the maps $\gamma\mapsto V_\gamma^{(1)}$ and $\gamma\mapsto V_\gamma^{(2)}$ are linear, $L(A_1,V_\gamma^{(1)}; \gamma\in\Gamma) \succ L(A_2,V_\gamma^{(2)}; \gamma\in\Gamma)$ iff $L(A_1,V_\gamma^{(1)}) \succ L(A_2,V_\gamma^{(2)})$ for all γ belonging to the convex hull of Γ.

Adaptation of the result to the models with nuisance parameters is straightforward.

6. COMPARISON OF LINEAR NORMAL MODELS

We shall say that a random vector X is subject to a linear normal model $N(A,V_\gamma; \gamma\in\Gamma)$ if X is normally distributed and subject to the linear model $L(A,V_\gamma; \gamma\in\Gamma)$. In the case of known covariance the symbol $N(A,V)$ will be used.

A linear normal model N_1 is said to be at least as informative as a linear normal model N_2 if to each finite-decision statistical problem and each decision rule δ_2 based on N_2 there exists a decision rule δ_1 based on N_1 such that the risk of δ_1 is uniformly not greater than the risk of δ_2.

The main result in this Section is

Theorem 8. (Hansen and Torgersen (1974)).

(i) A linear normal model $N(A_1, I_{n_1})$ is at least as informative as a linear normal model $N(A_2, I_{n_2})$ iff $A_1^*A_1 - A_2^*A_2$ is n.n.d.

(ii) A linear normal model $N(A_1, \gamma I_{n_1}; \gamma>0)$ is at least as informative as a linear normal model $N(A_2, \gamma I_{n_2}; \gamma>0)$ iff $A_1^*A_1 - A_2^*A_2$ is n.n.d. and $n_1 \geq n_2 + \mathrm{rank}(A_1^*A_1 - A_2^*A_2)$.

Remark 5. The part (i) of the Theorem has been proved earlier by Kiefer (1959).

Torgersen (1980) extends the Theorem for arbitrary models with known covariances in

Theorem 9.

(i) Model $N(A_1, V_1)$ is at least as informative as a model $N(A_2, V_2)$ iff $L(A_1, V_1) \succ L(A_2, V_2)$.

(ii) Model $N(A_1, \gamma V_1; \gamma>0)$ is at least as informative as model $N(A_2, \gamma V_2; \gamma>0)$ iff $L(A_1, V_1) \succ L(A_2, V_2)$ and $\mathrm{rank}(V_1) - \mathrm{rank}(V_2) \geq \mathrm{rank}[P(A_1^*V_1^-A_1 - A_2^*V_2^-A_2)P]$, where P is the orthogonal projector on the orthogonal complement of $A_1^*[N(V_1)]$.

Return to Theorem 4 to see that this is an interesting complement to Theorem 9(ii).

REFERENCES

EHRENFELD, S. (1955). Complete class theorem in experimental design. Third Berkeley Symp. Math. Statist. Prob. 1 69-75.

HANSEN, O.H. and TORGERSEN, E.N. (1974). Comparison of linear normal experiments. Ann. Statist. 2 367-373.

KIEFER, J. (1959). Optimum experimental designs. J. R. Statist. Soc. B. 21 272-319.

RAO, C.R. (1973). Linear Statistical Inference and its Applications. 2nd ed. Wiley, New York.

STĘPNIAK, C. (1977). On comparison of linear models. VII Coll. Agro-Biometr. 395-401 (In Polish).

STĘPNIAK, C. (1981). Comparison of linear models having singular covariances. Unpublished manuscript.

STĘPNIAK, C. (1982a). Optimal allocation of observations in one-way random normal model. Ann. Inst. Statist. Math. 34 175-180.

STĘPNIAK, C. (1982b). Optimal allocation of units in experimental design with hierarchical and cross classification. Submitted to Ann. Inst. Statist. Math.

STĘPNIAK, C. and TORGERSEN, E. (1981). Comparison of linear models with partially known covariances with respect to unbiased estimation. Scand. J. Statist. 8 183-184.

SWENSEN, A.R. (1980). Deficiencies between linear normal experiments. Ann. Statist. 5 1142-1155.

TORGERSEN, E. (1980). Orderings of linear models. Statist. Res. Report. Univ. of Oslo.

WANG, S. G. and WU, C.F. (1981). Comparison of general linear experiments. Unpublished manuscript.

Proc. of the 3rd Pannonian Symp. on Math. Stat., Visegrád, Hungary 1982
J. Mogyoródi, I. Vincze, W. Wertz, eds

HOMOMORPHISMS OF RENEWAL SEQUENCES

G.J. SZÉKELY

Department of Probability Theory,
Eötvös Loránd University
Budapest, Hungary

SUMMARY.
Renewal sequences form a commutative semigroup under multiplication, as do the positive integers, and a decomposition into the product of irreducible and anti-irreducible elements is also valid. We pose the question, how far can the parallel be carried. This is closely related to the problem of existence of nontrivial homomorphisms from this semigroup into divisible and non-divisible algebraic structures.

INTRODUCTION

First we recall the definition of renewal sequences. Let X be a non-negative valued random variable, and let $X_1, X_2, \ldots$ be independent copies of it. Denote $S_0 = 0$, $S_1 = X_1$, $S_2=X_1+X_2, \ldots$ The sequence $u=(u_0,u_1,u_2,\ldots)$ is a renewal sequence if there exists an X such that $u_n=\Pr(S_j=n$ for some $j)$, $n=0,1,2,\ldots$. Very little generality is lost if we restrict ourselves to aperiodic renewal sequences defined by the condition g.c.d.$\{n:u_n>0\}=1$. In the following every renewal sequence is supposed to be aperiodic and the set of all these sequences will be denoted by U.

The starting point of the algebraic theory of renewal sequences is the observation that U is a commutative semigroup if the operation $\otimes$ is the pointwise multiplication:

$$(u' \otimes u'')_n = u'_n u''_n \qquad n=0,1,2,...$$

(see Daley(1965), Kingman(1968), (1972)).

Let U be given the topology which it inherits as a subspace of the product space $[0,1]^\infty$ of all sequences $(x_n : n \geq 0)$ with $0 \leq x_n \leq 1$. Then U is a compact metrisable space and the multiplication $\otimes$ is continuous to this topology, i.e. U is a topological semigroup. The problem of the existence of nontrivial continuous homomorphisms from certain subsemigroups of U into the additive topological group of real numbers is closely related to the problem of decomposition of renewal sequences into irreducible and anti-irreducible factors (see Kendall(1967), (1968), Davidson(1968),(1969). The paper of Davidson(1968) contains the proof of the non-existence of continuous homomorphism from the complete U into the additive group of real numbers. In the First PSMS, I proved the existence of (non-continuous) homomorphism from U into the real numbers (Székely(1981)). Now we show why the problem of existence of (non-trivial) homomorphisms from U to the additive group of integer numbers Z is important.

PRIMES AND HOMOMORPHISMS

Write $u|v$ if there exists an u' $(u,v,u' \in U)$ such that $v=uu'$. A renewal sequence π is called a prime if $\pi|uv$ implies either $\pi|u$ or $\pi|v$.

<u>THEOREM.</u> If π is a prime renewal sequence, then

$$\varphi(u) = \max\{n : \pi^n | u\}$$

is a homomorphism from U to Z.

<u>PROOF.</u> The problem is caused by the fact that

(i) $u=\pi^k u'$, $v=\pi^m v'$ $\pi \nmid u', \pi \nmid v'$ and

(ii) $\pi^{k+m+1} \mid \pi^{k+m} u'v'$

does not seem to be an obvious contradiction.

Write $u \prec v$ if every 0 of u is also a 0 of v. It is obvious that $u|v$ implies $u \prec v$. Now call u aggressive if $u \prec v$ also implies $u|v$ (cf. Ruzsa-Székely(1982)).

Lemma 1. If a prime renewal sequence π is not aggressive then φ is a homomorphism.

Proof of Lemma 1. Let π be a prime renewal sequence such that $\pi \prec u$ but $\pi \nmid u$ for some $u \in U$. If $\varphi(u_1)=k$, $\varphi(u_2)=m$ $(u_1, u_2 \in U)$ then $u_1=\pi^k u_1'$, $u_2=\pi^m u_2'$ where $\pi \nmid u_1' \in U$, $\pi \nmid u_2' \in U$. Now $u_1 u_2 = \pi^{k+m} u_1' u_2' = \pi^s u_3$ $(s=k+m, u_3=u_1'u_2')$ where $\pi \nmid u_3$ by the prime property. We show that $\pi^{s+1} \nmid \pi^s u_3$, otherwise $\pi^s u_3 = \pi^{s+1} u_3'$ for some $u_3' \in U$ and then $\pi^s (u_3 - \pi u_3')=0$, but $\pi \prec u$ thus $u(u_3-\pi u_3')=0$ that is $u_3 u = u u_3' \pi$ i.e. $\pi | u_3 u$ though $\pi \nmid u_3$ and $\pi \nmid u$ which is a contradiction.

Lemma 2. If π is a prime renewal sequence then it is not aggressive.

Proof of Lemma 2. Denote by n_0 the largest zero of π($n_0<\infty$ for $\pi_n>0$ if n is large enough). Now let $u_m=0 (m=1,2,\dots,n_0-1)$ and $u_{n_0}>0$. Then $\pi \nmid u$. Let $v_{n_0}=0$ and $v_{n_0+1}>\pi_{n_0+1}$ (<1). Then $\pi \prec uv$ (since $u_n v_n=0$ if $n \le n_0$) but $\pi \nmid v$. Now by the prime property of π we get $\pi \nmid uv$ i.e. π is not aggressive.

Lemma 1 and Lemma 2 obviously give the THEOREM.

A FURTHER PROBLEM

(i) The problem of existence of nontrivial homomorphisms from U to non-divisible algebraic structures is closely related to the following problem which is also interesting in itself.

We have mentioned that the product of renewal sequences is always a renewal sequence. What about their quotient? Obviously, the quotient is not always a renewal sequence. According to the Erdős-Feller-Pollard theorem every renewal sequence has a limit, i.e. $\lim_{n\to\infty} u_n$ always exists. Now a reasonable conjecture is the following. If x_n is an arbitrary positive sequence such that $0<\lim x_n<\infty$ exists then one can always find two renewal sequences u and v such that $u_n x_n = v_n$ $n=0,1,2,\ldots$.

(ii) The problem of existence of prime renewal sequences is still unsolved and our Theorem could help in proving their nonexistence by showing the nonexistence of homomorphisms from U to Z. (For the idea see Ruzsa-Székely(1982)).

REFERENCES

DALEY, D.J. (1965): On a class of renewal functions, Proc. Camb. Phil. Soc. 61, 519-526.

DAVIDSON, R. (1968): Arithmetic and other properties of certain Delphic semigroups: I-II, Z. Wahrscheinlichkeitstheorie verw. Gebiete 10, 120-145 and 146-172.

DAVIDSON, R. (1969): More Delphic theory and practice, Z. Wahrscheinlichkeitstheorie verw. Gebiete 13, 191-203.

KENDALL, D.G. (1967): Renewal sequences and their arithmetic, In: Symposium on Probability Methods in Analysis, pp.174-175, Lecture Notes in Math. 31, Springer, Berlin.

KENDALL, D.G. (1968): Delphic semigroups, infinitely divisible regenerative phenomena, and the arithmetic of p-functions, Z. Wahrscheinlichkeitstheorie verw. Gebiete 9, 163-195

KINGMAN, J.F.C. (1968): An approach to the study of Markov processes, J. Roy. Statist. Soc. B, 28, 417-447.

KINGMAN, J.F.C. (1972): Regenerative phenomena, Wiley, London.

RUZSA, I.Z.-SZÉKELY, G.J. (1982): Irreducible and prime distributions, In: Probability Measures on Groups (Proceedings of the Sixth Conference held at Oberwolfach, June 28-July 4, 1981), pp. 354-361. Lecture Notes in Math. 928, Springer, Berlin.

SZÉKELY, G.J. (1981): Extensions of partial homomorphisms in probability theory, In: The First Pannonian Symp. on Math. Stat. (Bad Tatzmannsdorf 1979) pp. 262-265. Lecture Notes in Statistics 8, Springer, Berlin.

Proc. of the 3rd Pannonian Symp.
on Math. Stat., Visegrád, Hungary 1982
J. Mogyoródi, I. Vincze, W. Wertz, eds

SINAI BILLIARD IN POTENTIAL FIELD (ABSOLUTE CONTINUITY)

A. VETIER

Technical University, Budapest, Hungary

1. SINAI BILLIARD IN POTENTIAL FIELD

A convex, open subset of the two-dimensional torus is called a scatterer if its boundary is a three times continuously differentiable curve with positive curvature. Let us have a finite number of scatterers, and let the billiard-table be the complement of the union of the scatterers. We suppose that the position of the scatterers guarantees that

1. the distance between any two scatterers is positive,

2. there exist $k_0>0$ and $\tau_{max}<\infty$ that any curve on the billiard-table with curvature less than k_0 cannot be longer than τ_{max}.

Let us consider a three times continuously differentiable potential on the torus. Under the influence of this potential a particle is moving on the billiard-table, and it is reflected elastically at the scatterers.

The state of the particle is determined by its place and velocity, so a state is a tangent vector to the billiard-table. Let M_h be the set of states in which the energy (potential energy plus kinetic energy) is equal to h. The motion of the particle defines some dynamics on the set of states: if the particle is in the state x at time 0 then $S^t x$ means its state at time t $/-\infty<t<\infty/$. According to the principle of conservation of energy, the sets M_h are invariant under the dynamics. So for each h we get a dynamical system $/M_h, \{S^t\}, \mu_h/$, where μ_h is the invariant measure on M_h induced by the Liouville measure. This dynamical system is called the Sinai billiard on the energy level h.

My result is the following.

Theorem. There exists an $h_0<\infty$ that if $h\geq h_0$ then every ergodic component of the Sinai billiard on the energy level h has positive measure.

The proof of the theorem (see [2]) consists of three main parts:

1. construction of the "stable and unstable fibers",
2. proof of the "absolute continuity",
3. application of the "Hopf method".

Knowing the proof of the ergodicity of the Sinai billiard without potential (see [1]), one can see that my result is the first half of the proof of the ergodicity of the Sinai billiard in potential field.

The construction of the stable and unstable fibers can be read in [3]. Now I am going to sketch the proof of the absolute continuity.

2. BILLIARDS WITH CONSTANT VELOCITY

Instead of having the torus and a potential on it let us restate the above mentioned notions in case of a general Riemannian surface without potential.

Let us consider a two-dimensional, three times continuously

differentiable, compact, closed, oriented Riemannian surface. A connected, open subset of the surface will be called a scatterer if its boundary is a three times continuously differentiable curve with positive curvature from inside. Let us have a finite number of scatterers. We call the complement of the union of the scatterers a billiard-table, which will be denoted by Q. The boundary of the billiard-table, denoted by ∂Q, consists of the union of the boundaries of the scatterers.

Let us imagine that a particle is moving on the billiard-table with unit velocity along geodetics, and at the boundary of the billiard-table it is reflected elastically by the scatterers. The state of the particle is determined by its place and velocity. So a state x means a place q and a direction v together: $x=(q,v)$. The state x can be identified with a unit vector of the tangent plane at the point q. The set of all states is called the phase-space, denoted by M.

The motion of the particle defines some dynamics $\{S^t\}_{-\infty<t<\infty}$ on the phase-space: if the particle is in the state x at time 0 then $S^t x$ means its state at time t $(-\infty<t<\infty)$. The states $S^t x$ $(-\infty<t<\infty)$ together constitute the trajectory of x. The curve on the billiard-table drawn by the particle will be called its path.

The Riemannian metric of the surface induces an area-measure μ_Q on the billiard-table. The set E^q of all unit vectors of the tangent plane at q can be obviously identified with the boundary of the unit circle in this tangent plane. The length-measure on the boundary of the circle corresponds to a measure σ^q on E_q by this identification. Since $M=\bigcup_{q\in Q} E^q$, the measure μ_Q and the measures σ^q $(q\in Q)$ together define a measure μ on M.

It is known that μ is invariant under the dynamics $\{S^t\}_{-\infty<t<\infty}$.

So we get a dynamical system $(M, \{S^t\}_{-\infty<t<\infty}, \mu)$, which we call a billiard with constant velocity. If, for example, the considered surface is the torus with the usual Riemannian metric then we get the well-known Sinai billiard.

It is shown in [3] that given a potential on the torus and an energy level h, one can define some Riemannian metric on the torus such that the billiard with constant velocity corresponding to this metric and the Sinai billiard on the energy level h have the same ergodic decomposition. This is why we study a billiard with constant velocity instead of the Sinai billiard in potential field.

3. ABSOLUTE CONTINUITY

3.1. The Riemannian metric on the billiard-table defines a Riemannian metric in the phase-space (see [3]), thus we can speak about the distance between any two states. We shall say that the states x and y belong to the same leaf if there exists such a t that the distance between $S^t x$ and $S^{t_0+t} y$ tends to 0 as $t\to\infty$. The notion of stable fiber was defined in [3]. It is obvious that if z belongs to the stable fiber of x then for every real value t_0 the states x and $S^{-t_0} z$ belong to the same leaf (Figure 1).

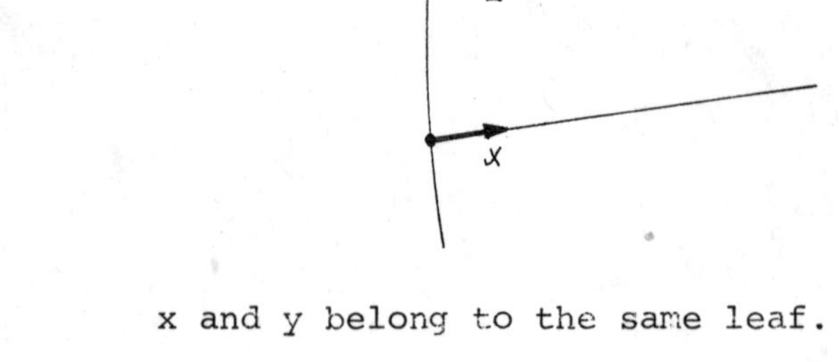

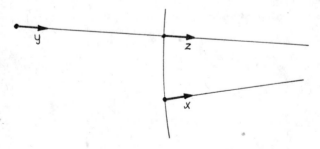

Figure 1. x and y belong to the same leaf.

3.2. Let $k(q)$ denote the curvature of the boundary of the billiard-table at q ($q\in\partial Q$), and let $k_{min} := \min_{q\in\partial Q} k(q)>0$.

Let g_i be an oriented curve of finite length on the billiard-table with curvature not less than k_{min} ($i=1,2$). Obviously there are two unit vectors orthogonal to the curve g_i and lying in the tangent plane at the distance s_i from the starting point of g. One of these vectors - as illustrated on Figure 2 - will be denoted by $x_i(s_i)$ ($i=1,2$). We suppose that $x_1(0)$ and $x_2(0)$ belong to the same leaf. Let Z_1 denote the set of values s_1 for which there exists such value s_2 that $x_1(s_1)$ and $x_2(s_2)$ belong to the same leaf. Obviously s_2 depends on s_1, which defines a function $s_2=\chi(s_1)$ on the set Z_1. Let $Z_2:=\{\chi(s_1):s_1\in Z_1\}$. We suppose that the sets Z_1 and Z_2 are dense at 0.

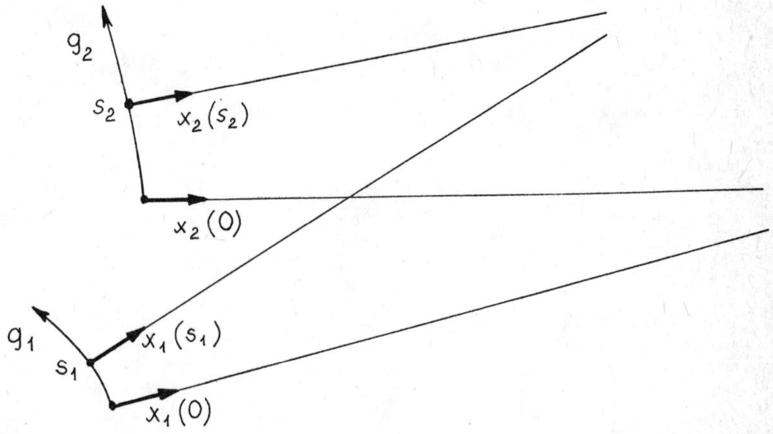

Figure 2. For $s_2=\chi(s_1)$, $x_1(s_1)$ and $x_2(s_2)$ belong to the same leaf.

One can see from [1] that the so called absolute continuity means the fulfilment of the following theorem.

Theorem. The limit

$$\lim_{\substack{z_1 \to 0 \\ z_1 \in Z_1}} \frac{\chi(z_1)}{z_1}$$

exists, and is positive.

The aim of this paper is to give a sketch of the proof of this Theorem. The proof is based on two lemmas. First we shall formulate these lemmas.

3.3. Let g be an oriented curve of finite length on the billiard-table with curvature not less than k_{min}. Let x(s) denote the unit vector - as illustrated on Figure 3 - orthogonal to g and lying in the tangent plane at the distance s from the starting point of g. Let us imagine that at time 0 particles start from each state x(s). For each time t≧0 the places of these particles define a curve on the billiard-table, denoted by $S^t g$ (Figure 3).

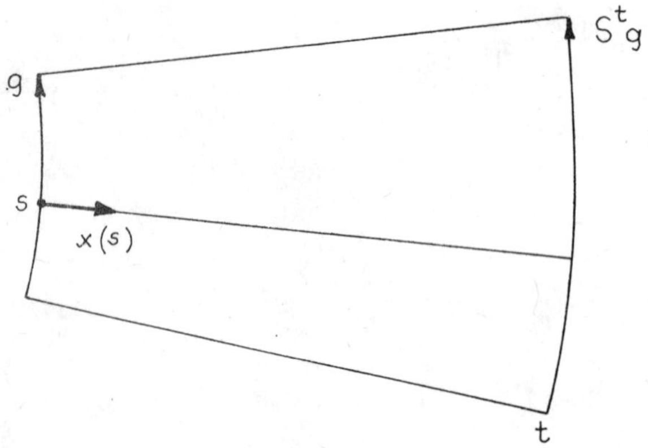

Figure 3. Definition of x(s) and $S^t g$

Let us choose and fix a value s_0, and let us consider the curvature of the curve $S^t g$ along the path of the state $x(s_0)=x$ as the function of the length-parameter t. We denote this curvature by $\varkappa_x(t)$. It is an elementary fact (see [3]) that the function $\varkappa_x(t)$ satisfies the following "differential equation with jumps":

(i) if t is not a moment of collision in the future of x then

$$\mathcal{H}_x'(t)=-K_x(t)-\mathcal{H}^2_x(t),$$

where $K_x(t)$ means the Gaussian curvature of the billiard-table at the point of this path with parameter-value t,

(ii) if t is a moment of collision in the future of x then

$$\mathcal{H}_x(t+0)-\mathcal{H}_x(t-0)=\frac{2k}{\sin\Psi}$$

where k means the curvature of the boundary of the billiard-table at the place of collision, and Ψ is the angle between the boundary of the billiard-table and the path of x at the place of the collision.

Let us notice that $\mathcal{H}_x(t)$ is uniquely determined by x and $\mathcal{H}_x(0)$.

3.4. The pair (x,ℓ) will be called a <u>state-curvature pair</u>, if $x \in M$ and $\ell \geq k_{min}$. Let a denote a state-curvature pair: $a=(x,\ell)$. Let us consider the solution of the differential equation with jumps mentioned above with initial condition $\mathcal{H}_x(0)=\ell$. The state-curvature pair $(S^t x, \mathcal{H}_x(t))$ will be denoted by $S^t a$ $(t>0)$. We introduce the notation $I(a,T):=\int_0^T \mathcal{H}_x(t)dt$ $(T>0)$.

<u>Lemma 1. If</u> $a_i=(x_i,\ell_i)$ <u>is a state-curvature pair</u> (i=1,2) <u>and the states</u> x_1 <u>and</u> x_2 <u>belong to the same leaf then the limit</u>

$$\Delta I(a_1,a_2):=\lim(I(a_1,T_1)-I(a_2,T_2))$$

<u>exists, provided that</u> T_1 <u>and</u> T_2 <u>tend to</u> ∞ <u>in such a way that</u> $S^{T_1}x_1$ <u>and</u> $S^{T_2}x_2$ <u>belong to the same stable fiber</u> (Figure 4).

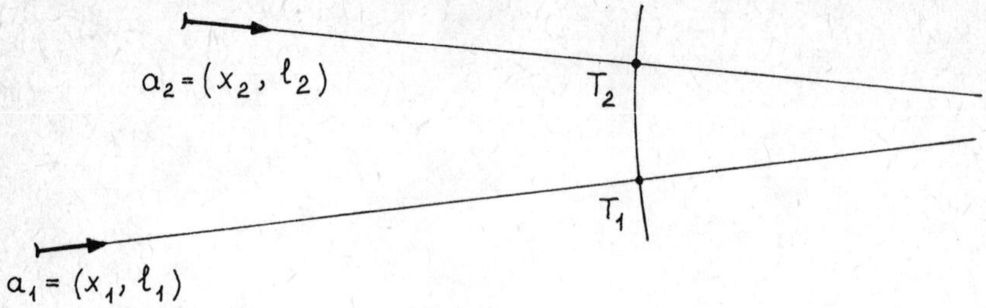

Figure 4. T_1 and T_2 tend to ∞ in such a way that $S^{T_1}x_1$ and $S^{T_2}x_2$ belong to the same stable fiber.

The proof of Lemma 1 can be found in [2]. Here we give only the key ideas of the proof.

It follows from Theorem 5.5 of [3] that the paths of the states x_1 and x_2 come closer and closer exponentially. Provided that t_1 and t_2 tend to ∞ in such a way that $S^{t_1}x_1$ and $S^{t_2}x_2$ belong to the same stable fiber, this fact implies that $|K_{x_1}(t_1)-K_{x_2}(t_2)|$ tends to 0 exponentially. This implies that $|\mathcal{H}_{x_1}(t_1)-\mathcal{H}_{x_2}(t_2)|$ tends to 0 exponentially, from which the statement of Lemma 1 easily follows.

3.5. In Section 3.3 we defined the state x(s) in connection with the curve g. Now let $\ell(s)$ denote the curvature of g at distance s from the starting point of g. Let us introduce the notation $a(s)=(x(s),\ell(s))$. The starting piece of g with length z will be denoted by g_z. It is a well-known fact that the length of $S^T g_z$ is equal to

$$\int_0^z \exp I(a(s),T)\,ds.$$

Thus there exists an $s=s(z)\in(0,z)$ such that

$$\text{length of } S^T g_z = z\cdot\exp I(a(s),T). \qquad (*)$$

3.6. The constant ω was defined by Theorem 3.1 in [3]. Corollary 3.7 of [3] implies that for a fixed value z there exists a unique T such that

$$\text{length of } S^T g_z = \exp[-\tfrac{\omega}{2}T].$$

Let $\mathcal{T}(z)$ denote this value of T. It is obvious that $\lim_{z\to 0}\mathcal{T}(z)=\infty$.

Lemma 2.

$$\lim_{z\to 0}\ \max_{\substack{0\le s\le z\\ 0\le T\le \mathcal{T}(z)}} |I(a(s),T)-I(a(0),T)|=0.$$

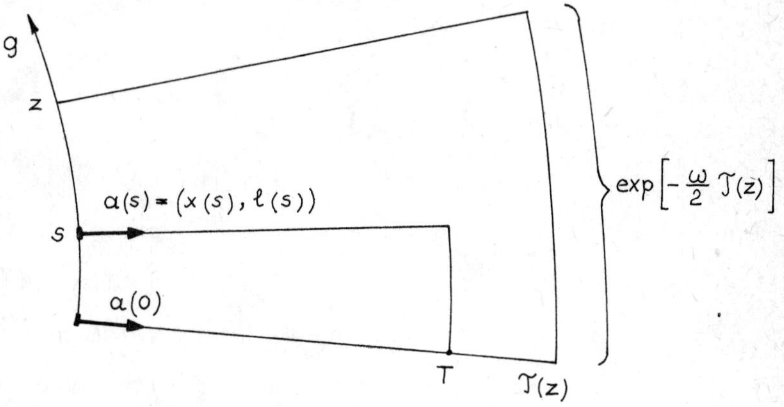

Figure 5. Illustration to Lemma 2.

The proof of Lemma 2 can be found in [2]. Here we give only a short explanation why Lemma 2 is true.

It follows from the definition of $\mathcal{T}(z)$ that for $t\in[0,\mathcal{T}(z)]$ the length of $S^t g$ is exponentially small compared to $\mathcal{T}(z)$. This implies that for $s\in[0,z]$ the pieces of the paths of the states $x(s)$ and $x(0)$ corresponding to the time interval $[0,\mathcal{T}(z)]$

are exponentially near to each other compared to $\mathcal{T}(z)$. Thus for $s\in[0,z]$ and $t\in[0,\mathcal{T}(z)]$, both $|K_{x(s)}(t)-K_{x(0)}(t)|$ and $|\mathcal{H}_{x(s)}(t)-\mathcal{H}_{x(0)}(t)|$ are exponentially small compared to $\mathcal{T}(z)$, from which the statement of Lemma 2 follows.

3.7. In this section, using Lemmas 1 and 2, we shall prove the Theorem (see Figure 6).

We define the state $x_i(s_i)$, the state-curvature pair $a_i(s_i)$ and the value $\mathcal{T}_i(z_i)$ in connection with the curve g_i $(i=1,2)$ as the corresponding objects were defined in connection with the curve g. Now let us choose the values $T_i=T_i(z_i)$ $(i=1,2)$ in such a way that

(i) the states $S^{T_1}x_1(0)$ and $S^{T_2}x_2(0)$ belong to the same stable fiber, and

(ii) $T_i\leq \mathcal{T}_i(z_i)$ $(i=1,2)$, and

(iii) $T_1=\mathcal{T}_1(z_1)$ or $T_2=\mathcal{T}_2(z_2)$.

Let us suppose for the sake of simplicity that $T_1=\mathcal{T}_1(z_1)$.

Theorem 5.5 of [3] implies that the distance between the points of the paths of the states $x_i(0)$ $(i=1,2)$ with parameter value T_i $(i=1,2)$ is smaller than $\text{const}\cdot\exp[-\omega T_1]$. Since in the definition of $\mathcal{T}_1(z_1)$ there is $\exp[-\frac{\omega}{2}T_1]$, which is "much greater" than $\text{const}\cdot\exp[-\omega T_1]$, the distance mentioned above is "small" compared to the length of the curves $S^{T_i}g_{i,z_i}$ $(i=1,2)$.

This implies that

$$\lim_{z_1\to 0}\frac{\text{length of } S^{T_2}g_{2,z_2}}{\text{length of } S^{T_1}g_{1,z_1}}=1. \qquad (**)$$

We get from formula (*) that

$$z_i=(\text{length of } S^{T_i}g_{i,z_i})\cdot\exp[-I(a_i(s_i),T_i] \quad (i=1,2).$$

So using (**), this gives the result

$$\lim_{\substack{z_1 \to 0 \\ z_1 \in Z_1}} \frac{z_2}{z_1} = \lim_{\substack{z_1 \to 0 \\ z_1 \in S_1}} \exp[I(a_1(s_1),T_1) - I(a_2(s_2),T_2)].$$

Now using Lemma 2, it follows that

$$= \lim_{\substack{z_1 \to 0 \\ z_1 \in Z_1}} \exp[I(a_1(0),T_1) - I(a_2(0),T_2)]$$

and, finally, using Lemma 1, we are led to the expression

$$= \exp \Delta I(a_1(0), a_2(0)),$$

which completes the proof of the Theorem.

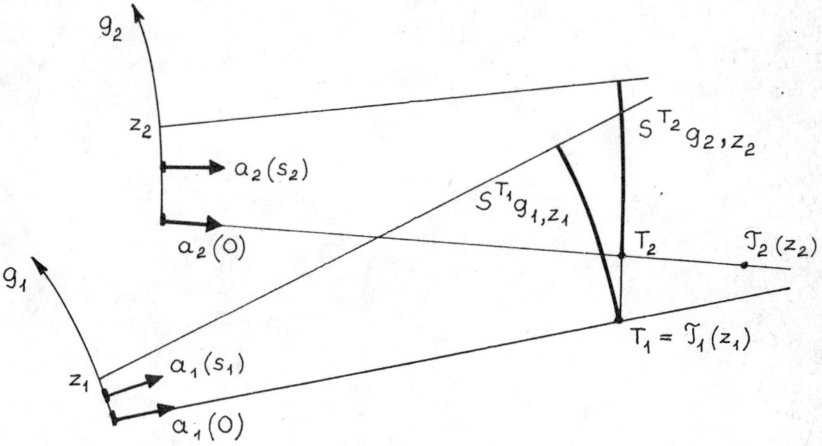

Figure 6. Illustration to the proof of the Theorem.

REFERENCES

[1] Ya. G. SINAI, Russ. Math. Survey, 25, 137-189 (1976), (in Russian)

[2] A. VETIER, Billiard on bent surfaces, thesis (1982), (in Hungarian)

[3] A. VETIER, Sinai billiard in potential field (Construction of stable and unstable fibers), in Proc.Coll.Limit Theorems, Veszprém, 1982 (ed. P. Révész)

Proc. of the 3rd Pannonian Symp. on Math. Stat., Visegrád, Hungary 1982
J. Mogyoródi, I. Vincze, W. Wertz, eds

CONTRIBUTION TO A CHARACTERIZATION PROBLEM

I. VINCZE

Mathematical Institute, Hungarian Academy of Sciences, Budapest, Hungary

1. INTRODUCTION

In [3] the following problem due to A. Rényi and the author was posed:

Let

$$f(t) = 1 + a_1 t + a_2 t^2 + \dots \quad , \quad a_i > 0 \tag{1.1}$$

be an entire function satisfying the relations

$$\int_0^{+\infty} \frac{a_n t^n}{f(t)} \, dt = 1 \ , \qquad n=0,1,2,\dots \quad . \tag{1.2}$$

The question is whether the function $f(t)=e^t$ is the unique solution of (1.1) and (1.2). If yes, this would give a joint characterization of the Poisson and the gamma distributions.

Multiplying (1.2) by λ^n and summarizing with respect to n we obtain the functional equation

$$\int_0^{+\infty} \frac{f(\lambda t)}{f(t)} \, dt = \frac{1}{1 - \lambda} \ , \quad 0 \le \lambda < 1 \tag{1.3}$$

which is equivalent to (1.2).

The first result in this direction is due to Hall and Williamson [2]; they proved that if $f(t)$ satisfies (1.1) and (1.3) then

$$\frac{f'(t)}{f(t)} = 1 + \frac{0(1)}{\log t} \qquad \text{as} \quad t \to \infty \ .$$

As a next step, Hayman and the author [4] sharpened this statement proving by elementary tools that for arbitrary $t>0$ and $A\geq 56$

$$\left|\frac{f'(t)}{f(t)} - 1\right| < A(1+t)^{-1/2}$$

holds; from this inequality the relation

$$e^{t-2A\sqrt{t}} < f(t) < e^{t+2A\sqrt{t}} \qquad (1.4)$$

follows. To obtain this result some lemmas were proved and the aim of the present note is to improve some of them.

From (1.2) - being the sequence a_n^{-1}, $n=0,1,2,\ldots$, a moment sequence - it follows that

$$a_n^2 > a_{n-1}a_{n+1} \quad , \qquad n=1,2,\ldots$$

Using the notation $r_n = a_{n-1}/a_n$ this inequality means that $r_n<r_{n+1}$. It can be seen that the term a_nt^n is the maximum term in the interval $r_n\leq t<r_{n+1}$. Being $f(t)$ an entire function, $r_n\to+\infty$ as $n\to\infty$. For the value of r_n in [4] the following inequality was obtained

$$n+2-2\sqrt{n+1}<r_n<n+1+2\sqrt{n}$$

which was the main tool for proving (1.4).

In [5] the authors pointed out the probabilistic feature of their considerations and also proved the following inequality

$$r_n\leq r_{n+1}\left(1 - \frac{1}{(n+1)^2}\right), \qquad n=0,1,2,\ldots \qquad (1.5)$$

It was remarked that for proving the conjecture $f(t) = e^t$ the inequality

$$r_n\leq r_{n+1}\left(1 - \frac{1}{n+1}\right)$$

would be needed.

Hall and the author [1] proved that the conjecture $f(t) = e^t$ is true when $f(t)^{-1}$ is completely monotone, i.e. its derivatives satisfy the relation

$$(-1)^n \frac{d^n}{dt^n} \left(\frac{1}{f(t)}\right) > 0 , \qquad t>0 \tag{1.6}$$

for n=1,2,... . This relation is trivial for n=1 and we shall prove in the next paragraph its validity for n=2, i.e. that

$$\frac{d^2}{dt^2} \left(\frac{1}{f(t)}\right) > 0 , \qquad t>0 . \tag{1.7}$$

Here we turn to some consequence of (1.7). According to a theorem of the author [6] the validity of the relation (1.6) for n=1,2,...,k involves that the sequence

$$\left(\frac{1}{n} + \frac{1}{k}\right) \frac{a_{n-1}}{a_n}$$

is monotonically increasing. This means for k=2 that

$$r_n < r_{n+1} \left(1 - \frac{2}{(n+1)(n+2)}\right) , \qquad n=1,2,... \tag{1.8}$$

holds.

Concerning the coefficients a_n in [4], relation $a_n \geq a_{n+1}$, $n \geq 2$ was justified. Now we prove the following

PROPOSITION 1.1

When f(t) satisfies (1.1) and (1.2) then the inequalities

$$\frac{3}{2} > a_1 > a_2 \geq a_3 \geq \ldots \quad \text{and} \quad \frac{3}{2} > a_1 > \frac{1}{2} \tag{1.9}$$

hold.

Proof: Denoting $f(t)^{-1}$ by $\varphi(t)$, (1.3) has the form

$$\int_0^{+\infty} f(\lambda t)\, \varphi(t) dt = \frac{1}{1-\lambda} , \qquad 0 \leq \lambda < 1 .$$

Introducing the variable $u=\lambda t$ we have

$$\int_0^{+\infty} f(u)\, \varphi\left(\frac{u}{\lambda}\right) du = \frac{\lambda}{1-\lambda} , \qquad 0 \leq \lambda < 1 .$$

Multiplying the first derivative by λ^2 and differentiating both sides once again we have

$$\int_0^{+\infty} f(u)\, u^2\, \varphi''\left(\frac{u}{\lambda}\right) du = \frac{2\lambda^3}{(1-\lambda)^3} .$$

Returning to the original variable t we get

$$\int_0^{+\infty} f(\lambda t)t^2\varphi''(t)dt = \frac{2}{(1-\lambda)^3},$$

which has the form

$$\sum_{n=0}^{\infty} \lambda^n \int_0^{+\infty} a_n t^{n+2}\varphi''(t)dt = \sum_{n=0}^{\infty} (n+1)(n+2)\lambda^n.$$

Comparing the coefficients on the two sides for n=0,1 we obtain

$$\int_0^{+\infty} t^2\varphi''(t)dt = 2, \quad a_1 \int_0^{+\infty} t^3\varphi''(t)dt = 6 .$$

Integrating by parts the expression $\int_0^{+\infty} t\varphi''(t)dt$ we obtain

$$\int_0^{+\infty} t\varphi''(t)dt = 1 ,$$

finally

$$\int_0^{+\infty} \varphi''(t)dt = \varphi'(+\infty) - \varphi'(0) = a_1 .$$

In this way the values a_1, 1, 2, $6/a_1$ form a moment sequence, consequently

$$1 < 2a_1 \quad \text{and} \quad 4 < \frac{6}{a_1} ,$$

from which

$$\frac{1}{2} < a_1 < \frac{3}{2}$$

follows. Applying (1.8) to n=1 we obtain $r_1 < \frac{3}{2} r_2$, on the other hand $r_1 = \frac{1}{a_1} > \frac{2}{3}$, thus $r_2 > 1$, consequently $a_2 = a_1/r_2 < a_1$, q.e.d.

In [4] the relation

$$\frac{f(t) - 1}{t \cdot f(t)} < \frac{C}{1+t}$$

with C=6 was proved. Now we improve the constant C giving

PROPOSITION 1.2

If f(t) satisfies (1.1) and (1.2) then

$$\frac{f(t) - 1}{t\, f(t)} < \frac{2}{1+t} .$$

Proof: Due to (1.9) we can write

$$(1+t)\frac{f(t)-1}{t} = a_1 + (a_1+a_2)t + (a_2+a_3)t^2 + $$
$$+ \ldots < 2+2a_1t+2a_2t^2+ \ldots = 2f(t).$$

Since in the proof of Theorem 3 in [4] the coefficient C=6 plays the role of a proportional factor, we claim

PROPOSITION 1.3

When f(t) satisfies (1.1) and (1.2) then

$$e^{t-2A\sqrt{t}} < f(t) < e^{t+2A\sqrt{t}}$$

with $A \geq 19$.

2. A LEMMA FOR POWER SERIES WITH POSITIVE COEFFICIENTS

LEMMA

Let $f(t)=a_0+a_1t+a_2t^2+\ldots$ be a power series with positive coefficients and with radius of convergence R>0. Whenever the relations $a_n^2>a_{n-1}a_{n+1}$, n=1,2,..., hold, the function $f(t)^{-1}$ is strictly convex in [0, +∞), i.e. the inequality

$$\frac{d^2}{dt^2}\left(\frac{1}{f(t)}\right) > 0, \qquad 0\leq t<R$$

holds.

Proof of the Lemma: As

$$\frac{d^2}{dt^2}\left(\frac{1}{f(t)}\right)= \frac{2f'(t)^2 - f(t)f''(t)}{f^3(t)},$$

it is sufficient to prove that the numerator is positive. By simple calculation we obtain

$$2f'^2(t)-f(t)f''(t) = 2\sum_{n=0}^{\infty} t^n\left[\sum_{k=0}^{n}(k+1)(n-k+1)a_{k+1}a_{n-k+1}\right] -$$

$$-\sum_{n=0}^{\infty} t^n\left[\sum_{k=0}^{n}(k+2)(k+1)a_{k+2}a_{n-k}\right] = \quad (2.1)$$

$$= \sum_{n=0}^{\infty} t^n\left\{-(n+1)(n+2)a_0a_{n+2} + \sum_{k=0}^{n}\left[2(k+1)(n-k+1)-k(k+1)\right]a_{k+1}a_{n-k+1}\right\} =$$

$$= \sum_{n=0} t^n\left[-(n+1)(n+2)a_0a_{n+2} + \sum_{k=0}^{n}(k+1)(2n-3k+2)a_{k+1}a_{n-k+1}\right].$$

Now we turn to odd and even powers separately. In both cases we prove that the coefficients of t^n are positive. For proving it let us observe that they have the following structure:

$$A_{n,o}a_oa_{n+2} + \sum_{k=0}^{n} A_{n,k+1}a_{k+1}a_{n-k+1} \quad ,$$

and a slight calculation shows that $\sum_{k=0}^{n+1} A_{n,k}=0$ for $n=0,1,2,\ldots$. Now we shall rearrange the terms in such a way that the positive products a_oa_{n+2}, a_1a_{n+1}, a_2a_n, ... form an increasing sequence, while the coefficients up to $k=i$ will be negative and for $k>i$ nonnegative. Our Lemma will be proved by applying the following simple

PROPOSITION 2.1

Let us have two sequences $A_1, A_2, \ldots, A_s$ and $B_1,B_2,\ldots,B_s$ with $A_j<0$ for $j=1,2,\ldots,i$ and $A_j\geq 0$ for $j=i+1,i+2,\ldots,s$, $\sum_{j=1}^{s} A_j=0$ and $0<B_1<B_2<\ldots<B_s$. Then

$$\sum_{j=1}^{s} A_jB_j>0 \quad .$$

The proof of this statement is contained in the following sequences of inequalities.

$$\sum_{j=1}^{s} A_jB_j = \sum_{j=1}^{i} A_jB_j + \sum_{j=i+1}^{s} A_jB_j > B_i \sum_{j=1}^{i} A_j + B_{i+1} \sum_{j=i+1}^{s} A_j =$$

$$= (B_{i+1}-B_i) \sum_{j=i+1}^{s} A_j > 0.$$

Now we turn to the proposed rearrangement.

The case n=2m. From the last expression of (2.1) we have for the coefficient of $t^n=t^{2m}$

$$-(2m+1)(2m+2)a_o a_{2m+2} + \sum_{k=0}^{2m} (k+1)(4m-3k+2)a_{k+1}a_{2m-k+1}.$$

The last sum will be divided into two parts: $\sum_{0}^{m} + \sum_{m+1}^{2m}$. In the second part the index k will be changed for $k'=2m-k$ where k' varies from 0 to m-1. Then we have

$$\sum_{k=0}^{m} (k+1)(4m-3k+2)a_{k+1}a_{2m-k+1} - \sum_{k=0}^{m-1} (2m-k+1)(2m-3k-2)a_{k+1}a_{2m-k+1}.$$

Now adding the first term, by easy calculations we have

$$-(2m+1)(2m+2)a_o a_{2m+2} + \sum_{k=0}^{m-1} [2m^2+6m+4-6(k-m)^2]a_{k+1}a_{2m-k+1} +$$

$$+ (m+1)(m+2)a_{m+1}^2 .$$

It is clear that the coefficients of the positive products $a_{k+1}a_{2m-k+1}$ are negative up to an index k=i and nonnegative for k>i. On the other hand, according to our assumption on the coefficients we have

$$a_{k+1}^2 > a_k a_{k+2}$$

$$a_{k+2}^2 > a_{k+1}a_{k+3}$$

$$\dots$$

$$a_{2m-k}^2 > a_{2m-k-1}a_{2m-k+1}$$

$$a_{2m-k+1}^2 > a_{2m-k}a_{2m-k+2}$$

from which the relation

$$a_{k+1}a_{2m-k+1} > a_k a_{2m-k+2}, \quad k=0,1,\dots,m$$

follows, completing the proof.

The case n=2m+1. Substituting the new notation into the coefficient of $t^n = t^{2m+1}$ we have

$$\sum_{k=0}^{2m+1} (k+1)(4m-3k+4)a_{k+1}a_{2m-3k+2} = \sum_{0}^{m} + \sum_{m+1}^{2m+1} .$$

Let $k'=2m-k+1$, then we have

$$\sum_{k=0}^{2m+1} = \sum_{k=0}^{m} (k+1)(4m-3k+4)a_{k+1}a_{2m-k+2} -$$

$$- \sum_{k=0}^{m}(2m-k+2)(2m-3k-1)a_{k+1}a_{2m-k+2} =$$

$$= \sum_{k=0}^{m} [2m^2+8m + \frac{15}{2} - 6(k-m-\frac{1}{2})^2] a_{k+1}a_{2m-k+2} .$$

Using the argumentation as in the case of n=2m we arrive to the proof of our Lemma.

REFERENCES

1 HALL, R. R. and VINCZE, I. (1981). On a simultaneous characterization of the Poisson law and the gamma distribution. In: Analytical Methods in Probability Theory (Oberwolfach, 1980), pp. 54-59, Lecture Note in Math. 861, Springer, Berlin.

2 HALL, R. R. and WILLIAMSON, J. H. (1976). On a certain functional equation. J. London Math. Soc.(2) 12, 133-136.

3 HAYMAN, W. K. (1967). Research Problems in Function Theory. Athlone, London.

4 HAYMAN, W.K. and VINCZE, I. (1978). A problem on entire functions. In: Complex Analysis and Its Applications, pp. 591-594, Nauka, Moscow.

5 HAYMAN, W.K. and VINCZE, I. (1979). Markov type inequalities and entire functions. In: Analytic Function Methods in Probability Theory, pp. 153-163, Coll. Math. Soc. J. Bolyai 21, North-Holland, Amsterdam.

6 VINCZE, I. (1980). On a certain class of multiple monotone and completely monotone functions. In: Functions, Series, Operators (Proc. Fejér-Riesz Colloq., Budapest, 1980), Coll. Math. Soc. J. Bolyai 35, North-Holland, Amsterdam. (To appear.)

Proc. of the 3rd Pannonian Symp. on Math. Stat., Visegrád, Hungary 1982
J. Mogyoródi, I. Vincze, W. Wertz, eds

BALANCE EQUATIONS AND Σ'-TYPE CONDITIONS FOR POINT PROCESSES

A. WAKOLBINGER, G. EDER

Universität Linz, Austria

INTRODUCTION

Motivated mainly by Nguyen and Zessin's paper [10] Integral and differential characterization of the Gibbs process, Matthes, Mecke and Warmuth (1979) gave a link between differential and integral conditional description of point processes. Recently, still more investigations in this direction also have been carried out by Kallenberg (1982).
Georgii (1975),(1979) found similar results as in [10] for canonical Gibbs processes. Finally, Georgii (1981) and the authors (1982) presented results corresponding to those of Matthes et al. for general point processes, but in the canonical rather than in the grand canonical picture.
The present paper is - in a unified language - a supplement to the results of Matthes et al. (1979), Georgii (1981), and the authors (1982).

We introduce (cf. [3]) balance equations of higher orders for point processes, and give a connection with reversibility w.r. to multiple jump generators, thus covering results of Glötzl ([4]).

Moreover, we introduce (cf. [9]) Σ'-type conditions of higher order for point processes, which are closely connected to balance equations, and show in a few examples how to construct conditional probabilities w.r. to suitable σ-fields from such equations.

NOTATION

We fix a Polish space X, and denote by $\mathfrak{X}$ the σ-algebra of Borel sets in X, and by $\mathfrak{B}$ the set of all $V\in\mathfrak{X}$ which are bounded. N will be the set of all integer-valued measures μ on X with $\mu(V)<\infty$ for all $V\in\mathfrak{B}$; for $B\in\mathfrak{X}$, we also write $\beta_B(\mu)$ instead of $\mu(B)$. Let $\mathcal{N}$ denote the σ-algebra on N generated by the mappings $\beta_V (V\in\mathfrak{B})$. For $B\in\mathfrak{X}$ we denote by $\mathcal{N}_B$ the σ-algebra generated by the mappings $\beta_V (V\subset B, V\in\mathfrak{B})$, and by ε_B the σ-algebra generated by the mappings $\beta_V (V\subset B^c, V\in\mathfrak{B})$ and β_B (where B^c denotes $X\backslash B$).

For $x\in X$, δ_x will denote point mass in x.

If $x=(x_1,\ldots,x_n)\in X^n$, we will also write δ_x instead of $\delta_{x_1}+\ldots+\delta_{x_n}$.

By a point process (which we will usually name P), we simply mean a probability measure on $(N,\mathcal{N})$.

1. CONDITIONAL DESCRIPTION OF POINT PROCESSES

Such a description can be

(i) of "integral type", i.e. via conditional probabilities w.r. to a certain family of sub-σ-algebras of $\mathcal{N}$, e.g. $P[\,.\,|\mathcal{N}_{V^c}]\ V\in\mathfrak{B})$, or

(ii) of "differential type", e.g. by ratios of the form

$$\frac{P(d(\mu+\delta_{x_1}+\ldots+\delta_{x_n}))}{P(d(\mu+\delta_{y_1}+\ldots+\delta_{y_n}))} \quad \text{with } m,n\geq 0 \text{ fixed e.g. } \frac{P(d(\mu+\delta_x))}{P(d\mu)}.$$

Of course, ratios as in (ii) are only formal and must be given a precise meaning.

First of all, one needs a measure which "differentially" corresponds to $P(d(\mu+\delta_{x_1}+\ldots+\delta_{x_m}))$. This is given by the so-called m-th order reduced Campbell measure (see, e.g. [7]) $C_P^{(m)}$ on $\mathcal{N}\times\mathfrak{X}^m$, defined by

$$C_P^{(m)}(A):=\int_N\int_{X^m}\mu^{(m)}(dx_1,\ldots,dx_m)1_A(\mu-\delta_{x_1}-\ldots-\delta_{x_m},x_1,\ldots,x_m)P(d\mu)$$

$$(A\in\mathcal{N}\times\mathfrak{X}^m)$$

with $\mu^{(m)}(dx_1,\ldots,dx_m)=\mu(dx_1)(\mu-\delta_{x_1})(dx_2)\ldots(\mu-\delta_{x_1}-\ldots-\delta_{x_{m-1}})(dx_m)$.

Hence, an equation prescribing a ratio like in (ii) for P might be of the form

$$C_P^{(m)}(d\mu,dx)\rho_{m,n}(\mu,x,dy)=C_P^{(n)}(d\mu,dy)\rho_{n,m}(\mu,y,dx) \tag{1}$$

where $\rho_{m,n}$ is a kernel from $N\times X^m$ to X^n

and $\rho_{n,m}$ is a kernel from $N\times X^n$ to X^m.

We call such an equation (in accordance with a definition of Georgii [3], who considered the case m=n=1) a (non-separated) balance equation for P of degree (m,n).

We now give an example where such an equation occurs.

2. REVERSIBILITY w.r. TO MULTIPLE JUMP GENERATORS

For $m,n\geq 0$, let $\rho_{m,n}$ be a kernel from $N\times X^m$ to X^n, such that for all $\mu\in N$, $x\in X^m$

(a) $\rho_{m,n}(\mu,x,\cdot)$ is a symmetric measure on $\mathfrak{X}^n$, and symmetric in x

(b) $\rho_{m,n}(\mu,x,D_x^{(n)})=0$, with $D_{(x_1,\ldots,x_m)}^{(n)}:=\{(y_1,\ldots,y_n)\in X^n\,|\,y_i=x_j$ for some $i,j\}$

(c) $\int_{X^{m+n}\setminus(V^c)^{m+n}}\mu^{(m)}(dx)\rho_{m,n}(\mu-\delta_x,x,dy)<\infty$ for all $V\in\mathcal{B}$

and let $\rho_{n,m}$ be a kernel from $N\times X^n$ to X^m with properties (a), (b),(c) (n and m interchanged).

If f is bounded and $\mathcal{N}_V$-measurable for some $V\in\mathcal{B}$ (we denote the set of all such f by $\mathcal{D}$), put

$$\mathcal{L}_{\{m,n\}}f(\mu):=\int\mu^{(m)}(dx)\rho_{m,n}(\mu-\delta_x,x,dy)[f(\mu-\delta_x+\delta_y)-f(\mu)]+$$

$$+\int\mu^{(n)}(dy)\rho_{n,m}(\mu-\delta_y,y,dx)[f(\mu-\delta_y+\delta_x)-f(\mu)]$$

(Note that (c) guarantees $\mathcal{L}_{\{m,n\}}f<\infty$ for $f\in\mathcal{D}$.)

Note that if some dynamic exists for which $\mathcal{L}_{\{m,n\}}$ is (an extension of) its generator, then $\rho_{m,n}(\mu,x,dy)$ gives the jump rate of the m-tuple x to the n-tuple in dy, given the surrounding μ. (The case $\{m,n\}=\{1,0\}$ corresponds to birth and death, the case $\{m,n\}=\{1,1\}$ corresponds to simple jumps.) We do not bother with existence questions, but rather define:

A point process P is called <u>$\mathcal{L}_{\{m,n\}}$-reversible</u> if, for all f and $g\in\mathcal{D}$, $\int f\,\mathcal{L}_{\{m,n\}}g\,dP=\int g\,\mathcal{L}_{\{m,n\}}f\,dP$.

<u>Theorem 1:</u> P is $\mathcal{L}_{\{m,n\}}$-reversible iff (1) holds.

<u>Proof</u>:

α) First, a straightforward calculation shows that for $f,g\in\mathcal{D}$

$$\int[f\,\mathcal{L}_{\{m,n\}}g-g\,\mathcal{L}_{\{m,n\}}f]dP=$$

$$=\begin{cases}\int[C_P^{(m)}(d\mu,dx)\rho_{m,n}(\mu,x,dy)-C_P^{(n)}(d\mu,dy)\rho_{n,m}(\mu,y,dx)]\cdot\\ \quad\cdot[f(\mu+\delta_x)g(\mu+\delta_y)-f(\mu+\delta_y)g(\mu+\delta_x)] & \text{if } m>n\\ 2\int C_P^{(n)}(d\mu,dx)\rho_{n,n}(\mu,x,dy)[f(\mu+\delta_x)g(\mu+\delta_y)-f(\mu+\delta_y)g(\mu+\delta_x)] & \text{if } m=n.\end{cases}$$

From this, the direction from the right to the left is already clear.

β) Secondly, take $B_1,\dots,B_m$, $C_1,\dots,C_n \in \mathcal{L}$ such that $\bigcup_{i=1}^{m} B_i$ and $\bigcup_{j=1}^{m} C_j$ are disjoint.

Let $B := \bigcup_{\pi \text{ perm. of } \{1,\dots,m\}} B_{\pi(1)} \times \dots \times B_{\pi(m)}$, and C be defined analogously. Take an $F \in \mathcal{N}$ which is $\mathcal{N}_V$-measurable for some $V \in \mathcal{L}$, and such that

$F \subset (\{\beta_{B_1}=k_1\} \cap \dots \cap \{\beta_{B_m}=k_m\} \cap \{\beta_{C_1}=l_1\} \cap \dots \cap \{\beta_{C_n}=l_n\})$ for some $k_1,\dots,k_m, l_1,\dots,l_n \in N_0$.

Then, put $f := 1_{\{\mu+\delta_x \mid \mu \in F, x \in B\}}$, and $g := 1_{\{\mu+\delta_y \mid \mu \in F, y \in C\}}$, and observe that

$$f(\mu+\delta_x)g(\mu+\delta_y) - f(\mu+\delta_y)g(\mu+\delta_x) =$$

$$= \begin{cases} 1_F(\mu)1_B(x)1_C(y) & \text{for } m>n \\ 1_F(\mu)[1_B(x)1_C(y) - 1_B(y)1_C(x) & \text{for } m=n. \end{cases}$$

From α) and β), and from the assumptions b) and c) on ρ_{mn} and ρ_{nm}, now also the direction from the left to the right in Theorem 1 follows by standard arguments.

<u>Remark 1:</u> Theorem 1 has been proved for the case $(m,n)=(1,0)$ by Glötzl [4] and for the case $(m,n)=(1,1)$ by Georgii [3] and Glötzl [4]. Using the same ideas as in the proof of Theorem 1, we are able to prove the following

<u>Proposition 1:</u> Let, for $K \in N$, and $0 \leq m+n \leq K$, ρ_{mn} and ρ_{nm} be kernels with properties a), b), c). Put $\mathcal{L}_K := \sum_{m+n \leq K} \mathcal{L}_{\{m,n\}}$ (this would correspond to a dynamic where jumps of various multiplicity can occur).

Then a point process P is $\mathcal{L}_K$-reversible iff it is $\mathcal{L}_{\{m,n\}}$-reversible for all m,n with $m+n \leq K$.

3. SEPARATED BALANCE EQUATIONS

In special cases, the variables x and y in equation (1) (section 1) may be "separated", i.e., one can find kernels σ_m (from N to X^m) and σ_n (from N to X^n) such that (1) is equivalent to

$$C_P^{(m)}(d\mu,dx)\sigma_n(\mu,dy)=C_P^{(n)}(d\mu,dy)\sigma_m(\mu,dx) \qquad (2)$$

(see Georgii [3] for a short discussion in the case m=n=1).

We treat only the special case in which there exists a measure λ on X (with $\lambda(X)>0$, $\lambda(V)<\infty$ for $V\in\mathcal{L}$) such that $\rho_{m,n}$ (resp. $\rho_{n,m}$) is absolutely continuous w.r. to λ^n (resp. λ^m) in its last component. Then (1) takes the form

$$C_P^{(m)}(d\mu,dx)\lambda^n(dy)c_{mn}(\mu,x,y)=C_P^{(n)}(d\mu,dy)\lambda^m(dx)c_{nm}(\mu,y,x). \qquad (3)$$

First we remark that then c_{mn} and c_{nm} obey a "cycle relation":

Lemma 1: Suppose that (3) holds. Then

$$c_{mn}(\mu,x,y)c_{nm}(\mu,y,z)c_{mn}(\mu,z,v)c_{nm}(\mu,v,x)= \qquad (4)$$

$$=c_{mn}(\mu,x,v)c_{nm}(\mu,v,z)c_{mn}(\mu,z,y)c_{nm}(\mu,y,x)$$

$$C_P^{(m)}\times\lambda^{m+2n} \quad \text{a.a. } (\mu,x,y,z,v).$$

Proof: Use (3) four times to exchange x with y, y with z, z with v and v with x.

Using (4), we are in the position to prove

Proposition 2: Put for $\xi\in X^m$, $\eta\in X^n$

$$f_m^{(\xi,\eta)}(\mu,x)=\frac{c_{mn}(\mu,x,\eta)}{c_{nm}(\mu,\eta,x)}\,c_{nm}(\mu,\eta,\xi),$$

$$f_n^{(\xi,\eta)}(\mu,y)=\frac{c_{nm}(\mu,y,\xi)}{c_{mn}(\mu,\xi,y)}\,c_{mn}(\mu,\xi,\eta).$$

a) Assume

$$\begin{cases} 0<c_{nm}<\infty & C_P^{(m)}\times\lambda^n \quad \text{a.e.} \\ 0<c_{mn}<\infty & C_P^{(n)}\times\lambda^m \quad \text{a.e.} \end{cases} \tag{5}$$

Then (3) is equivalent to

$$C_P^{(m)}(d\mu,dx)f_n^{(\xi,\eta)}(\mu,y)\lambda^n(dy)=C_P^{(n)}(d\mu,dy)f_m^{(\xi,\eta)}(\mu,x)\lambda^m(dx) \tag{6}$$

$$\lambda^m\times\lambda^n \quad \text{a.a.} \quad (\xi,\eta).$$

b) Assume that (4) and (5) hold for all μ,x,y,z,v.

Then (3) holds iff (6) holds for one (all) pairs ξ,η.

The proof of b) is easy; the proof of a) is somewhat technical but rather straightforward, and will be omitted.

4. Σ'-TYPE CONDITIONS OF HIGHER ORDERS

Motivated by the considerations above, we give some regularity conditions for point processes, generalizing definitions and results of Matthes, Mecke, Warmuth [9], Georgii [3], Rauchenschwandtner [11] and the authors [13].

Definition: For $m\geq n$ a point process P is said to be a $\Sigma^{(m,n)}$-process if, for all $V\in\mathcal{B}$, $C_P^{(m)}[\cdot\times V^m]\ll C_P^{(n)}[\cdot\times V^n]$.

Remark 2:

a) Note that condition $\Sigma^{(1,0)}$ is the same as condition

(Σ') $\quad C_P^{(1)}[\cdot\times V]\ll P \quad$ for all $V\in\mathcal{B}$.

This condition was introduced by Matthes et al. [9] (for further results in this direction, see Rauchenschwandtner [11] and Kallenberg).

b) Furthermore, note that each process trivially is a $\Sigma^{(n,n)}$-process ($n\in\mathbb{N}$).

To give an intuitive understanding of conditions $\Sigma^{(m,n)}$, we state the following simple

Lemma 2: Let P be a point process, fix $n\in\mathbb{N}$ and $A\in\mathcal{N}\times\mathfrak{X}^n$. Then $C_P^{(n)}(A)=0$ iff $P(\{\mu+\delta_{x_1}+\ldots+\delta_{x_n}:(\mu,x_1,\ldots,x_n)\in A\}=0$.

The proof follows from the definition of $C_P^{(n)}$.

Corollary: A point process P is $\Sigma^{(m,n)}$ iff for all $F\in\mathcal{N}$,

$$P[\{\mu+\delta_x \mid \mu\in F, x\in X^n\}]=0 \text{ implies } P[\{\mu+\delta_y \mid \mu\in F, y\in X^m\}]=0.$$

For the case $(m,n)=(1,0)$, this is a characterization of Σ'-processes noticed by Rauchenschwandtner [11] (see also [12]): "by adding points, you cannot get allowed configurations from forbidden ones".

Of course, condition $\Sigma^{(1,0)}$ is stronger than all other conditions $\Sigma^{(m,n)}$. An example of a $\Sigma^{(2,0)}$-process, concentrated on infinite configurations, which is not $\Sigma^{(1,0)}$, is as follows:

Let $X=\mathbb{Z}$, and $(M_n)_{n\in\mathbb{N}}$ be a numbering of the finite subsets of $\mathbb{Z}$ with even cardinality.

Let, for $n\in\mathbb{N}$, μ_n be given by

$$\mu_n(\{k\}) = \begin{cases} 1 & \text{if } k\notin M_n \\ 0 & \text{if } k\in M_n \end{cases}$$

and define the point process P by $P(\{\mu_n\})=\frac{1}{2^n}$.

Then, using the above Corollary, it is immediate that P is $\Sigma^{(2,0)}$ but not $\Sigma^{(1,0)}$. Also, the following remark, which we will need below, is a direct consequence of this corollary:

Remark 3: Condition $\Sigma^{(2,1)}$ implies condition $\Sigma^{(n,1)}$.

A link between balance equations of degree (m,n) and property $\Sigma^{(m,n)}$ is given by the following

Theorem 2: A point process P is a $\Sigma^{(m,n)}$-process iff there exist kernels σ_m from N to X^m and σ_n from N to X^n with

$$0<\sigma_n(\mu,V^n)<\infty \quad C_P^{(m)}[\cdot\times V^m] \text{ a.a. } \mu \quad \text{for all } V\in\mathcal{L}, \tag{7}$$

and such that (2) holds.

Proof: " $\Leftarrow$ ": We have to show that, for all $V\in\mathcal{B}$, $C_P^{(m)}[\cdot\times V^m]\ll$ $\ll C_P^{(n)}[\cdot\times V^n]$. Now take $V\in\mathcal{B}$ and $F\in\mathcal{N}$ with $C_P^{(n)}[F\times V^n]=0$. Then (2) implies $\int_F\int_{V^m} C_P^{(m)}(d\mu,dy)\sigma_n(\mu,V^n)=0$. From this, together with (7), there follows the assertion.

" $\Rightarrow$ ": Let Q be a σ-finite measure on $\mathcal{N}$, and σ_m and σ_n be kernels from N to X^m resp. X^n such that

$$C_P^{(m)}(d\mu,dx)=Q(d\mu)\sigma_m(\mu,dx) \text{ and } C_P^{(n)}(d\mu,dy)=Q(d\mu)\sigma_n(\mu,dy).$$

(The existence of such Q, σ_m, σ_n follows from the fact that $C_P^{(m)}[\cdot\times X^m]$ as well as $C_P^{(n)}[\cdot\times X^n]$ are countable sums of σ-finite measures, hence dominated by some σ-finite measure Q, w.r. to which both $C_P^{(m)}$ and $C_P^{(n)}$ then can be disintegrated; for a general argument see A.3.3 in the forthcoming edition of [6]). From $C_P^{(n)}[\cdot\times V^n]=\sigma_n(\mu,V^n)Q(d\mu)$ there follows $\sigma_n(\mu,V^n)>0$ $C_P^{(n)}[\cdot\times V^n]$ a.a. μ, and hence, by $\Sigma^{(m,n)}$, we obtain (7). (2), however, is immediate.

Remark 4:

a) For (m,n)=(1,0), Theorem 2 is a rephrasing of Theorem 1.1 in Matthes et al. [9] (see also Kallenberg [5]):

Property Σ′ is equivalent to existence of a (P a.a. determined) kernel η_P from N to X with $\eta_P(\mu,V)<\infty$ for $V\in\mathcal{B}$, $\mu\in N$, such that $C_P^{(1)}(d\mu,dx)=\eta_P(\mu,dx)P(d\mu)$.

Such a kernel is called Papangelou kernel in [9].

b) A kernel σ(not identical zero) obeying (2) (with m=n=1) is called conditionally balancing P by Georgii [3]. Theorem 2 states the existence of such kernels for arbitrary point processes. (This observation also follows directly from A.3.3 in the forthcoming edition of [6], and also appears in Kallenberg [7]).

c) A kernel σ_n satisfying (2) and (7) (with m=n) is called <u>n-th order balancing kernel of P</u> in [13]; its existence also is guaranteed by Theorem 2.

([13], Proposition 1.5 yields that σ_n satisfies (2) and (7) iff there exists a σ-finite measure Q on $\mathcal{N}$ with $C_P^{(n)}(d\mu,dx)=$ $=Q(d\mu)\sigma_n(\mu,dx)$).

5. BALANCE EQUATIONS AND CONDITIONAL PROBABILITIES

Matthes et al. [9] give an important link between the Papangelou kernel η_P of a Σ'-process P and its conditional probabilities $P[\cdot|\mathcal{N}_{V^c}]$:

<u>Theorem 3</u> ([9], Satz 2.1): Let η_P be a version of the Papangelou kernel of a point process P.

Then, for $V\in\mathcal{L}$ and $F\in\mathcal{N}_V$,

$$P[F|\mathcal{N}_{V^c}](\mu)=$$

$$=P[\beta_V=0|\mathcal{N}_{V^c}](\mu)\cdot(1_F(\sigma)+\sum_{n=1}^{\infty}\frac{1}{n!}\int_{V^n}1_F(\delta_{x_1}+\ldots+\delta_{x_n})\eta_P(\mu_{V^c},dx_1)\ldots$$

$$\ldots\eta_P(\mu_{V^c}+\delta_{x_1}+\ldots+\delta_{x_{n-1}},dx_n)\qquad P\text{ a.a. }\mu. \qquad (8)$$

(If we put in (8) F=N, then the l.h.s. equals one P a.e., hence we can identify $P[\beta_V=0|\mathcal{N}_{V^c}](\mu)$ in terms of η_P and obtain Theorem 2.2 of [9].)

The following theorems are in the same abstract framework as [9]):

<u>Theorem 4:</u> (Georgii [2],[3]; and [13], Proposition 1.8)

Let P be a point process, and, for $n\in\mathbb{N}$, σ_n be an n-th order balancing kernel of P (see Remark 4 c)). Then, for $V\in\mathcal{L}$ and $F\in\mathcal{N}_V$

$P[F|\varepsilon_V](\mu)=$

$$= \frac{1}{\sigma_{\mu(V)}(\mu_{V^c}, V^{\mu(V)})} \int_{V^{\mu(V)}} 1_F(\delta_{x_1}+\ldots+\delta_{x_{\mu(V)}})\sigma_{\mu(V)}(\mu_{V^c}, dx_1,\ldots,dx_{\mu(V)})$$

P a.a. μ.

To illustrate the connections between higher order $\Sigma^{(m,n)}$ conditions and conditional probabilities, we introduce, for $m>n$, and $V\in\mathcal{B}$ the σ-algebra $\mathcal{A}_V^{(m,n)}:=\sigma(\mathcal{N}_{V^c},\{\beta_V=0\},\ldots,\{\beta_V=n-1\}, \{\beta_V\equiv 0 \bmod m-n\},\ldots\{\beta_V\equiv m-n-1 \bmod m-n\})$.

For example, $\mathcal{A}_V^{(1,0)}=\mathcal{N}_{V^c}$; $\mathcal{A}_V^{(2,0)}=\sigma(\mathcal{N}_{V^c},\{\beta_V\equiv 0 \bmod 2\})$; $\mathcal{A}_V^{(2,1)}=\sigma(\mathcal{N}_{V^c},\{\beta_V=0\})$.

We claim that kernels σ_m,σ_n as in Theorem 2 are "the right ones" for computing $P[\cdot|\mathcal{A}_V^{(m,n)}]$, and illustrate this for the cases (2,0) and (2,1).

Theorem 5:

a) Let P be a $\Sigma^{(2,0)}$-process, η be a kernel from N to X^2 such that $C_P^{(2)}(d\mu,dx,dy)=P(d\mu)\eta(\mu,dx,dy)$, and σ be a first order balancing kernel of P. Then, for $V\in\mathcal{B}$ and $F\in\mathcal{N}_V$,

$$P[F|\mathcal{A}_V^{(2,0)}](\mu)=P[\beta_V=0|\mathcal{A}_V^{(2,0)}](\mu)[1+\sum_{n=1}^{\infty}\frac{1}{2n!}\int_{V^{2n}} 1_F(\delta_{x_1}+\ldots+\delta_{x_{2n}}).$$

$$.\eta(\mu_{V^c},dx_1,dx_2)\ldots\eta(\mu_{V^c}+\delta_{x_1}+\delta_{x_{2n-2}},dx_{2n-1},dx_{2n})]$$

$$+P[\beta_V=1|\mathcal{A}_V^{(2,0)}\}(\mu)[\frac{1}{\sigma(\mu_{V^c},V)}(\int_V \sigma(\mu_{V^c},dx)1_F(\delta_x)$$

$$+\sum_{n=1}^{\infty}\frac{1}{(2n+1)!}\int_{V^{2n+1}}\sigma(\mu_{V^c},dx_1)\eta(\mu_{V^c}+\delta_{x_1},dx_2,dx_3)\ldots$$

$$\ldots\eta(\mu_{V^c}+\delta_{x_1}+\ldots+\delta_{x_{2n-1}},dx_{2n},dx_{2n+1})1_F(\delta_{x_1}+\delta_{x_{2n+1}})]$$

P a.a. μ.

b) Let P be a $\Sigma^{(2,1)}$-process, and, for all $m \in \mathbb{N}$, σ_m be a kernel from N to X^m such that (9) and (10) hold (see Remark 5 a))

$$C_P^{(n)}(d\mu,dx)\sigma_1(\mu,dy)=C_P^{(1)}(d\mu,dy)\sigma_n(\mu,dx) \tag{9}$$

$$\sigma_1(\mu,V)>0 \quad C_P^{(n)}[\cdot\times V^n] \quad \text{a.a. } \mu. \tag{10}$$

Then, for all $V \in \mathcal{L}$ and $F \in \mathcal{N}_V$,

$$P[F|\mathcal{A}_V^{(2,1)}](\mu)=1_F(0)1_{\{\beta_{V=0}\}}(\mu)+ \tag{11}$$

$$+P[\beta_V=1|\mathcal{A}_V^{(2,1)}](\mu)\,\frac{1}{\sigma_1(\mu_{V^c},V)}\sum_{n=1}^{\infty}\frac{1}{n!}\int_{V^n}1_F(\delta_{x_1}+\ldots+\delta x_n)\sigma_n(\mu_{V^c},dx_1,\ldots,dx_n)$$

P a.a. μ.

The proof of Theorem 5 follows the same line as that of Theorems 3 and 4.

Remark 5:

a) In Theorem 5 b), the existence of kernels σ_n obeying (9) and (10) follows, using Remark 3, as in the proof of Theorem 2.

b) If in Theorem 5 a), $\eta(\mu,dx,dy)=h(\mu,x,y)\lambda(dx)\lambda(dy)$ for some $h>0$, then for $\lambda\times\lambda$ a.a. (ξ,η),

$$\sigma^{(\xi,\eta)}(\mu,dx):=\frac{h(\mu+\delta_\xi,x,\eta)}{h(\mu+\delta_x,\xi,\eta)}\lambda(dx) \text{ is a balancing kernel for P.}$$

c) As a "constructive remark" to Theorem 5 b), we note:

If $C_P^{(2)}(d\mu,dx_1,dx_2)\lambda(dy)f(\mu,y)=C_P^{!}(d\mu,dy)\lambda^2(dx_1,dx_2)k(\mu,x_1,x_2)$ with $f>0$ then, for $\lambda^{\mathbb{N}}$-a.a. $(z_1,z_2,\ldots,)$, (11) holds with

$\sigma_1(\mu,dx):=f(\mu,x)\,\lambda(dx)$, $\sigma_2(\mu,dx_1,dx_2):=k(\mu,x_1,x_2)\lambda^2(dx_1,dx_2)$,

$$\sigma_n^{(z_1,\ldots,z_{n-2})}(\mu,dx_1,\ldots,dx_n):=$$

$$:=\frac{k(\mu,x_1,z_1)k(\mu+\delta_{x_1},x_2,z_2)\ldots k(\mu+\delta_{x_1}+\ldots\delta_{x_{n-2}},x_{n-1},x_n)}{f(\mu+\delta_{x_1},z_1)f(\mu+\delta_{x_1}+\delta_{x_2},z_2)\ldots f(\mu+\delta_{x_1}+\ldots+\delta_{x_{n-2}},z_{n-2})}\lambda^n(dx_1,\ldots,dx_n).$$

ACKNOWLEDGEMENT:

We thank P. Weiss and R. Takacs for discussions, and J. Fritz for a valuable suggestion.

REFERENCES

[1] GEORGII, H.O.: Canonical and grand canonical Gibbs states for continuum systems. Commun. math. Phys. 48, 31-51 (1975).

[2] GEORGII, H.O.: Canonical Gibbs measures. Lecture Notes in Mathematics 760, Springer, 1979.

[3] GEORGII, H.O.: Equilibria for particle motions: conditionally balanced point random fields; to appear in: Proc. of the Conf. on Exchangeability in Prob. and Statistics, Rome 1981; North Holland.

[4] GLÖTZL, E.: Time reversible and Gibbsian point processes I, Math. Nachrichten 102, 217-222 (1981), II, to appear in Math. Nachrichten.

[5] KALLENBERG, O.: On conditional intensities of point processes. Z. Wahrscheinlichkeitstheorie verw. Gebiete 41, 205-220 (1978).

[6] KALLENBERG, O.: Random measures. Akademie-Verlag, Berlin 1975 and Academic Press, London-New York 1976.

[7] KALLENBERG, O.: Conditioning in point processes. Preprint, Göteborg, 1982.

[8] MATTHES, K., KERSTAN, J., MECKE, J.: Infinitely divisible point processes, Wiley, New York 1978. Russian edition: Nauka, Moskva, 1982.

[9] MATTHES, K., MECKE, J., WARMUTH, W.: Bemerkungen zu einer Arbeit von Nguyen Xuan Xanh und Hans Zessin, Math. Nachrichten 88, 117-127 (1979).

[10] NGUYEN X.X., ZESSIN, H.: Integral and differential characterizations of the Gibbs process. Math. Nachrichten 88, 105-115 (1979).

[11] RAUCHENSCHWANDTNER, B.: Gibbsprozesse und Papangeloukerne. VWGÖ, Wien 1980.

[12] RAUCHENSCHWANDTNER, B., WAKOLBINGER, A.: Some aspects of the Papangelou kernel, in: Point processes and queuing problems, pp. 325-336, North Holland, 1981.

[13] WAKOLBINGER, A., EDER, G.: A condition Σ_λ^c for point processes. Preprint, Linz, 1982.

Proc. of the 3rd Pannonian Symp. on Math. Stat., Visegrád, Hungary 1982
J. Mogyoródi, I. Vincze, W. Wertz, eds

SOME REMARKS ON THE ESTIMATION OF PARAMETERS IN SINGLE- AND MULTIVARIATE MIXED MODELS

M.T. WESOŁOWSKA-JANCZAREK

Institute of Applied Mathematics, Agricultural University, Lublin, Poland

1. INTRODUCTION

The problem of estimation of variance components in one-variable random and mixed models is generalized to estimation of covariance matrices in the multivariate case.

A simple way of this generalization is presented here and a generalization of the Searle's Method (see Searle (1968)) in this way is also given.

The Searle's Method can be used for the mixed models with interactions between fixed and random effects. This method is some modification of Henderson's Method 2.

Three Henderson's Methods of estimation of the variance components were described for the first time by Henderson (1953).

2. DEFINITIONS AND PRELIMINARIES

The following definitions and relations will be used in this paper.

Definition 2.1. The Kronecker product of order p $\underline{A} \underset{p}{\otimes} \underline{B}$ of a matrix $\underset{m,n}{\underline{A}} = \left[\underset{p,p}{\underline{A}_{ij}}\right]$ and $\underset{r,s}{\underline{B}} = \left[\underset{p,p}{\underline{B}_{kl}}\right]$ where m=tp, n=up, r=vp, s=wp, $\underline{A}_{ij}$ and $\underline{B}_{kl}$ are the ij-th and kł-th submatrices of size pxp of $\underline{A}$

and $\underline{B}$ respectively is the matrix $\underline{F}$ of order vptxupw given by

$$\underline{F} = \underline{A} \underset{p}{\otimes} \underline{B} = [\underline{A}_{ij}\underline{B}_{k\ell}] \tag{2.1}$$

$(i=1,\ldots,t;\ j=1,\ldots,u;\ k=1,\ldots,v;\ \ell=1,\ldots,w)$.

When p=1 we have the usual Kronecker product of matrices.

The properties of the Kronecker product of order p are the following:

If $\underline{A}$, $\underline{B}$ and $\underline{C}$ are matrices, $\underline{I}_p$ a pxp unit matrix and a is a number, then we have

(1) $a(\underset{m_1p,n_1p}{\underline{A}} \underset{p}{\otimes} \underset{m_2p,n_2p}{\underline{B}}) = a\underline{A} \underset{p}{\otimes} \underline{B} = \underline{A} \underset{p}{\otimes} a\underline{B}$;

(2) $(\underset{m_1p,n_1p}{\underline{A}} \underset{p}{\otimes} \underset{m_2p,n_2p}{\underline{B}}) \underset{p}{\otimes} \underset{m_3p,n_3p}{\underline{C}} = \underline{A} \underset{p}{\otimes} (\underline{B} \underset{p}{\otimes} \underline{C})$;

(3) $(\underset{m_1p,n_1p}{\underline{A}} + \underset{m_1p,n_1p}{\underline{B}}) \underset{p}{\otimes} \underset{m_2p,n_2p}{\underline{C}} = (\underline{A} \underset{p}{\otimes} \underline{C}) + (\underline{B} \underset{p}{\otimes} \underline{C})$;

(4) $\underset{m_1p,n_1p}{\underline{A}} \underset{p}{\otimes}(\underset{m_2p,n_2p}{\underline{B}} + \underset{m_2p,n_2p}{\underline{C}}) = (\underline{A} \underset{p}{\otimes} \underline{B}) + (\underline{A} \underset{p}{\otimes} \underline{C})$;

(5) $(\underset{m_1,n_1}{\underline{A}} \otimes \underset{m_2p,n_2p}{\underline{B}}) \underset{p}{\otimes} \underset{m_3p,n_3p}{\underline{C}} = \underline{A} \otimes (\underline{B} \underset{p}{\otimes} \underline{C})$.

If $m_2 = n_2 = m_3 = n_3 = 1$, then

$\underline{A} \otimes \underline{B} \underset{p}{\otimes} \underline{C} = \underline{A} \otimes \underline{BC}$,

(6) $\underset{np,np}{\underline{A}} \underset{p}{\otimes} \underset{p,p}{\underline{B}} = \underline{A}(\underline{I}_n \otimes \underline{B})$,

(7) $[\underset{mn,mn}{\underline{A}} (\underset{n,n}{\underline{B}} \otimes \underset{p,p}{\underline{C}})] \underset{p}{\otimes} \underset{p,p}{\underline{D}} = \underline{A}[(\underline{B} \otimes \underline{C}) \underset{p}{\otimes} \underline{D}] = \underline{A}(\underline{B} \otimes \underline{CD})$.

<u>Definition 2.2</u>. The general trace of order p of an mpxmp matrix $\underline{G} = \left[\underset{m,m}{\underline{G}_{ij}}\right]$, i,j=1,...,p, is defined as a pxp matrix $\underline{T}$ such that the ij-th element of $\underline{T}$ is the trace of $\underline{G}_{ij}$, the ij-th submatrix of size mxm of $\underline{G}$. The general trace of $\underline{G}$ is denoted by

$$\underset{p,p}{\underline{T}} = \mathrm{tr}_p\, \underline{G} = [\mathrm{tr}\, \underline{G}_{ij}]. \tag{2.2}$$

The properties of the general trace of order p are the following:

(1) $\mathrm{tr}_p(\underset{mp,mp}{\underline{A}} + \underset{mp,mp}{\underline{B}}) = \mathrm{tr}_p\,\underline{A} + \mathrm{tr}_p\,\underline{B}$,

(2) $\mathrm{tr}_p[(\underset{n,n}{\underline{A}} \otimes \underset{p,p}{\underline{B}}) \underset{p}{\otimes} \underset{p,p}{\underline{C}}] = \underline{A} \otimes \mathrm{tr}\,\underline{BC}$,

(3) $\mathrm{tr}_p[(\underset{n,n}{\underline{A}} \otimes \underset{p,p}{\underline{B}}) \underset{p}{\otimes} \underline{I}_p] = \mathrm{tr}_p[(\underline{A} \otimes \underline{I}_p) \underset{p}{\otimes} \underline{B}] = \underline{A}\,\mathrm{tr}\,\underline{B}$.

A proof of the (1) - (5) properties of the Kronecker product of order p and (1) - (2) of the general trace of order p we can find in Mikołajczak (1976). The remaining properties follow from the previous ones and from the definitions.

The following operator turning a column vector into a matrix will be used.

Let $\underline{y} = [\underline{y}_1', \underline{y}_2', \ldots, \underline{y}_p']'$ is an np×1 vector and $\underline{y}_j$ is an n×1 subvector for each j=1,...,p. Then $\underline{Y}$ is an n×p matrix of the form

$$\underline{Y} = R_p(\underline{y}) = [\underline{y}_1, \underline{y}_2, \ldots, \underline{y}_p] = \underline{y}_j \qquad (2.3)$$

and its properties are the following:

(1) $R_p(\underline{y} + \underline{z}) = R_p(\underline{y}) + R_p(\underline{z})$,

(2) $R_p[(\underline{I}_p \otimes \underset{n,n}{\underline{A}})\,\underline{y}] = \underline{A}\,R_p(\underline{y})$.

For the matrix given by (2.3) a generalized quadratic form will be defined.

Definition 2.3. An expression

$$\underline{Y}'\underline{A}\,\underline{Y} = [\underline{y}_j'\underline{A}\underline{y}_{j*}], \qquad j,j^*=1,\ldots,p \qquad (2.4)$$

where $\underline{A}$ is an n×n symmetrical matrix and $\underline{y}_j$, $\underline{y}_{j*}$ are vectors defined above, we call a generalized quadratic form of order p.

If $\underline{A} = \underline{I}_n$, where $\underline{I}_n$ is an n×n unit matrix, we will get a generalized sum of squares. That is a form

$$\underline{Y}'\underline{Y} = [\underline{y}_j'\underline{y}_{j*}], \quad j,j*=1,\ldots,p. \tag{2.5}$$

Theorem 2.1. Let $\underline{y}$ be an npx1 vector of random variables with the covariance matrix $\underline{\Sigma}_{\underline{y}} = [\underline{V}_{jj*}]$, where $\underline{V}_{jj*}$ is a matrix of covariance for the subvectors $\underline{y}_j$ and $\underline{y}_{j*}$, then

$$\varepsilon(\underline{Y}'\underline{A}\ \underline{Y}) = tr_p(\underline{\Sigma}_{\underline{y}} \underset{n}{\otimes} \underline{A}) + \varepsilon(\underline{Y}')\ \underline{A}\varepsilon(\underline{Y}), \tag{2.6}$$

where $\underline{Y}'\underline{A}\ \underline{Y}$ is given by (2.4) and $\varepsilon(\underline{Y})$ is an expected value of $\underline{Y}$.

The proof of this theorem is given by Mikołajczak (1976).

Definition 2.4. $\underline{A}^-$ is a generalized inverse of a matrix $\underline{A}$ when the following conditions are satisfied

(1) $\underline{A}\ \underline{A}^-\underline{A} = \underline{A}$,

(2) $\underline{A}^-\underline{A}$ is an idempotent matrix,

(3) $r(\underline{A}) = r(\underline{A}^-\underline{A})$ where $r(\underline{A})$ is the rank of $\underline{A}$.

3. MODELS AND ASSUMPTIONS

The general linear model in one-variable or multivariate case can be written as

$$\underline{y} = \underline{X}\ \underline{\beta} + \underline{e}. \tag{3.1}$$

a) One-variable case. $\underline{y}$ is an nx1 vector of observations, $\underline{X}$ is an nxt known matrix of an experimental design, $\underline{\beta}$ is a tx1 vector of fixed and random parameters and $\underline{e}$ is an nx1 vector of random errors with $\varepsilon(\underline{e}) = \underline{0}$ and $\underline{\Sigma}_{\underline{e}} = \sigma_e^2\underline{I}_n$.

In (3.1) $\underline{\beta}$ will be partitioned as $\underline{\beta}' = [\mu, \underline{\beta}_f', \underline{\beta}_r'] = [\mu, \underline{\beta}_f', \underline{\beta}_1', \ldots, \underline{\beta}_{k-1}']$, where μ is a mean, $\underline{\beta}_f$ is a vector of q fixed parameters, $\underline{\beta}_r$ is an sx1 vector of random parameters, $\underline{\beta}_j$ is s_jx1 subvector of $\underline{\beta}_r$ for every j=1,...,k-1, $s = \sum_{j=1}^{k-1} s_j$ and k is a number of random subvectors together with the vector of errors($\underline{\beta}_k = \underline{e}$ and s_k=n). The design matrix will be partitioned respectively $\underline{X} = [\underline{1}_n | \underline{X}_f | \underline{X}_r] = [\underline{1}_n | \underline{X}_f | \underline{X}_1 | \dots | \underline{X}_{k-1}]$ where $\underline{1}_n$ is a column vector of n once.

Then (3.1) is of the form

$$\underline{y} = \underline{1}_n \mu + \underline{X}_f \underline{\beta}_f + \underline{X}_r \underline{\beta}_r + \underline{e} \tag{3.2}$$

with the following assumptions:

$$\varepsilon(\underline{y}) = \underline{1}_n \mu + \underline{X}_f \underline{\beta}_f \tag{3.3}$$

and a covariance matrix of $\underline{y}$

$$\underline{\Sigma}_{\underline{y}} = \sum_{j=1}^{k-1} \underline{X}_j \, \underline{\Sigma}_{\underline{\beta}_j} \, \underline{X}_j' + \sigma_e^2 \underline{I}_n \, . \tag{3.4}$$

But if $\underline{\Sigma}_{\underline{\beta}_j} = \sigma_j^2 \underline{I}_{s_j}$ and if we denote $\underline{X}_k = \underline{I}_n$ and $\sigma_e^2 = \sigma_k^2$, then

$$\underline{\Sigma}_{\underline{y}} = \sum_{j=1}^{k} \sigma_k^2 \, \underline{X}_j \underline{X}_j' \, . \tag{3.5}$$

b) <u>Multivariate case</u>. $\underline{y}$ is an np×1 vector of n observations for p variables, $\underline{X}$ is an np×tp design matrix, $\underline{\beta}$ is tp×1 vector of parameters and $\underline{e}$ is an np×1 vector of error under the same assumption as in the one-variable case.

Again in (3.1) $\underline{\beta}$ and $\underline{X}$ will be partitioned as $\underline{\beta} = [\underline{\mu}, \underline{\beta}_f, \underline{\beta}_r]$ and $\underline{X} = [\underline{I}_p \otimes \underline{1}_n | \underline{I}_p \otimes \underline{X}_f | \underline{I}_p \otimes \underline{X}_r]$ where $\underline{1}_n$, $\underline{X}_f$, and $\underline{X}_r$ are the same as in one-variable case, $\underline{\beta}_r$ and $\underline{X}_r$ will be partitioned as previous, $\underline{\mu}$ is a vector of p means, $\underline{\beta}_f$ is a qp×1 vector of fixed parameters and $\underline{\beta}_r$ is an sp×1 vector of random parameters.

Then the model (3.1) can be written as

$$\underline{y} = (\underline{I}_p \otimes \underline{1}_n) \, \underline{\mu} + (\underline{I}_p \otimes \underline{X}_f) \, \underline{\beta}_f + (\underline{I}_p \otimes \underline{X}_r) \, \underline{\beta}_r + \underline{e} \, . \tag{3.6}$$

For the further considerations model (3.6) will be written in the more general form

$$\underline{y} = (\underline{I}_p \otimes \underline{1}_n) \, \underline{\mu} + (\underline{I}_p \otimes \underline{X}_f) \, \underline{\beta}_f + (\underline{I}_p \otimes \underline{U}) \, \underline{\delta} \tag{3.7}$$

where a random error is connected with other random parameters.

In (3.7) $\underline{\mu} = [\mu_1, \dots, \mu_p]'$, $\underline{X}_f$ and $\underline{U} = [\underline{X}_r | \underline{I}_n]$ where $\underline{X}_r = [\underline{X}_1 | \dots | \underline{X}_{k-1}]$ and $\underline{X}_k = \underline{I}_n$ are known matrices of design and the submatrices $\underline{X}_j$ (j=1,...,k-1) of size $n \times s_j$ are adequate to a given experimental design. Finally $\underset{wp,1}{\underline{\delta}'} = [\underline{\delta}'_1, \dots, \underline{\delta}'_p]$ where $\underline{\delta}'_ł = [\underline{\xi}_1^{ł'}, \dots, \underline{\xi}_k^{ł'}]$ and every $\underline{\xi}_j^ł$ (j=1,...,k) is a vector of s_j random effects for the ł-th variable ($\underline{\xi}_k^ł = \underline{e}^ł$) $w = \sum_{j=1}^{k} s_j = s+n$ and t=1+q+s is a number of parameters for each of p variables.

In model (3.7) we have the following assumptions

$$\varepsilon(\underline{y}) = (\underline{I}_p \otimes \underline{1}_n)\underline{\mu} + (\underline{I}_p \otimes \underline{X}_f)\underline{\beta}_f \;,\; \varepsilon(\underline{\delta}) = \underline{0}\;, \tag{3.8}$$

$$\underline{\Sigma}_{\delta} = \varepsilon(\underline{\delta}\,\underline{\delta}') = [\underline{\Sigma}_{\underline{\delta}_ł\underline{\delta}_{ł^*}}] = [\mathrm{diag}(\sigma_1^{łł^*}\underline{I}_{s_1}, \dots, \sigma_k^{łł^*}\underline{I}_{s_k})] \tag{3.9}$$

for ł,ł*=1,...,p with

$$\varepsilon(\xi_{ji}^ł \xi_{ji'}^ł) = \begin{cases} \sigma_j^{łł^*} & \text{if } i=i' \\ 0 & \text{if } i \neq i' \end{cases} \tag{3.10}$$

for each i,i'=1,...,s_j, where $\underline{\xi}_j^ł = [\xi_{j1}^ł, \dots, \xi_{js_j}^ł]'$, j=1,...,k.

Moreover, for a given vector $\underline{\xi}_j$, with $\varepsilon(\underline{\xi}_j) = \underline{0}$, the covariance matrix for p variables will be

$$\underset{p,p}{\underline{\Sigma}_{\underline{\xi}_j}} = \underline{\Sigma}_j = [\sigma_j^{łł^*}],\; j=1,\dots,k,\; ł,ł^*=1,\dots,p. \tag{3.11}$$

For these assumptions the covariance matrix of $\underline{y}$ is of the form (see Oktaba and Jagiełło (1977))

$$\underline{\Sigma}_{\underline{y}} = \sum_{j=1}^{k} (\underline{\Sigma}_j \otimes \underline{X}_j\underline{X}'_j). \tag{3.12}$$

However, if the random effects are correlated, that is $\varepsilon(\xi_{ji}^ł \xi_{ji'}^{ł^*}) \neq 0$ for each ł,ł* and i≠i' then

$$\underline{\Sigma}_{\underline{y}} = \sum_{j=1}^{k} [(\underline{I}_p \otimes \underline{X}_j)\underline{\Sigma}_j^*(\underline{I}_p \otimes \underline{X}_j)'], \tag{3.13}$$

where $\underline{\Sigma}_j^*$ is a covariance matrix of $\underline{\xi}_j$ of size $ps_j \times ps_j$.

We notice that if (3.10) is true, then

$$\underline{\Sigma}_j^* = \underline{\Sigma}_j \otimes \underline{I}_{s_j} \tag{3.14}$$

and $\underline{\Sigma}_{\underline{y}}$ is as (3.12) instead of (3.13).

4. A SIMPLE WAY OF GENERALIZATION OF THE VARIANCE COMPONENTS ESTIMATION TO P VARIABLES

In one-variable mixed models the estimators of variance components are a solution of equations that are taken from equating of expected values of some quadratic forms to their calculated values. The expected values of quadratic forms are linear functions of the unknown variance components σ_j^2, that we want to estimate.

In multivariate models the components of variance σ_j^2 must be replaced by the covariance matrices $\underline{\Sigma}_j$. Its estimators will be obtained from the same equations as the previous, but the quadratic forms must be replaced by suitable generalized quadratic forms. Their expected values can be obtained from (2.6) and properties (1)-(7) of the Kronecker product of order p and (1)-(3) of the general trace of order p and the relation $R_p(\underline{y})$ given in section 2.

5. ESTIMATION OF COVARIANCE MATRICES

A. Balanced data

We consider the model (3.6) such that the number of fixed and random effects in each group are established, and the assumptions are given in section 3.

Using the relation $R_p(\underline{y})$, the vector of observations $\underline{y}$ will be rearranged into a matrix $\underline{Y}$ and then generalized quadratic forms can be formed as in analysis of variance. Their expected values can be obtained using (2.6) and the suitable system of equations can also be formed.

These equations are the same as the equations for one-variable case where the components of variance are replaced by the proper covariance matrices for the respective groups of parameters of p-variables.

The similar results for the multivariate random models can be found in Ahrens (1972) and (1977).

B. Unbalanced data

The estimators of components of variance for the models (3.6) with unequal numbers of observations in subclasses can be obtained by Henderson's Methods 2 and 3.

In particular, if there is an interaction between fixed and random effects in the model, Searle's Method 4 given by Searle (1968) is applied. This method is some modification of Henderson's Method 2.

Generalization of Henderson's Method 2 to p variables can be found in Wesołowska-Janczarek (1980) and of Henderson's Method 3 in Wesołowska-Janczarek (1982).

In the following section, generalization of Searle's Method 4 will be presented.

6. SEARLE'S METHOD IN ONE- AND P-VARIABLE CASE

A. One-variable case

We consider the model (3.2). At first estimators of fixed effects with normal equations will be obtained under the temporary assumptions $\hat{\mu}=0$ and random effects do not exist. Then the estimators of fixed effects from the equation

$$\underline{X}_f'\underline{X}_f\hat{\underline{\beta}}_f = \underline{X}_f'\underline{y} \tag{6.1}$$

will be of the form

$$\hat{\underline{\beta}}_f = (\underline{X}_f'\underline{X}_f)^- \underline{X}_f'\underline{y} = \underline{L}\underline{y} \tag{6.2}$$

where

$$\underline{L} = (\underline{X}_f'\underline{X}_f)^- \underline{X}_f \tag{6.3}$$

is a general inverse matrix of $\underline{X}_f$ under the condition

$$\underline{X}_f\underline{L}\underline{X}_f = \underline{X}_f. \tag{6.4}$$

Vector (6.2) will be used to adjust the vector of observations in the way

$$\underline{z} = \underline{y} - \underline{X}_f\hat{\underline{\beta}}_f = (\underline{I}_n - \underline{X}_f\underline{L})\underline{y} = \underline{W}\underline{y} \tag{6.5}$$

where

$$\underline{W} = \underline{I}_n - \underline{X}_f\underline{L} = \underline{I}_n - \underline{X}_f(\underline{X}_f'\underline{X}_f)^-\underline{X}_f' . \tag{6.6}$$

But by (3.2) and (6.4) we have

$$\underline{z} = (\underline{I}_n - \underline{X}_f\underline{L})\underline{1}_n\underline{\mu} + (\underline{X}_r - \underline{X}_f\underline{L}\underline{X}_r)\underline{\beta}_r + (\underline{I}_n - \underline{X}_f\underline{L})\underline{e} \tag{6.7}$$

and $(\underline{I}_n - \underline{X}_f\underline{L})\underline{1}_n = \underline{0}$ because $\underline{X}_f\underline{1}_n = q\underline{1}_n$ and $\underline{X}_f'\underline{1}_n = \frac{1}{q}\,\underline{X}_f'\underline{X}_f\underline{1}_n$.

Then (6.7) will be of the form

$$\underline{z} = \underline{W}[\underline{X}_r\underline{\beta}_r + \underline{e}] \tag{6.8}$$

and on the other hand we have (6.5).

It is to be mentioned that the matrix $\underline{W}$ is symmetrical, idempotent and invariant with the choice of $(\underline{X}_f'\underline{X}_f)^-$.

At last Henderson's Method 1 will be applied to the model (6.8), modified only by absence of a general mean in this model. The partition of the random vector $\underline{\beta}_r$ and the matrix $\underline{X}_r$ given in section 3a) will be used and $\underline{z}'\underline{z}$ and reducts $R(\hat{\underline{\beta}}_h) = \underline{z}'\underline{X}_h(\underline{X}_h'\underline{X}_h)^{-1}\underline{X}_h'\underline{z}$ for h=1,...,k-1 will be appointed.

Next the expected values of these reducts as linear functions of unknown components of variance will be found. The estimators of these components are the solution of the system of k equations obtained by equating these expected values to its calculated values.

These expected values here are of the form (see Searle (1968))

$$\varepsilon(\underline{z}'\underline{z}) = \varepsilon(\underline{y}'\underline{W}\,\underline{y}) = \mathrm{tr}(\underline{WX}_r \,\underline{\Sigma}_{\underline{\beta}_r}\, \underline{X}_r') + \sigma_e^2 \mathrm{tr}(\underline{W}) \tag{6.9}$$

$$\varepsilon[R(\hat{\underline{\beta}}_h)] = \mathrm{tr}[\underline{WX}_h(\underline{X}_h'\underline{X}_h)^{-1}\underline{X}_h'\underline{WX}_r \,\underline{\Sigma}_{\underline{\beta}_r}\, \underline{X}_r'] +$$

$$+ \sigma_e^2\{r(\underline{X}_h) + \mathrm{tr}[\underline{X}_f(\underline{X}_f'\underline{X}_f)^-\underline{X}_f'\underline{X}_h(\underline{X}_h'\underline{X}_h)^{-1}\underline{X}_h']\}$$

for each $h=1,\ldots,k-1$.

B. p-variable case

We now consider the model (3.7) under the assumptions (3.8) and (3.13). In the same way as previously, from the equation

$$(\underline{I}_p \otimes \underline{X}_f'\underline{X}_f)\hat{\underline{\beta}}_f = (\underline{I}_p \otimes \underline{X}_f')\underline{y} \tag{6.10}$$

we have

$$\hat{\underline{\beta}}_f = (\underline{I}_p \otimes \underline{X}_f'\underline{X}_f)^-(\underline{I}_p \otimes \underline{X}_f')\underline{y} = [\underline{I}_p \otimes (\underline{X}_f'\underline{X}_f)^-\underline{X}_f']\underline{y} \tag{6.11}$$

and

$$\underline{L} = \underline{I}_p \otimes (\underline{X}_f'\underline{X}_f)^-\underline{X}_f' \,. \tag{6.12}$$

Then we transform the mixed model into a random one in the following way

$$\underline{z} = \underline{y} - (\underline{I}_p \otimes \underline{X}_f)\hat{\underline{\beta}}_f = [\underline{I}_{np} - (\underline{I}_p \otimes \underline{X}_f)\underline{L}]\underline{y} \tag{6.13}$$

and denote

$$\underline{W}_1 = \underline{I}_{np} - (\underline{I}_p \otimes \underline{X}_f)\underline{L} = \underline{I}_p \otimes [\underline{I}_n - \underline{X}_f(\underline{X}_f'\underline{X}_f)^-\underline{X}_f'] = \underline{I}_p \otimes \underline{W} \tag{6.14}$$

where $\underline{W}$ is the same as in (6.6).

It is easy to see that the matrix $\underline{W}_1$ is symmetrical, idempotent and invariant also with the choice of $(\underline{X}_f'\underline{X}_f)^-$.

But on the other hand we notice that

$$\underline{z} = \underline{W}_1[(\underline{I}_p \otimes \underline{X}_r)\underline{\beta}_r + \underline{e}] \tag{6.15}$$

and for this model we apply Henderson's Method 1 (see Wesołowska-Janczarek (1981)).

To this purpose we rearrange the vector $\underline{z}$ into a matrix

$\underline{Z}$ using (2.3) and then we obtain

$$\underline{Z} = \underline{WX}_r R_p(\underline{\beta}_r) + \underline{W}R_p(\underline{e}) \quad . \tag{6.16}$$

For this matrix we form $\underline{Z}'\underline{Z}$ and the generalized reducts $R(\hat{\underline{\beta}}_h) = \underline{Z}'[\underline{X}_h(\underline{X}_h'\underline{X}_h)^{-1}\underline{X}_h']\underline{Z}$ for each $h=1,\ldots,k-1$.

We notice that

$$\varepsilon(\underline{Z}) = \underline{0} \tag{6.17}$$

and with regard to (3.13) covariance matrix of the random vector $\underline{z}$ is of the form

$$\underline{\Sigma}_{\underline{z}} = \sum_{j=1}^{k} (\underline{I}_p \otimes \underline{WX}_j)\, \underline{\Sigma}_j^*(\underline{I}_p \otimes \underline{X}_j'\underline{W}) \quad . \tag{6.18}$$

The expected values of $\underline{Z}'\underline{Z}$ and every $R(\hat{\underline{\beta}}_h)$ obtained by (2.6) are the following:

$$\varepsilon(\underline{Z}'\underline{Z}) = tr_p\{\sum_{j=1}^{k-1}(\underline{I}_p \otimes \underline{WX}_j)\underline{\Sigma}_j^*(\underline{I}_p \otimes \underline{X}_j'\underline{W})\} + tr_p\{(\underline{I}_p \otimes \underline{W})\underline{\Sigma}_e^*(\underline{I}_p \otimes \underline{W})\},$$

$$\varepsilon[R(\hat{\underline{\beta}}_h)] = tr_p\{\sum_{j=1}^{k-1}[(\underline{I}_p \otimes \underline{WX}_j)\underline{\Sigma}_j^*(\underline{I}_p \otimes \underline{X}_j'\underline{WX}_h(\underline{X}_h'\underline{X}_h)^{-1}\underline{X}_h')]\} + \tag{6.19}$$

$$+ tr_p\{(\underline{I}_p \otimes \underline{W})\underline{\Sigma}_e^*(\underline{I}_p \otimes \underline{WX}_h(\underline{X}_h'\underline{X}_h)^{-1}\underline{X}_h')\}$$

for $h=1,\ldots,k-1$.

The formulas (6.19) will be simplified when the effects in the model (3.7) are uncorrelated. Then according to (3.14) formulas (6.19) are of the form

$$\varepsilon(\underline{Z}'\underline{Z}) = \sum_{j=1}^{k-1}\{\underline{\Sigma}_j\, tr[\underline{W}(\underline{X}_j\underline{X}_j')]\} + \underline{\Sigma}_e\, tr(\underline{W}), \tag{6.20}$$

$$\varepsilon[R(\hat{\underline{\beta}}_h)] = \sum_{j=1}^{k-1}\{\underline{\Sigma}_j \cdot tr[\underline{W}(\underline{X}_j\underline{X}_j')\underline{WX}_h(\underline{X}_h'\underline{X}_h)^{-1}\underline{X}_h')\} +$$

$$+ \underline{\Sigma}_e \cdot tr[\underline{WX}_h(\underline{X}_h'\underline{X}_h)^{-1}\underline{X}_h'] \qquad \text{for } h=1,\ldots,k-1.$$

The system of equations can be obtained equating these expected values to their calculated values. The estimators of unknown covariance matrices are its solution.

Finally, if we compare (6.9) with (6.19) or (6.20) it is easy to see that in case of uncorrelated effects the formulas (6.9) and (6.20) are the same, and in the remaining cases suitable matrices are only Kronecker multiplied by matrix $\underline{I}_p$.

REFERENCES

AHRENS, H., (1972). Das Schätzen multivariater Varianzkomponenten. Biometrische Zeitschrift 14, 357-366.

AHRENS, H., (1977). On an Invariance Property for First- and Second-order Moments of Estimated Variance-covariance Components. Biometrical Journal 19, 485-496.

HENDERSON, C.R., (1953). Estimation of variance and covariance components.Biometrics 9, 226-252.

MIKOŁAJCZAK, J., (1976). Metody wyznaczania wartości oczekiwanych macierzy średnich kwadratów i iloczynów w wielozmiennych modelach losowych i mieszanych oraz zastosowania tych modeli w biologii i w przemyśle. Unpublished doctor dissertation.

SEARLE, S.R., (1968). Another look at Henderson's methods of estimating variance components. Biometrics 24, 749-778.

THOMPSON, R., (1973). The estimation of variance and covariance components with an application when records are subject to culling. Biometrics 29, 527-550.

OKTABA, W. and JAGIEŁŁO, G., (1977). Estimation of covariance matrices by Henderson's Method One in multivariate case. Siódme Colloquium Metodologiczne z Agrobiometrii, Puławy, 1, 8-44.

WESOŁOWSKA-JANCZAREK, M.T., (1980). Estimation of covariance matrix in random and mixed multivariate models. Studia Sci. Math. Hungarica (in press).

WESOŁOWSKA-JANCZAREK, M.T., (1982). Estimation of covariance matrices in unbalanced random and mixed multivariate models. Biometrical Journal (in press).

Proc. of the 3rd Pannonian Symp.
on Math. Stat., Visegrád, Hungary 1982
J. Mogyoródi, I. Vincze, W. Wertz, eds

ON IRREDUCIBLE MEASURES

A. ZEMPLÉNI

Eötvös Loránd University, Budapest, Hungary

1. INTRODUCTION

A famous decomposition theorem of Hinčin states [1] that every probability distribution function can be decomposed to the convolution of an infinitely divisible distribution without any irreducible component and a finite or at most countable convolution of irreducible distributions.
This decomposition is known not to be unique. One can easily construct an example such that

$$F_1 * F_2 = G_1 * G_2 * G_3,$$

where all distribution functions are irreducible.
The following problem is due to G.J. Székely:
If k,n are arbitrary positive integers, is it always possible to construct irreducible distribution functions F_1 and F_2 such that

$$F_1^{*k} = F_2^{*n} . \qquad (1)$$

Below we give irreducible distribution functions for the case k=2, n=3 for which (1) holds, and for the general case we give an affirmative answer to a somewhat weaker form of it, namely we prove the theorem: For arbitrary integers k, n≥2 there exist

σ-finite irreducible measures μ, ν, such that

$$\mu^n = \nu^k.$$

(Here μ^n and ν^k marks the convolution powers of μ and ν.)

2. THE CASE k=2, n=3

Let F_1 and F_2 be concentrated to the set of nonnegative integers. Then instead of F_i we can use their generating functions g_i.

If F_i is concentrated to a finite subset of $\mathbb{N}$, then g_i is a polynomial, and $g_1^2 = g_2^3$ implies the existence of a polynomial q such that $g_1 = q^3$ and $g_2 = q^2$. F_i must be irreducible, so q cannot have only positive coefficients.

Let $\tilde{q} := x^4+x^3-0.45x^2+x+1$ and $q := \frac{\tilde{q}}{3.55}$.

This q is suitable for us, because q^2 and q^3 have only positive coefficients and it can be easily computed that none of them can be decomposed into nontrivial product of two polynomials having only nonnegative coefficients.

Let $g_1 = q^3$ and $g_2 = q^2$. Then g_i $(i=1,2)$ are generating functions of irreducible distributions, for which

$$F_1^{*2} = F_2^{*3}$$

obviously holds.

It is clear that if $k|n$ $(k\neq n)$ then F_i cannot be concentrated to a finite subset of $\mathbb{N}$, since in this case $F_1^{*k} = F_2^{*n}$ implies that $g_1 = g_2^m$, where $m = \frac{n}{k}$, and so F_1 cannot be irreducible.

3. DECOMPOSITION OF Z

For $A, B \subset \mathbb{Z}$ let $A+B=\{z\in Z: z=x+y,$ where $x\in A, y\in B\}$.

If the measures μ and ν are concentrated to the subsets A_n and A_k of Z, respectively, then the values of $\nu^k(\{z\})$ and $\mu^n(\{z\})$ are determined by the decompositions of $z\in\mathbb{Z}$ into the sum of k terms from A_k and into the sum of n terms from A_n, respectively. The sets A_k and A_n will be given so, that they do not have too much decompositions of this type.

The construction of the sets A_k and A_n is analogous, so we give the definition of A_i for an arbitrary $i \geq 2$ integer at once.

$$A_i := \{\dots, b_m^{(i)}, b_{m-1}^{(i)}, \dots, b_1^{(i)}, 0, a_1^{(i)}, \dots, a_m^{(i)}, \dots\},$$

where the recursion

$$a_0^{(i)} = 0,\ b_0^{(i)} = 0,\ a_1^{(i)} = 1$$

$$b_m^{(i)} = -(i-1)a_m^{(i)} - m \tag{2}$$

$$a_{m+1}^{(i)} = -(i-1)b_m^{(i)} + m + 1 \tag{3}$$

$(m=1,2,\dots)$ determines the sequences $(a_m^{(i)})_{m=0}^{\infty}$ and $(b_m^{(i)})_{m=0}^{\infty}$.

In a considerable part of this section we fix an $i \geq 2$, and omit the superscript i.

The expressions (2) (and (3)) give a decomposition of the number $-m$ (and $m+1$) into the sum of i terms from A_i. These decompositions we shall call regular decompositions. So $A_i + \dots + A_i = Z$ (where on the left hand side are i terms).

From the recursion it is clear that

$$\dots < b_m < b_{m-1} < \dots < b_1 < 0 < a_1 < \dots < a_m < \dots$$

and

$$a_m < |b_m| < a_{m+1} \quad (m \geq 1).$$

A_i is an irreducible set since it is not bounded and if it were reducible then there would exist two subsets of A_i: $(c_k)_{k=1}^{\infty}$ and $(d_k)_{k=1}^{\infty}$ such that $c_k - d_k = c \neq 0$, but in our case from the recursion we get: $a_{m+1} = (i-1)^2 a_m + (i-1)m + m + 1 = (i-1)^2 a_m + im + 1$.

Hence $a_{m+1} - a_m = [(i-1)^2 - 1]a_m + im + 1 \to +\infty$ $(m \to +\infty)$, and in the same way

$$b_{m+1} - b_m \to -\infty \, (m \to +\infty).$$

The sequences forming A_i have big gaps, namely (2) and (3) imply that

$$(i-1)a_m + b_m < 0 < a_m + (i-1)\, b_{m-1} . \qquad (4)$$

<u>Lemma 1:</u> Let $t\in\mathbb{Z}$. Suppose that

$$t = \sum_{\alpha=1}^{m_1} b_{j_\alpha} + \sum_{\beta=1}^{m_2} a_{\ell_\beta} \qquad (5)$$

where $m_1 + m_2 \le i$ and $j_\alpha \ge j_{\alpha-1}$, $\ell_\beta \ge \ell_{\beta-1}$ for arbitrary α, β $(1<\alpha\le m_1,\ 1<\beta\le m_2)$.

Suppose that $a_j \le t < a_{j+1}$ (the case $b_m < t \le b_{m-1}$ can be handled similarly). If (5) is not the regular decomposition of t, then

a) $\ell_{m_2} \le j+1$ and $j_{m_1} \le j$

b) in the case $i>2$ $\ell_{m_2} \ge j$ also holds,

i.e. then there are only two possibilities:

$\ell_{m_2} = j$ and $j_{m_1} \le j-1$

or $\ell_{m_2} = j+1$ and $j_{m_1} \le j+1$.

<u>Proof</u>.

a) $t \ge 0$ implies that in (5) the second sum is not empty. Suppose indirectly that $\ell^* := \ell_{m_2} > j+1$.

In view of (4) it is obvious that we can get nonnegative number in (5) only when $j_{m_1} \le \ell^* - 1$. The minimum of these numbers is $a_{\ell^*} + (i-1)b_{\ell^*-1}$, but this is the regular decomposition of the number ℓ^*, so the smallest number in question is $a_{\ell^*} + (i-2)b_{\ell^*-1} + b_{\ell^*-2}$. But $a_{\ell^*} + (i-2)b_{\ell^*-1} + b_{\ell^*-2} =$
$= -b_{\ell^*-1} + \ell^* + b_{\ell^*-2} > (i-1)a_{\ell^*-1} + \ell^* + b_{\ell^*-2} > a_{\ell^*-1} \ge a_{j+1} > t$,
thus t cannot have a decomposition like this.

b) Let now $i>2$. Then $\ell_{m_2} < j$ cannot hold, because then $m_2 < i$ implies that the greatest decomposable number is $ia_{\ell_{m_2}}$, for which $ia_{\ell_{m_2}} < (i-1)^2 a_{\ell_{m_2}} < t$ (we have used $i>2$ in the first inequality).

So really only the cases

$$\ell_{m_2} = j \text{ or } \ell_{m_2} = j+1$$

are possible, and then - in view of (4) - $j_{m_1} \leq j-1$ or $j_{m_1} \leq j$, respectively.

Remark 1. The decompositions appearing in the lemma are called irregular decompositions.

Lemma 1 stated that if $a_j \leq t < a_{j+1}$, then the possible irregular decompositions of t contain only $b_j, b_{j-1}, \ldots, 0, \ldots, a_{j+1}$.

At the same time the regular decomposition of t contains b_{t-1} and a_t, and (for $t \geq 3$) $j+1 < t$ is obvious.

Lemma 2. If $t = \sum_{j=1}^{m} c_j$ ($c_j \in A_i$, when $j=1,\ldots,m$), and $m \leq i-1$, then this decomposition is unique.

Proof. The statement in the case i=2 is meaningless.

For the case i>2 applying Lemma 1 with the choice $m_1+m_2=m$ we get (with the notations above) that only $\ell_{m_2}=j$ or $\ell_{m_2}=j+1$ is possible.

In the case $\ell_{m_2}=j$ the greatest decomposable number is ma_j, on the other hand in the case $\ell_{m_2}=j+1$ the smallest decomposable number is $a_{j+1} + (m-1)b_j$.

$$a_{j+1} + (m-1)b_j > (m-i)b_j > (m-i+1)b_j + (i-1)a_j \geq ma_j$$

(because of $m-i+1 \leq 0$). So in the different manners there can be decomposed only different numbers.

Remark 2. The proof of Lemma 2 at the same time gives an algorithm for the decomposition of $t \in Z$ into the sum of m terms (where $m \leq i-1$): when

$$a_j + (m-1)\, b_{j-1} \leq t \leq ma_j \qquad (6)$$

for some j, then the greatest positive term in the decomposition will be a_j. (If there exist no such j, i.e. for some j

$$ma_j < t < a_{j+1} + (m-1)b_j$$

then t surely has no decomposition like that.)

Now we have to write (6) for $t_1 := t - a_j$ with m-1 instead of m, etc. (The last step can be the m-th one. In this case (6) gives the condition $t_{m-1} = a_j$.)

t has a desired decomposition if and only if, when all of the m steps can be done.

In the next lemma we prove that for i>2 in the "neighbourhood" of an arbitrary number there exists such a number, which has only regular decomposition.

Lemma 3. Suppose i>2, and let $t \in Z$. Suppose that t has an irregular decomposition in which the number of zeros is exactly m.

a) if m>1, then t+m+1 has not any irregular decomposition.

b) if $m \le 1$, then one of the numbers t+1, t+2, t+3 has no irregular decomposition.

Proof. First we prove that t+m+1 has no irregular decomposition with the same type (i.e. using the notations of Lemma 1: having the same value of ℓ_{m_2}) as the decomposition of t, which appears in our lemma.

t+m+1 would have such a decomposition if and only if, when $t+m+1-a_{\ell_{m_2}}$ were possible to decompose for a sum of at most i-1 terms, where the greatest positive term is not greater than $a_{\ell_{m_2}}$. Because of the conditions of our lemma $t' = t+m-a_{\ell_{m_2}}$ has such a decomposition. This gives that for this t′ the algorithm can be applied, and for the t'_{i-2} got in the last step we have $t'_{i-2} \in A_i$. Applying the same algorithm for $t'' = t+m+1-a_{\ell_{m_2}}$, then there exists a first j_o, for which $t'_{j_o} = (m-j_o)a_{\alpha_o}$.

This implies that for $j \leq j_0$ $t_j'' = t_j' + 1$ holds, and thus

$$(m-j_0)a_{\alpha_0} < t_{j_0}'' = t_{j_0}' + 1 < a_{\alpha_0+1} + (m-j_0-1)b_{\alpha_0}$$

($\alpha_0 > 0$ because in the decomposition of t+m there are no zeros). So in this step the algorithm for t'' cannot be continued thus t'' has not any such decomposition.

Since $i > 2$ we have $t_{j_0}' + 2 < a_{t_{j_0}} + (m-j_0 - 1)\, a_{t_{j_0}}$ thus t+m+2 also has no such decomposition.

To prove a) we have to show that t+m+1 has no decomposition having the other type (i.e. when $a_j \leq t < a_{j+1}$, then in the case $\ell_{m_2}=j$, for the decomposition of t+m+1 it remains to examine only $\ell_{m_2'}=j+1$ and respectively for $\ell_{m_2}=j+1$ only $\ell_{m_2'}=j$).

Suppose first that in the decomposition of t $\ell_{m_2}=j$ holds. Then

$$t \leq (i-2)a_j = -b_j - j - a_j = a_{j+1} + (i-2)b_j - (2j+1) - a_j =$$

$$= a_{j+1} + (i-2)b_j + b_{j-1} + (i-2)b_j - (2j+1)$$

$(i-2)b_{j-1} - (2j+1) + m + 1 < 0$ is obvious, so

$$t+m+1 < a_{j+1} + (k-2)b_j + b_{j-1}$$

where the right hand side is the smallest decomposable number with $\ell_{m_2'}=j+1$.

When in the decomposition of t $\ell_{m_2}=j+1$ holds, then

$$t \geqslant a_{j+1} + (i-3)b_i = j+1-2b_j = (2i-2)a_j + 3m+1 > ia_j$$

hence $t+m+1 > ia_j$ is obvious.

The proof of statement b):

Suppose $\ell_{m_2}=j$ and first let m=1.

Then according to the statement above t+2 and t+3 may have only irregular decomposition with $\ell_{m_2}=j+1$. But

$$a_{j+1} + (i-2)b_j = -b_j + j + 1 = (i-1)a_j + 2j + 1 > t+m+1$$

so in this decomposition there are not any zeros, thus t+m+2=t+3 cannot be decomposed either with $\ell_{m_2}=j+1$, or with $\ell_{m_2}=j$.

Now let m=0.

Then t+1 must have a decomposition with $\ell_{m_2}=j+1$, in which there is at least one 0 (otherwise t+2 cannot be decomposed). This decomposition can only be

$$t+1=a_{j+1}+(i-2)b_j \tag{7}$$

since $a_{j+1}+(i-3)b_j+b_{j-1}=j+1-2b_j+b_{j-1}=3j+1+2(i-1)a_j+b_{j-1}>ia_j$ (we have used $i>2$).

(7) implies that $t+1=-b_j+j+1=(i-1)a_j+2j+1$.

We know that t has an irregular decomposition with $\ell_{m_2}=j$, so $2j\in A_i$. Hence $2j+3\notin A_i$, thus t+3 has neither a decomposition with $\ell_{m_2}=j$, nor with $\ell_{m_2}=j+1$.

The case $\ell_{m_2}=j+1$ we can prove in the same way.

<u>Corollary</u>. Let $2<i<n$ and $j\in\mathbb{N}$. Then one of the numbers $j, j+1,\dots,j+n-1$ has no irregular decomposition.

<u>Proof</u>. If j has no irregular decomposition, then we are ready. Otherwise we can apply Lemma 3:

a) If $m\geq 2$ in the decomposition of j, then $j+m+1\leq j+1<j+n$ has no irregular decomposition.

b) If $m\leq 1$, then one of $j+1$, $j+2$, $j+3$ is suitable $(3\leq n-1)$.

Further we deal with irregular decompositions from A_k and A_n at the same time.

<u>Remark 3</u>. The corollary can be formulated as follows:

Let $2<k<n$. Then denoting $\{0,1,\dots,n-1\}$ by U_n, to every $a_j^{(n)}\in A_n$ there exist a $u\in U_n$, such that $a_j^{(n)}+u$ has no irregular decomposition from A_k. On the other hand from the definition of U_n is clear that $a_j^{(n)}+u$ has an irregular decomposition from A_n. This cannot be proved for k=2, but a somewhat weaker form is true:

Lemma 3'. Let $m_0 \in \mathbb{N}$ fixed. Then there exists a finite set $U_{n,2}$ such that for j large enough it is possible to choose a $u(j) \in U_{n,2}$ such that a) $a_j^{(n)} + u(j)$ has irregular decomposition from A_n

b) $a_j^{(n)} + u(j)$ has only such irregular decomposition from A_2 which has the form $a_\alpha^{(2)} + b_\beta^{(2)}$, where $\beta \geq m$.

Proof. It is obvious that the recursion (2), (3) for k=2 implies that $A_2 = \{\ldots, -m^2-m, -(m-1)^2-(m-1), \ldots, -2, 0, 1, \ldots, m^2, \ldots\}$.

An arbitrary $a' \in N$ may have two different types of decomposition from A_2 : 1) $a = \alpha'^2 + \beta'^2 (\alpha', \beta' \geq 0)$

2) $a = \alpha'^2 - \beta'^2 - \beta'$ $(\alpha', \beta' > 0)$.

A decomposition like 1) cannot be exist, when $a' \equiv 3 \pmod 4$.

Suppose that a' has a decomposition like 2) for some $\beta' < m_0$.

Then if a' is large enough, a' has exactly one such decomposition and for every $c \in N$ it can be given a natural N_c, such that for $a' > N_c$ none of the numbers

$$a+1, \ldots, a+c$$

have a decomposition like 2) with $\alpha \neq \alpha'$ and $\beta < m_0$ (where α' appears in the decomposition of a').

This is true, because let α' the minimal α, appearing in the decompositions of a' like 2) with $\beta' < m_0$:

$$a' = \alpha'^2 - \beta'^2 - \beta, \text{ hence } \alpha'^2 > a'. \text{ Thus}$$

$$(\alpha'+1)^2 - m_0^2 > a' + 2\sqrt{a'} - m_0^2 > a' + c \text{ if } a' \text{ is large enough.}$$

So really there are no other decompositions of a' of this type.

The next decomposable number with $\alpha_1 = \alpha' + 1$ is

$$(\alpha'+1)^2 - m_0^2 > a' + c$$

and the largest decomposable number with $\alpha_1 = \alpha' - 1$ is

$$(\alpha'-1)^2 - 1 < \alpha'^2 - m_0^2 - c - 1 < a'.$$

From the facts, mentioned above, can be proved, that for n>7

$U_{n,2}:=\{-5,\dots,-2,0,1,\dots,7\}$

will be suitable: it is clear that for each $u\in U_{n,2}$ u has a decomposition into a sum of k-1 terms from A_n, and this implies a).

To prove b) for example let $a_j^{(n)}-5\equiv 3\ (4)$.

Then none of $a_j^{(n)}-5$, $a_j^{(n)}+3$, $a_j^{(n)}+7$ have an irregular decomposition from A_2 like 1).

It can be seen from the statement above that if both a_j+3 and a_j-5 would have a decomposition like 2) with $\beta<m_0$, only when

$a_j^{(n)}-5=\alpha^2-4^2-4$

were for some α. Then $a_j^{(n)}+3=\alpha^2-3^2-3$ would stay, but then $a_j^{(n)}+7$ has no such irregular decomposition.

In the same way can be handled the other cases for

$u_{j(1)}, u_{j(2)}, u_{j(3)}\in U_{n,2}$ for which $a_j^{(n)}+u_{j(i)}\equiv 3\ (4)$.

Analogously for n≤7 is possible to construct a set

$U_{n,2}=\{u_1,\dots,u_{12}\}$ such that

(i) to every $0\le i\le 3$, $i\in\mathbb{N}$ there exist $u_{i_1}<u_{i_2}<u_{i_3}$ $U_{n,2}$

such that $u_{i_j}\equiv i\ (4)$ $(j=1,\dots,3)$

and $u_{i_3}-u_{i_2}$ is large enough with respect to $u_{i_2}-u_{i_1}$

(this means that there is no $\beta_3<\beta_2$ such that

$u_{i_3}-u_{i_2}=\beta_2^2+\beta_2-\beta_3^2-\beta_3$ where $u_{i_2}-u_{i_1}=\beta_1^2+\beta_1-\beta_2^2-\beta_2$)

(ii) $u_j=\sum_{\alpha=1}^{n-1} c_\alpha$ $(c_\alpha\in A_n,\ j=1,\dots,12)$.

Then choosing c in such a way that $c>\max_{0\le i\le 3}(u_{i_3}-u_{i_2})$ hold, in the same way as above, we can prove that this $U_{n,2}$ satisfies the conditions of our lemma.

The following lemma is analogous to Corollary 1, only the role of k and n are changed:

Lemma 4. Let $k<n$. Then there exists a finite set $U_{k,n} \subseteq Z$, such that for j large enough it is possible to choose a $u \in U_{k,n}$ such that a) a_j+u has an irregular decomposition from A_k

b) a_j+u has no irregular decomposition from A_n.

Proof. In order to satisfy condition a) $U_{k,n}$ should contain only numbers, having a decomposition into the sum of k-1 terms from A_k. The following two statements play an important role in the proof:

1) if $2<k$, then for $s \in \mathbb{N}$ there exists a $u \in N$ such that

$$s \le u < 2s \text{ and } u = \sum_{i=1}^{k-1} c_i, \text{ where } c_i \in A_k.$$

To the proof of this statement suppose $a_\alpha^{(k)} \le s < a_{\alpha+1}^{(k)}$, and cut the interval $[a_\alpha^{(k)}, a_{\alpha+1}^{(k)})$ into sub-intervals:

$$[a_\alpha^{(k)}, 2a_\alpha^{(k)}), \dots, [(k-2)a_\alpha^{(k)}, (k-1)a_\alpha^{(k)}),$$

$$[(k-1)a_\alpha^{(k)}, a_{\alpha+1}^{(k)}+(k-2)b_\alpha^{(k)}),$$

$$[a_{\alpha+1}^{(k)}+(k-2)b_\alpha^{(k)}, a_{\alpha+1}^{(k)}+(k-3)b_\alpha^{(k)}), \dots, [a_{\alpha+1}^{(k)}+b_\alpha^{(k)}, a_{\alpha+1}^{(k)}).$$

It can be computed easily that to each sub-interval $[c_j, d_j)$ appearing above $\frac{d_j}{c_j} \le 2$ ($s>0$ implies $\alpha>0$, so $c_j>0$) holds, except the case $k=3$, $\alpha=1$: then $[2a_1, a_2+b_1)=[2,5)$, and so in any case we can choose as u the greater endpoint of the subinterval containing s in the interior, and when s is an endpoint, then $u=s$ can be chosen. (The description above gives a decomposition into the sum of k-1 terms for the endpoints of the subintervals.)

2) In the case $k>3$ for arbitrary $\eta>0$ there exist an N_η, such that for every $v \in \mathbb{N}$, $v>N_\eta$ belonging to the interval $(v, (3+\eta)v)$ there exist a u, such that $u = \sum_{i=1}^{k-2} c_i$ ($c_i \in A_k$).

To the proof we cut the interval $[a_\alpha^{(k)}, a_{\alpha+1}^{(k)})$ into subintervals like above: $[a_\alpha^{(k)}, 2a_\alpha^{(k)}), \dots, [(k-3)a_\alpha^{(k)}, (k-2)a_\alpha^{(k)}),$

$$[(k-2)a_\alpha^{(k)}, a_{\alpha+1}^{(k)}+(k-3)b_\alpha^{(k)}),$$

$$[a_{\alpha+1}^{(k)}+(k-3)b_\alpha^{(k)}), \dots, [a_{\alpha+1}^{(k)}+b_\alpha^{(k)}, a_{\alpha+1}^{(k)}).$$

Only one subinterval increased comparing with the case in statement 1), so it is enough to deal only with this one.

$$\frac{a_{\alpha+1}^{(k)} + (k-3)b_\alpha^{(k)}}{(k-2)a_\alpha^{(k)}} = \frac{-2b_\alpha^{(k)} + \alpha + 1}{(k-2)a_\alpha^{(k)}} =$$

$$= \frac{2(k-1)a_\alpha^{(k)} + 3\alpha + 1}{(k-2)a_\alpha^{(k)}} \leq 3 + \frac{3\alpha+1}{(k-2)a_\alpha^{(k)}} .$$

But from inequality $a_\alpha \leq v < a_{\alpha+1}$ follows that if v tends to infinity, then so does α thus when v is large enough then we can write the following estimate:

$$3 + \frac{3\alpha+1}{(k-2)a_\alpha^{(k)}} \leq 3 + \eta .$$

From our previous consideration it can be seen that there exists a suitable u (i.e. $u \in A_3$) in the interval $(v, (4+\eta)v)$ also in the case k=3.

In the proof of the Lemma 4 we distinguish the cases k=2, k=3 and k>3.

First observe that we can suppose that $a_j^{(k)}$ has an irregular decomposition from A_n (when it bias not, then for this $a_j^{(k)}$, $0 \in U_{k,n}$ is suitable):

$$a_j^{(k)} = \sum_{\alpha=1}^{n-\ell-m} c_\alpha + \sum_{\beta=1}^{\ell} 1 + \sum_{\gamma=1}^{m} 0$$

$(c_\alpha \in A_n,\ c \neq 0,1,\ c_1 \geq c_\alpha$ and let $a_{\ell*}^{(n)} = c_1)$.

In all cases we shall distinguish the subcases: $m \leq 1$ and $m \geq 2$.

First let k>3 and consider the subcase $m \leq 1$.

Then according to Lemma 3 b) one of the numbers $a_j^{(k)}+1, \ldots, a_j^{(k)}+3$ cannot be decomposed irregularly from A_n, so for this subcase $1,2,3 \in U_{k,n}$ is suitable $(k-1 \geq 3)$.

Now let $m \geq 2$. $a_j^{(k)}+m$ has an irregular decomposition from A_n.

Since m≥2 we can determine that the next number with irregular decomposition is the

$$\sum_{\alpha=1}^{n-1-m} c_\alpha + a_2^{(n)} + (\ell+m-1)b_1^{(n)} = a_j^{(k)} - 1 + n^2 - n + 2 - (\ell+m-1)n. \qquad (8)$$

Namely in the proof of Lemma 3 we have seen that in case m≥2 for large enough j none of $a_j^{(k)}+m+1,\dots,a_j^{(k)}+m+c$ have irregular decomposition with $\ell_{m_2} \neq \ell^*$, where c is a constant, to be fixed later. Then in the decomposition of $a_j^{(k)}+m+c$ the greatest term is the same $a_{\ell^*}^{(n)}$ as in the decomposition of $a_j^{(k)}+m$. This implies that to get the decomposition for $a_{j^*}^{(k)}+m+c$ we can apply the algorithm in remark 2 for $a_j^{(k)}+m+c-a_{\ell^*}^{(n)}$.

We look for the smallest c>0, for which the algorithm works. This c is not smaller than

$$a_{j'+1}^{(n)} + (m'-1)b_{j'}^{(n)} - m'a_{j'}^{(n)} \qquad (9)$$

where the algorithm for a_j+m+1 stops in the (i-1-m')-th step. But in the decomposition of $a_j^{(k)} + m - a_{\ell^*}^{(n)}$ the number with the smallest absolute value is 1, so in (9) j'=1, m'=m+1, thus really

$$a_j^{(k)} - \ell + a_2^{(n)} - (\ell+m-1)b_1^{(n)} \qquad (10)$$

is the next, decomposable number after $a_j^{(k)} + m$. (If the algorithm stops for $a_j^{(k)} + m + c - a_\alpha^{(n)}$ at some other step earlier, then

$$c = a_{j^*+1}^{(n)} + (m^*-1)b_{j^*}^{(n)} - m^* a_{j^*}^{(n)} \geq 2j^*+1+(n-1-m^*)a_{j^*}^{(n)} > a_2^{(n)}$$

would have held, which is impossible because of in (10) $c<a_2$; the last inequality holds because m*<m implies that m*<n-1.)

According to the statement 1) there exists an u in the interval [m+1, 2m+2) which is decomposable into a sum of k-1 terms from A_k.

So we are ready, if

$$-\ell+n^2-n+2-(\ell+m-1)n\geq 2m+2. \quad (11)$$

(Then we have to put to each $m: 3\leq m<n$ a $u(m)$ into $U_{k,n}$.)

Observe that $\ell+m\leq n-2$ implies (11):

namely $-\ell+n^2-n+2-(\ell+m-1)n\geq -\ell+n^2-n+2-(n-3)n=-\ell+2n+2\geq m+n+4>2m+2$.

So it remains to examine the subcase $\ell+m=n-1$.

In this case according to (8) $a_j-\ell+n+2$ can be decomposed irregularly. After this the following numbers have irregular decomposition:

$$a_j^{(k)}-\ell+2n+2(=c_1+a_2^{(k)}+(n-3)b_1^{(n)}),\ a_j^{(k)}-\ell+2n+3,$$
$$a_j^{(k)}-\ell+3n+2,\ a_j^{(k)}-\ell+3n+3,\ a_j^{(k)}-\ell+3n+4,$$
$$a_j^{(k)}-\ell+4n+2,\ldots,\ a_j^{(k)}-\ell+4n+5 \text{ etc.}$$

(Let $c>n+5$ and then we can prove it in the same way as above.)

According to the statement 2) in the interval $m,-\ell+3n+2$ there exist a $u=\sum_{i=1}^{k-2} d_i$ $(d_i\in A_k)$. (Because of $3n+2-\ell=2n+3+m>3m$, so the choice $\eta=\frac{1}{n-1}$ is good.)

Then $k<n$ implies that one of the numbers

$$a_j^{(k)}+u,\ a_j^{(k)}+u+1,\ a_j^{(k)}+u-b_1^{(k)}=a_j^{(k)}+u-k$$

does not coincide with any of the above listed decomposable numbers. Now we are ready with the case $k>3$ (because this last statement implies that now to each m we have to put into $U_{k,n}$ 3 numbers, independently from j).

The proof for $k=3$:

We do not need to prove the statement for the subcase $2\leq m$, $m+1\leq n-2$, because in this case we didn't use the condition $k>3$ above.

If $m\leq 1$, then the interesting case is, when both $a_j^{(k)}+1$ and $a_j^{(k)}+2$ can be decomposed irregularly from A_n. From the proof of Lemma 3 it can be seen that then not only $a_j^{(k)}+3$ but $a_j^{(k)}+a_2^{(k)}=a_j^{(k)}+8$ cannot have irregular decomposition - if j is large enough. If $m+\ell=n-1$ then the remark after the statement 2) implies that in the interval $(m,-\ell+4n+1)$ there exists $a_r^{(k)}\in A_k$. It appears difficult only when both $a_j^{(k)}+a_r^{(k)}$ and $a_j^{(k)}+a_r^{(k)}+1$ are among the decomposable numbers listed above. But then, for $n>4$, $a_j^{(k)}+a_r^{(k)}-3$ cannot be decomposed from A_n.

And finally, if $n=4$ then, in the case $m\geq 2$, $a_2=8$ is in the interval $(m,-\ell+2n+2)$ so $a_j^{(k)}+8$ has no irregular decomposition. If $m=1$ then $\ell=2$ implies that $a_j^{(k)}-\ell+2n+2=a_j^{(k)}+8$ has an irregular decomposition, but

$$a_j^{(k)} + a_2^{(k)} + b_1^{(k)} = a_j^{(k)}+5$$

has none.

So it remains only the case k=2.

In this case the statement 1) is essentially true:

$$\frac{a_s^{(2)}}{a_{s-1}^{(2)}} = \frac{s^2}{(s-1)^2} \leq 2 \quad \text{if} \quad s\geq 2+\sqrt{2}, \text{ i.e. if } s\geq 4.$$

So we have almost proved our assertion in the subcase $m\geq 2$, $\ell+m\leq n-2$, since in the interval $[m+1, 2m+2)$ there exists a quadratic number, if $m>1$. But then also there are not any numbers in the interval $[m+1, m+n+4)$ with irregular decomposition from A_n, and in this interval obviously exists quadratic number. According to the subcase $\ell+m=n-1$ we have seen that in the interval $(a_j^k-\ell+n+2, a_j^{(k)}-\ell+2n+2)$ there are not any numbers with irregular decomposition.

It is easy to see that in the interval $(-\ell+n+2, -\ell+2n+2)$ there is a quadratic number if $n>7$, and on the other hand this holds also in case $n\leq 7$ excepting the cases: $n=7$ and $\ell=0$, $n=5$ and $\ell=3$,

$n=4$ and $\ell=1$ or $\ell=2$, $n=3$ and $\ell=0$ or $\ell=1$. For these cases we can prove the lemma knowing the irregularly decomposable numbers from A_n - similarly to the proof of above.

Finally, for $m\leq 1$ applying the same idea which we have used in the proof of Lemma 3 it can be seen that at least one of the numbers

$$a_j^{(2)}+1,\ a_j^{(2)},\ a_j^{(2)}-2(=a_j^{(2)}-b_1^{(2)}),$$
$$a_j^{(2)}+4(=a_j^{(2)}+a_2^{(2)}),\ a_j^{(2)}+9(=a_j^{(2)}+a_2^{(2)}) \qquad (12)$$

cannot be decomposed irregularly.

In all cases we obtained a finite number of u_r, which are independent from j, so we have got the desired set $U_{k,n}$ for all cases.

4. THE CONSTRUCTION OF THE MEASURES

Let us denote $\nu(\{a_\ell^{(k)}\})$ by ν_ℓ, $\nu(\{b_\ell^{(k)}\})$ by $\nu_{-\ell}$,

$\mu(\{a_\ell^{(n)}\})$ by μ_ℓ, $\mu(\{b_\ell^{(n)}\})$ by $\mu_{-\ell}$.

These values will be given by induction according to ℓ in such way that every ν_r, μ_r will be positive ($r\in\mathbb{Z}$) and

$$\nu^k(\{r\}) = \mu^n(\{r\}) \qquad (13)$$

will hold for $|r|<\ell$. This equation means that $\mu^n=\nu^k$.

From the Lemma 1 a) it can be seen that 0 can be decomposed only regularly. So

$$\nu^k(\{0\}) = \mu^n(\{0\})$$

is equivalent to $\nu_o^k=\mu_o^n$.

Hence we can determine ν_o when μ_o is an arbitrary given positive number.

Now suppose by the induction that ν_r, μ_r are already determined positive numbers, when $|r|<\ell_o$, such that the equality (13) holds in case $|r|<\ell_o$.

The equation (13) in case $r=\ell_o$ is

$$\mu^n(\{\ell_o\}) = \nu^k(\{\ell_o\}) = k\nu_{-\ell_o}^{k-1}\cdot \nu_{\ell_o}+S_{\ell_o} \qquad (14)$$

where on the right hand side the first term is given by the regular decomposition of ℓ_o, and the second term contains the contributions of the possible irregular decompositions.

Observe that invoking Lemma 1 in the irregular decompositions the terms $b_{\ell_o-1},\ldots, a_{\ell_o}$ can appear, so the only non-determined number in (14) is ν_{ℓ_o}. The same holds for $\mu^n(\{\ell_o\})$, so μ_{ℓ_o} and ν_{ℓ_o} can be determined to be positive and to fulfill (14). Doing the same for $-\ell_o$ we get $\nu_{-\ell_o}$ and $\mu_{-\ell_o}$, so by induction we can determine the measures μ,ν which are concentrated to the sets A_n and A_k respectively, and so these are irreducible, and for which $\mu^n = \nu^k$ holds.

<u>Theorem</u>. For the above constructed μ,ν we have $\nu(A_k)=\mu(A_n)=+\infty$.

<u>Proof</u>. First suppose that $2<k<n$. Suppose indirectly that $\nu(A_k)<\infty$, this implies $\mu(A_n)<\infty$.

Then for an arbitrary $\delta>0$ there exists n_o and m_o, such that $\nu_n<\delta$ when $|n|>n_o$ and $\mu_m<\delta$ when $|m|>m_o$. Let $k<n$,

$\varepsilon_1 := \min\limits_{u\in U_{k,n}}(\nu^{k-1}(\{u\}))$ ($U_{k,n}$ is the set, constructed in Lemma 4).

$\varepsilon_2 := \min\limits_{0\le i\le n-1} \mu^{n-1}(\{i\})$.

$$\text{Let } j>\max(n_o,m_o)-\min_{u\in U_{k,n}}(u) \qquad (15)$$

and so great that it would be possible to apply Lemma 4.

Let $t:=a_j^{(k)}+u$ (this is the u which is given by Lemma 4), then $n\mu_t\,\mu_{-t+1}^{n-1} = \mu^n(\{t\})$ because of the Lemma 4 t has no irregular decomposition from A_n. The inequality

$$\nu^k(\{t\})\ge\nu_i\varepsilon_1$$

is obvious. But (15) implies that $\mu_{-t+1}<\delta$, hence

$$\mu_t \geq \frac{\nu_j \varepsilon_1}{n \cdot \delta^{n-1}} \geq \nu_j \text{ when } \delta^{n-1} \leq \frac{\varepsilon_1}{n} . \qquad (16)$$

According to Lemma 3 there exists an i: $0 \leq i \leq n-1$ such that $a^{(n)}_{\mu_t}+i$ has no irregular decomposition from A_k, thus similarly to the case above:

$$k\nu_{a^{(n)}_{\mu_t}+i}\, \nu^{k-1}_{-(a^{(n)}_{\mu_t}+i)+1} \geq \mu_t \cdot \varepsilon_2 \text{ hence (because } a^{(n)}_{\mu_t}>t>n_0)$$

$$\nu_{a^{(n)}_{\mu_t}+i} \geq \frac{\mu_t \varepsilon_2}{k\delta^{k-1}} \geq \mu_t \geq \nu_j \text{ when } \delta^{k-1} \leq \frac{\varepsilon_2}{k} . \qquad (17)$$

If we choose such a δ, for which $\delta \leq \frac{\min(\varepsilon_1,\varepsilon_2)}{n}$ holds then δ satisfies (16) and (17).

Summing up we have got that to every ν_j (where j is great enough) there exists an $\ell>j$ such that $\nu_\ell \geq \nu_j$. But this contradicts our assumption on A_k.

In the case k=2 the method of the proof is the same as above. The first half of it can be done in the same way: now we also get a t, for which $\mu_t \geq \frac{\nu_j \varepsilon_1}{n^{n-1}}$.

Fix an $\eta \leq \frac{1}{4}$ and let n_0 be large enough to assure that $\nu_{n_0}+\nu_{-n_0}+\nu_{n_0+1}+\nu_{-(n_0+1)}+\ldots+\nu_{n_0+N}+\nu_{-n_0-N}<\eta$ for an arbitrary $N \in \mathbb{N}$.

Suppose that j is such a large number that we can apply Lemma 3' for $m_0=n_0$. This gives for $a^{(n)}_{\mu_t}$ that

$$2\nu_{a^{(n)}_{\mu_t}+u} \cdot \nu_{[-(a^{(n)}_{\mu_t}+u)+\ell]} + S_{a^{(n)}_{\mu_t}+u} \geq n\mu_t \varepsilon_2' \qquad (18)$$

(where $\varepsilon_2' = \min\limits_{u \in U_{n,2}} (\mu^{n-1}(u))$). Since according to Lemma 3' in S appear only such terms, which come from a decomposition having the form

$$a_{\mu_t}^{(n)} + u = \alpha^2 - \beta^2 - \beta \text{ where } \beta \geq n_0.$$

Then $a_{\mu_t}^{(n)} + u > a_j^{(2)}$ implies $\alpha \geq j$. Suppose that $\nu_\ell \leq \nu_j$ when $\ell > j$.

In this case $S_{a_{\mu_t}^{(n)}+u} \leq 2\nu_j \eta$.

So from (18) we get (because $a_{\mu_t}^{(n)} + u > n_0$)

$$\nu_{a_{\mu_t}^{(n)}+u} \geq \frac{n\mu_t \varepsilon_2' - 2\nu_j \cdot \eta}{2\eta} \geq \nu_j \left(\frac{\varepsilon_1 \varepsilon_2'}{n^2 \delta^{n-1} 2\eta} - 1 \right) \geq \nu_j ,$$

when $\delta^{n-1} \leq \frac{\varepsilon_1 \varepsilon_2'}{n^2}$.

Thus for every j which is large enough there exists an $\ell > j$ such that $\nu_\ell > \nu_j$ which is in contradiction with $\nu(A_2) < \infty$.

Remark 4. The nonuniqueness of Hinčin's decomposition theorem is enlightened from another point of view in [2].

5. REFERENCES

1 LINNIK, Ju. V. - OSTROWSKI, I. V.: Decompositions of Random Variables and Vectors (Moscow, 1972 in Russian).

2 RUZSA, I. Z. - SZÉKELY, G. J.: No distribution is prime (to appear).

SUBJECT INDEX

This index contains key-words for the individual papers and is intended to help the reader to find results of his interest. Key-words are indexed only once per article.